Thaddeus Mann

Spermatophores

Development, Structure,
Biochemical Attributes and Role
in the Transfer of Spermatozoa

With 50 Figures

Springer-Verlag Berlin
Heidelberg GmbH 1984

Zoophysiology   Volume 15

Coordinating Editor: D. S. Farner

Editors:
B. Heinrich K. Johansen H. Langer
G. Neuweiler D. J. Randall

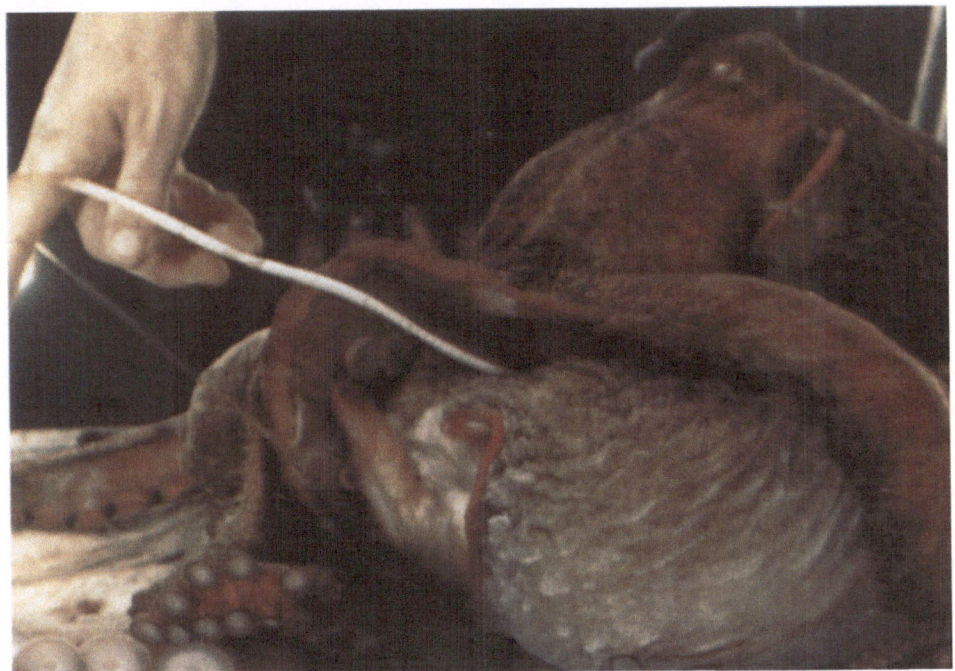

The 1-m-long spermatophore of the Giant Octopus of North Pacific, *Octopus dofleini martini,* being pulled out manually from animals copulating in a sea-water-filled observation tank. The upper, white portion of the spermatophore contains the sperm rope with spermatozoa, and the lower, thin portion contains the ejaculatory apparatus; the junction between the two portions is located in the region pressed against the forefinger. *Top right* the head of the male, in profile; *bottom right* ventral aspect of the female's head. (Mann et al. 1969)

Thaddeus Mann

Spermatophores

Development, Structure,
Biochemical Attributes and Role
in the Transfer of Spermatozoa

With 50 Figures

Springer-Verlag Berlin
Heidelberg GmbH 1984

Professor Dr. THADDEUS MANN
1, Courtney Way
Cambridge CB4 2EE
England

The front cover illustrates the spermatophore of cuttlefish, as first seen by Swammerdam and described by him in Biblia Naturae

ISBN 978-3-642-82310-7   ISBN 978-3-642-82308-4 (eBook)
DOI 10.1007/978-3-642-82308-4

Library of Congress Cataloging in Publication Data. Mann, Thaddeus, 1908 – Spermato-phores: development, structure, biochemical attributes, and role in the transfer of spermatozoa. (Zoophysiology; v. 15) Bibliography: p. Includes indexes. 1. Spermato-phores. 2. Spermatozoa. I. Title. II. Series. QP255.M25 1984 592.01662 84-14181

2131/3130-543210

Preface

Physiology and biochemistry of male reproductive function and semen became the main area of my research in 1944, after my attention was finally diverted from animal cells in general, to mammalian spermatozoa specifically. Ever since, the interest has remained largely focussed on reproductive problems in mammals, the work continuing mostly at the University of Cambridge, where I was privileged to hold also the Marshall-Walton Professorship in Physiology of Reproduction. This work led to the publication of three books, *The Biochemistry of Semen* (Methuen 1954), *The Biochemistry of Semen and of the Male Reproductive Tract* (Methuen 1964) and lately, in co-authorship with my wife, Dr. Cecilia Lutwak-Mann, *Male Reproductive Function and Semen – Themes and Trends in Physiology, Biochemistry and Investigative Andrology* (Springer-Verlag 1981).

In 1960, thanks to the Lalor Foundation, I was able to avail myself for the first time of a chance to visit the Marine Biological Laboratory at Woods Hole and there to take part in a study of reproduction in marine animals. Ever since, first as Visiting Professor of Biology at the State University of Florida, and later as the Walker Ames Professor and frequent visitor to the Department of Zoology at the University of Washington in Seattle, it has been my good fortune to sustain this new interest and to pursue it further. The spermatophores of cephalopod molluscs aroused my special curiosity, not least because it puzzled me greatly why in contrast to all mammals, a male squid or octopus does not ejaculate semen in a liquid state, but chooses instead to pack the spermatozoa into the tube-shaped capsule of a spermatophore, before transferring them to the female. Apart from the common American squid, *Loligo pealii,* and the common octopus, *Octopus vulgaris,* extensive use was made of the giant octopus of the North Pacific, *Octopus dofleini martini.* This creature's giant spermatosphores, each 1 m long, became the main object of a collaborative study with Dr. Arthur Martin, Dr. John Thiersch and several other colleagues. The study continued for many happy years and led to a number of new findings. We were able to demonstrate that the spermatophoric plasma, that is, the fluid which surrounds the spermatozoa inside the spermatophore, bears close chemical similarity to mammalian epididymal plasma. We found that, unlike the spermatozoa of mammals, those of the octopus are unable to metabolize fructose to L(+)lactic acid, but instead survive at the expense of their glycogen, which they break down to D(−)lactic acid. We have shown that the so-called spermatophoric or ejaculatory reaction which makes possible the release of spermatozoa from the spermatophore depends on influx of seawater and osmoregulation as the main driving force, but in the course of that reaction, not merely salt and water, but also foreign chemicals can enter the interior of the spermatophore.

Three centuries have passed since Swammerdam made the discovery of the first spermatophore in *Sepia*, and over the years that followed similar sperm-encompassing devices were shown to occur in phyla other than Mollusca, including Platyhelminthes, Aschelminthes, Phoronida, Annelida and many Arthropoda, such as Onychophora, Myriapoda, Insecta, Crustacea and Arachnida. They were also shown to be present in Chaetognatha, Pogonophora and occasionally, in Vertebrata, such as certain fishes and salamanders. Yet, as I became painfully aware during the pursuit of my own inquiries, no one seems to have made an attempt to collect and update information about the various phyla so as to present it in the form of a single treatise. Hopefully, the present monograph, which includes references to about 770 publications, will meet that need even though it deals predominantly with problems of function and adaptation to environment. I have tried deliberately to shy away from phylogenetic speculations and have left aside the various, sometimes rather wild, evolutional theories (which I distrust). I have made no attempt to tinker with problems of taxonomy (about which I know little), and as regards terminology, I admit that apart from Lord Rothschild's (1965) Classification of Living Animals, I relied principally, partly for historical reasons, on the terms used by authors of the original publications.

Those readers who might feel that I have accorded preferential treatment to cephalopod spermatophores, I would like to remind of a comment made by William Hoyle (1907) during his presidential address to the Zoological Section of the British Association for the Advancement of Science, on the subject of reproduction in Cephalopoda:

"The impression left upon my mind by a score of Presidential Addresses to this Section, which it has been my privilege to hear, is that the speaker who treats of the subject matter of his own researches has the best prospect of making his remarks interesting and profitable to his audience."

"What I have ventured to lay before you are a few fruits of the little garden plot in whose culture I have been privileged to take a humble share."

"The plot I have tried to cultivate has been a very small one, and I have had but little leisure to peep over the fence and see what my neighbours were doing."

The task of writing the monograph, which occupied a great deal of my time for about 2 years, could not have been accomplished without the support of the Royal Society, the help of Prof. A. Labhart and Mrs. A. Pfau during the search for early literature at the University of Zurich, and the generous assistance of those colleagues who either read parts of the text or gave permission to reproduce various figures: Drs. R. A. Brandon, J. Martan, L. D. Russell and E. J. Zalisko (Carbondale, Illinois), H. Breucker (Hamburg), R. Dallai (Siena), K. G. Davey (York University, Ontario), H. G. Drecktrah (University of Wisconsin Oshkosh), C. Erséus (Stockholm), W. H. Fahrenbach (Oregon Primate Research Center), D. S. Farner (Seattle), B. Feldman-Muhsam (Jerusalem), G. E. Gregory (Rothamsted Experimental Station), W. Grewe (Anstalt Helgoland), R. Hartmann (Cologne), A. W. Martin (Seattle), W. G. Robison (National Institutes of Health, Bethesda), V. Storch (Heidelberg), P. Talbot and M. J. Kooda-Cisco (Riverside, California), H. E. Vistorin (Waiblingen, FRG), W. Westheide (Osnabrück), P. Weygoldt (Freiburg) and R. L. Zimmer (Harvard University). I take also great pleasure in expressing my thanks to Mrs. Carmen Frankl for drawing the figures, Mrs. Jennifer Constable for typing the manuscript, and Mr. Michael Jackson for patient guidance during the preparation of the monograph for the Springer-Verlag. Indeed, the publishers and editors alike have gone far beyond normal obligations, and their assistance is gratefully acknowledged.

Cambridge, August 1984   THADDEUS MANN

Contents

Chapter 4. *Annelida*

Chapter 5. *Onychophora and Myriapoda*

Chapter 6. *Insecta*

Chapter 1

General Considerations

In contrast to most vertebrates, which discharge semen in a fluid state, that is, as suspensions of free spermatozoa in seminal plasma, in many invertebrates, but only a very few vertebrates, the spermatozoa undergo aggregation into all sorts of conglomerates or packages, some loosely, others tightly wrapped, before they are passed on to the females. In most instances the wrappers consist of male accessory secretions which in the course of enfolding the spermatozoa have undergone gelation, coagulation or similar forms of hardening and solidification. The resulting sperm-encompassing devices can range in appearance from simple, undifferentiated mucous coats to structurally complex, multilayered, often tough and highly elastic tunics or capsules.

Spermatophores are of the latter, that is, the encapsulated type, quite distinct from all those loose agregates that are represented by so-called sperm-bundles, spermatodesms or spermatozeugmata. Depending on texture and moulding properties of the encompassing material, a spermatophore may take the shape of a "thread", "tube", "worm", "seed", "pear", "balloon", "flask", "disk", and other forms. Similarly variable can be its size, ranging from the 0.15-mm-long and 0.04-mm-wide spermatophore of *Diarthrodes cystoecus* (a small copepod, less than 1 mm long, which inhabits marine algae) to the 1-m-long and 1-cm-wide spermatophore of *Octopus dofleini* (the giant octopus, up to 50 kg in weight, in the Pacific). Equally variable can be the number of spermatozoa squeezed into a single spermatophore: from one or two in some crustaceans, to 10^8–10^{10} in the giant octopus. As regards colour, in most instances the spermatophores are of a milky white appearance, conferred upon them by the sperm mass visible through the transparent capsule, but occasionally they may be, partly at least, of a yellow, orange, purplish-red or brown colour.

1.1 Beginnings

Jan Swammerdam, 1637–1685, the first to take notice of the *white stylet-like bodies (albi quidam styluli)* present in the male reproductive tract of cuttlefish, described and illustrated these "minute parts" in his *Biblia Naturae,* (published posthumously in Holland, 1737–1738; and as *The Book of Nature* in England 1758). Except for a few details, Swammerdam's *"microscopical view of one of these minute parts"* (Fig. 1) differs little from the present-day low-power microscopic image of a *Sepia spermatophore;* "its hinder part, loose and transparent" corresponding to the region which is filled mainly by *spermatophoric plasma;* the "white

1

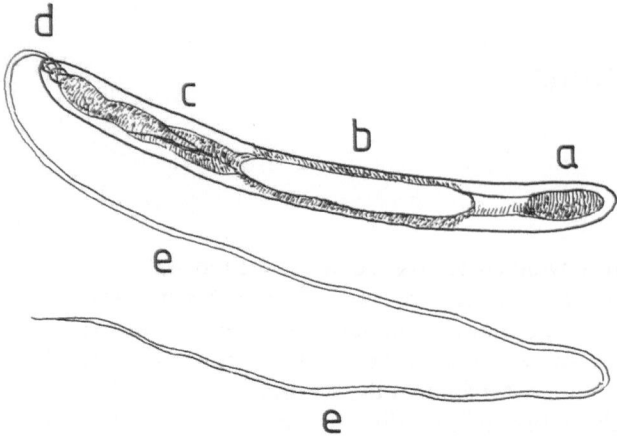

Fig. 1. The spermatophore of cuttlefish, *Sepia maris,* as shown by Swammerdam in *Biblia Naturae* (legend *left*) and *The Book of Nature* (legend *right*)

Quidam istorum Stylulorum sub microsco pio conspectus

A microscopical view of one of these parts

a Portio ejus posterior, libera, perlucida

a Its hinder part, loose and transparent

b Substantia alba, intus conclusa, quae per aquam intro penetrantem excutitur

b A white substance enclosed in the said part, and which is forced out of it by the water that penetrates it

c Locus, quo Stylulus anterius, aequae ac posterius, translucidus est instar vitri

c Places where it is transparent at each end

d Ejusdem elegantes convolutiones in extremo anteriore

d Beautiful windings of the same on its fore extremity

e, e Subtile ejus Filamentum, quod, instar staminis Bombycini, in aere durescit

e, e Its fine, delicate, or slender filament, which hardens in the open air like the Silk-Worm's tread

substance enclosed" representing the *mass of spermatozoa* packed into a *sperm rope;* the transparent "fore extremity" depicting the *ejaculatory apparatus,* including its coiled portion in the form of "beautiful windings"; and the "slender filament, which hardens in the open air like the Silk-Worm's thread" being the filamentous *cap thread.*

Swammerdam's remarkable sense of perception is the more admirable since in addition to depicting fairly accurately the stylet-like "minute parts" which in due course, became known as *sperm-carrying machines* or *spermato-phores,* he was quick to notice a phenomenon to which we would now refer as the spermatophoric reaction, that is, the discharge of the sperm mass in the form of a "little white body which, on its escape, rolls and curls itself up in a serpentine manner."

"This surprising little body, when viewed with the microscope, looks like a very white Earthworm, divided into a great many exceeding small rings; and if left in the water for some time, it expands and grows bigger by degrees, by the water it imbibes, which makes me imagine, that the water may possibly be the cause of that wonderful motion observable in these parts, on their being put into it."

2

The discovery of spermatophores in the Calamary (squid), reported by John Tuberville Needham (1745) in his *New microscopical discoveries,* marked the next important step in early developments. *Milt-vessels* was the name that Needham bestowed upon the tubular structures which he found stored in one of the squid's male accessory organs: the *milt-bag;* but during the century that followed, most investigators preferred to call them *Needhamia expulsatoria,* and to refer to the storage organ as *Needham's bag, sac* or *pouch.* Explosive indeed, was the response of a squid spermatophore to contact with water, as perceived by Needham (p. 47) in the *Action of the Milt-Vessels upon Several repeated Experiments:*

"Tho' many of the Milt-Vessels, when they are ripe for Action, and disengaged from that glutinous Matter which surrounds them while they are in the Milt-Bag, will act immediately in the open Air, for which perhaps the slightest Pressure during Extraction may be sufficient, yet the Generality of them will not only bear a translation to the Object-Plate, and lie quiet for Observation, but also require a Drop of Water to moisten the upper Extremity of the inclosing Case, before they begin to Operate.
Upon Application of this, the Extremity begins to evolute, and unfold itself, and the two slender Ligaments, which emerge out of the Case turn and twist themselves in various Direction,"
thereafter
"the inferior Part of the Apparatus which contains the Semen, extends itself in Length proportionably, with a Motion at the same Time upwards, which may be perceived by an increase of the Vacuity at the Bottom of the Case,"
and finally,
"the Semen flows out of the Cup, consisting of small opake Globules swimming in a sort of serous Matter, just in the same Form, and without any Appearance of Life, as I had seen it before, when diffused at large in the Milt-Bag."

Not absolutely certain about the presence of spermatozoa (Animalcules) in the Semen, Needham (p. 56) preferred

"To conclude, if I had ever seen the supposed Animalcules in the Semen of any living Creature, I could perhaps be able to determine with some Certainty, whether they were really *living* Creatures, or might possibly be nothing more than immensely less Machines analogous to these Milt-Vessels, which may be only in Large, what those are in Miniature. For by Mr. Levenhoeck's calculation of the Number and Size of the Animalcules in the Semen of a Cod-fish, one Million of them would scarce equal one of these Milt-Vessels ...".

Much less cautious and critical than Needham were some of the investigators who followed in his footsteps and tried to guess accurately what the true nature of Swammerdam's *tubes à ressort* (Denys-Monfort 1802) and the *fameuses anguilles de Needham* (Cuvier 1805, 1817) might be. In spite of Denys-Monfort's assurances to have positively identified *zoospermes* in Needham's milt-vessels, to Wagner (1834–1835) the "so-called Needham's bodies" were just some sort of parasitic "intestinal worm", most probably an *Echinorhynchus.* Likewise, Delle Chiaje (1822–1830; Vol. 4, 1829) felt that they are nothing else but true helminthes; the one in cuttlefish he identified as *Scolex dibothrius,* and that in octopus as a special species of *Monostomes* ("Monostoma del polpo"). Siebold (1836 a, b), whose observations on sperm "cysts", "bundles", and "bunches" *(Samenbläschen, Bündel, Büschel)* contributed so much to the early knowledge of sperm-aggregates in invertebrates (insects, flukes, and thorny-headed worms such as *Echinorhynchus),* while agreeing that the cephalopodan Swammerdam's tubes *(Swammerdamsche Röhrchen)* serve the purpose of reproduction, tended to view these tubes not as sperm containers but as individual, highly developed spermatozoa: *höher entwickelte Spermatozoen.* As late as 1839, Philippi, though by then convinced that the semen-machines *(Samenmaschinen)* of octopus are in fact, "neither intestinal worms nor independent animals of any kind, but must be considered as most wonderful, purposely organized and unique semen-containers or machines," still felt uncertain whether the small bodies inside these containers should be regarded as true spermatozoa *(Samenthierchen).*

A year later, the name *Spermatophore* was introduced for the first time into the literature on cephalopodan reproduction. It was proposed by Milne Edwards (1840) in a brief communication addressed to the *Annales des sciences naturelles:*

> „Ainsi, ces corps que Cuvier appelle les *fameux filamens machines de Needham,* ne sont ni des animalcules spermatiques, ni des vers parasites, mais des instrumens de fécondation tels que je n'en connais pas encore d'exemple dans le règne animal; nous proposons de les appeler des *Spermatophores,* et je ne puis mieux les comparer qu'aux grains de pollen qui renferment aussi les corpuscles fécondateurs, et qui éclatent de même, pour s'en décharger, lorsqu'ils sont parvenus de l'appareil mâle, sur l'organe femelle de la fleur."

Two years later, in another, more detailed paper published in the same journal (of which Milne Edwards was the editor) he presented a full account of his collaborative study with Peters, *Sur les Spermatophores des Céphalopodes,* of five species: Calmar commun (*Loligo vulgaris*), Seiche officinale (*Sepia officinalis*), Elédon musqué (*Eledone moschata*), Poulpe à longs bras (*Octopus macropus*), and Poulpe commun (*Octopus vulgaris*). Some years later, Duvernoy (1853) proposed another name, that of *spermaphore,* but in the end, *spermatophore* was the term that became universally accepted, not merely by malacologists but by those that discovered and investigated similar sperm-encompassing devices in arthropods and other phyla.

The beginnings of the history of arthropodan spermatophores are uncertain. It has been suggested (Whitman 1891) that Swammerdam's remarks in *Biblia Naturae* about the many peculiar "regular particles" which he found in the reproductive tract of hermit crabs should be looked upon as the first published record of spermatophores being present in an arthropod, but convincing evidence of the occurrence of *Samenschläuche* in *Pagurus* became available only much later, when Kölliker (1841) published the results of his extensive investigations on spermatozoa of invertebrate animals. By that time there were other indications that spermatophores occur in arthropods.

Ray Lankester (1871), the first to discover spermatophores in two species of *Tubifex* (Oligochaeta), became so impressed by the structural and functional characteristics of spermatophores that he acclaimed them as "an example of a kind of organization elsewhere without parallel," and remarked:

> "Did we know of a number of free unicellular organisms after complete development becoming fixed together by a cement to form a secondary organism capable of locomotion and possibly of nutrition, we should have a parallel to the spermatophores; as it is, they are, I believe, the only examples of the building up of an organ or quasi-organism by agglomeration instead of histogenesis."

By the end of the XIXth century, after a great deal had already been learned about the structure of spermatophores in a large number of invertebrate animals, it became also abundantly clear that even within the same order, there can be species that manufacture spermatophores, and others that depend on the use of liquid semen, and furthermore, that spermatophores differ so much in size, shape, and structure between species that in many instances they can be used as reliable, species-specific taxonomic markers. This phenomenon, which is particularly well expressed amongst insects, led Petersen (1907) to conclude that in butterflies and moths at any rate, "the diversity in the appearance of spermatophores proves once more that the features of genital organs provide us with the most trust-worthy indicator for recognizing species-characteristics"; a statement that brings to mind an earlier one about spermatozoa (Wagner and Leuckhart 1852), "that one may often safely venture to infer from the specific shape of these elements the systemic position and the name of the animals investigated".

1.2 Definitions

The amazingly great differences of shape and size amongst spermatophores pre-cluded the development of a nomenclature that could be uniformly applied to dif-ferent parts of the spermatophore structure in all animals. However, by way of a representative example, below are given several definitions commonly used for describing major parts of the spermatophore in the American common squid, *Loligo pealii,* a species that over many years served frequently as a model upon which morphological studies of spermatophores in other animals were based.

Essentially, the spermatophore of *Loligo pealii,* as schematically represented in Fig. 2, is composed of two main parts: the *body,* about 1 cm long, and the thread-like appendage, or *cap thread* which, when fully unwound, would measure several centimetres in length. The thinner (anterior) part of the body contains the *ejaculatory apparatus* consisting of the *ejaculatory tube* and the bulbous *cement body.* Most of the internal space in the wider (posterior) part of the body is oc-cupied by the mass of tightly packed spermatozoa, except for the pocket at the

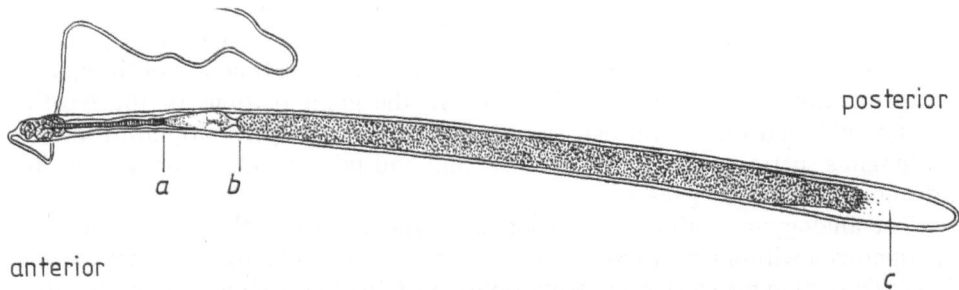

Fig. 2. Diagram of spermatophore in the squid, *Loligo pealii.* The cap thread extends from the anterior extremity. The anterior part of the body contains the ejaculatory apparatus, and the posterior part is filled largely by the mass of spermatozoa (stripped area), *a* junction between the two parts of the ejac-ulatory apparatus, the thin ejaculatory tube and the bulbous cement body; *b* junction between the ce-ment body and the sperm mass; *c* the pocket at the posterior end of the spermatophore, filled with fluid (spermatophoric plasma). (Austin et al. 1964)

posterior end, which is filled with the highly viscous fluid, the *spermatophoric plasma*. Together, the spermatozoa and spermatophoric plasma constitute the *spermatophoric semen*. The body of the spermatophore is bounded by amorphous *membranes* and these in turn are surrounded by *tunics* (for further details see Chap. 3.4).

The complex set of events culminating in the discharge of spermatophoric semen is usually called a *spermatophoric reaction* or *ejaculatory reaction*. In the common squid the spermatophoric reaction takes only a few seconds, but in other animals, such as the giant octopus for example, it can extend over a period of hours. Although the sequence of events associated with the spermatophoric reaction varies between species, in most instances several definitive stages can be clearly distinguished. In the giant octopus, for example (Mann et al. 1970), where because of its slow progress the spermatophoric reaction can be followed in a leisurely fashion, the main stages have been defined as (1) *extrusion of the ejaculatory apparatus,* (2) *advance of the sperm rope* (the rope-shaped tubule packed with spermatozoa), and (3) formation of the *spermatophoric bladder,* a bulbous structure formed by the evaginated tunic, into which the whole sperm mass is propelled. When some time later the spermatophoric bladder has ruptured, the spermatozoa are finally set free.

1.3 Spermatophore as Repository and Transport Vehicle for Spermatozoa. Certain Similarity to Epididymis

Whatever its size, shape or other gross-morphological features may be, the spermatophore fulfils in the first place the role of a storage container and transport vehicle for spermatozoa. In many species, all parts of this vehicle are manufactured and totally assembled inside the male accessory organs of reproduction so that the final appearance of the spermatophore is wholly determined before it is delivered by the male. This happens in cephalopod molluscs, including cuttlefish, squid, and octopus. In these animals, the male reproductive tract not only manufactures the spermatophores, but is capable of storing them for long periods of time, in a manner that reminds one of the sperm-storing function of the epididymis towards mammalian spermatozoa. In the giant octopus of the North Pacific we found on several occasions perfectly formed spermatophores inside Needham's spermatophoric sac of males that had been caught some months in advance of the breeding season.

The analogy to epididymis extends much further. As in the epididymis, the spermatozoa within a spermatophore appear to be mostly immotile. Belonoschkin (1929 a, b) in his work on spermatophores of *Octopus vulgaris,* compared the dormant state of the spermatozoa to that of a "winter sleep" *(Winterschlaf)*. Our experience has been that spermatozoa freshly collected from spermatophores of *Octopus dofleini martini* (Chap. 3) are either immotile or exhibit only sluggish movements, even at the peak of the breeding season. These spermatozoa acquire their motility only gradually after they have been separated from the spermatophoric plasma (which inhibits sperm motility) and then suspended in either seawater or isotonic saline solution (about 2.7% NaCl). Such suspensions, however, must be kept at a relatively low temperature around 8°–10 °C, which corresponds to that of the natural habitat of the animal in the North Pacific (Mann et al. 1970). Unlike those of mammals, the spermatozoa of octopus become less motile at higher temperatures, and at around 35 °C their ability to move is irretrievably lost (Brooks et al. 1971).

The similarity between spermatophores and epididymis is to some extent also reflected in certain common biochemical characteristics of spermatophoric and epididymal plasma. In the giant octopus, the spermatophoric plasma obtained by centrifugation of spermatophoric semen resembles epididymal plasma in its high content of organic as opposed to inorganic matter, the relatively high concentration of potassium, and above all, the presence of large amounts of glycoproteins, glycopeptides, glycerylphosphorylcholine, carnitine, and certain highly active enzymes such as glycosidases (Brooks et al. 1974 a, b; Mann 1975; Mann et al. 1970, 1973).

Equally significant is the pattern of the biochemical changes associated with the processes of formation and maturation of spermatophores in the male tract of the giant octopus. During the passage of semen from the testis to the spermatophoric sac, the concentration of the extracellular electrolytes gradually declines, and at the same time the inorganic constituents are largely replaced by organic substances (Mann et al. 1981 b). Analogous events occur in the mammalian epididymis: as testicular semen, rich in salt but poor in organic matter, enters the epididymis, the dilute testicular plasma is turned into the much more viscous and concentrated epididymal plasma, and at the same time electrolytes are replaced by glycoproteins, glycerylphosphorylcholine, carnitine, and other organic substances (reviewed by Mann and Lutwak-Mann 1981). Finally, like the epididymis, the spermatophore provides, in some species at any rate, a site where the germ cells themselves undergo maturation. As we shall see later there are some animals in which at the time of encapsulation into spermatophores the germ cells are still immature, resembling more closely spermatids than spermatozoa.

1.4 Direct and Indirect Insemination Routes

All of the above outlined considerations apply to animals in which the process of spermatophore assembly is both initiated and completed within the male reproductive tract. There are, however, a great many invertebrates in which the blending of the accessory secretions into the spermatophore's "capsule" occurs not prior to but during copulation. This happens when different parts of the spermatophore are sequentially secreted and ejaculated by the male, so that the final appearance of the completely assembled spermatophore is determined only later, mainly by the shape of the internal lumen of the female reproductive tract (vagina or bursa copulatrix).

Furthermore, the direct route of insemination does not constitute the sole method of introducing spermatophores into females. Many animals use an indirect route for transfer of spermatozoa. In some instances of this kind, instead of being inserted into the genital opening, the spermatophore is placed by the male elsewhere on the female's body, but positioned so that it can be easily picked up by the female and transferred to her genital opening. There are other instances where a spermatophore is deposited by the male on some sort of a *substratum*, a leaf perhaps, or a piece of rock, or in some cases a special web, before it is picked up by the female and transferred to her genital opening. Other animals again do not use the genital route at all. In some of them, a spermatophore is initially attached to the skin of the female, and there the spermatozoa are liberated and afterwards injected into her skin, which enables them to inseminate the female by the method of so-called *hypodermic impregnation*.

The substrate-mediated indirect method of spermatophore transfer is particularly characteristic of many terrestrial arthropods. Schaller (1965), whose studies contributed greatly to the present knowledge of the mechanisms employed in this type of transfer, proposed that a special scale, *A-B-C*, should be used to define the degree of bodily contact between the sexual partners.

A. No mating: Males and females act independently, both in space and time, e.g. Oribatei (except *Collohmannia*), Symphyla Pselaphognatha, Collembola Arthropleona (except *Podura*); *A1*, the males add to the spermatophore a signalling device that guides the females towards it, e.g. *Polyxenus* (Pselaphognatha).

B. Only one mate is active: Males select sexually mature females and deposit spermatophores in their presence; the females either take no notice of their "mates" (*Dicyrtomina*, Collembola Symphypleona) or are pushed to the spermatophore (*Podura*).

C. Mating: Two active partners are required for sperm transfer, e.g. Scorpionida, Pedipalpida, Pseudoscorpionida, Arrenuridae (water mites), *Collohmannia* (Oribatei), Chilopoda, Thysanura, *Sminthurides aquaticus* (Collembola Arthropleona); *C1*, the partners do not establish any physical contact, but "correspond" by means of chemical long-distance stimuli and "mating parade" ("dancing acts"), e.g. *Chelifer* (Pseudoscorpionida), *Sarax, Admetus* (Pedipalpida); *C2*, the males construct a special signalling web (Chilopoda: *Scolopendra* and *Lithobius;* Thysanura: *Lepisma*) or filament (Thysanura: *Machilis*) on which they deposit the sperm drop, to be collected by the females, in some cases (*Machilis*) the male pushing the female towards the drop; *C3*, the partners grasp each other firmly [Scorpionida, *Thelyphonus* (Pedipalpida), *Dendrochernes* (Pseudoscorpionida), *Smithurides aquaticus*], or the male is closely linked to the female (Arrenuridae), or the female clings to the male [*Schizomus* (Pedipalpida)].

1.5 Role of Spermatophore in Fertilization and Nutrition

The most obvious way in which spermatophores help "to augment the likelihood of fertilization" (Swedmark 1964) is by minimizing sperm losses during insemination. Of some importance must also be their role in protecting the spermatozoa from a hostile environment, particularly along the route of indirect insemination after a spermatophore has been deposited on a substratum or extruded into water. Equally important is probably the role that the wall of the spermatophore performs in increasing the chances of sperm survival within the female genital tract, as for instance in those crustaceans in which spermatophores are known to preserve structural integrity for long periods, namely, until the time of spawning and fertilization (Chap. 7). At the same time, one must not forget that the outer wall of the spermatophore does not represent an absolute barrier to the entry of foreign chemicals. An influx of such substances from seawater into spermatophores undergoing spermatophoric reaction has been demonstrated in the octopus (Mann et al. 1981 b).

Of undoubted importance in augmenting the chances of fertilization are also the aphrodisiac and gustatory properties of some spermatophores. Under conditions where the spermatophores are initially deposited on a substrate, the aphrodisiacs, acting as pheromones, can play a key role in attracting the female's attention. As regards the gustatory, that is, the nutritional role of spermatophores, this can be equally important as follows from many observations, especially on insects (Chap. 6). In some insects, a single spermatophore accounts for as much as one-third of the male's total body weight, and under such circumstances, it constitutes an obvious male investment of great nutritional value to the female in terms of calories, protein and other nutrients, including, as we shall see later, certain specific stimulants of oogenesis and oviposition. Small wonder that on numerous occasions female insects can be observed to devour spermatophores soon after the spermatozoa had been discharged in the genital tracts. An interesting example is provided by the post-copulatory behaviour of female katydids (Orthoptera, Tettigoniidae). Available evidence indicates that when given the choice between two singing males of different weight, the female always mates with the larger male producing a heavier spermatophore which she then eats (Gwynne 1982, 1984).

Equally obvious is why the male's ability to turn out normal spermatophores must be adversely affected by underfeeding or malnutrition. The male calanoid copepod *Diaptomus clavipes,* for example, responds to food shortages by reducing drastically the output of spermatophores (Cooney and Gehrs 1980). Feeding of D-glucoascorbic acid or D-araboascorbic acid to males of the moth *Spodoptera littoralis* induces sterility owing to malformation of spermatophores (Navon and Levinson 1976; Navon and Marcus 1982; further details in Chap. 6).

Apart from providing aphrodisiacs and nutrients, spermatophores can sometimes also act as sex-repellents which effectively dissuade males from approaching a female that has already been inseminated. This particular anti-aphrodisiac property of spermatophores is closely allied to similar attributes of certain male accessory secretions, those in particular that coagulate after deposition in the female tract and form so-called mating or copulatory plugs. Pheromonal anti-aphrodisiac sex repellents have been described in a number of spermatophore-carrying lepidopteran insects (Chap. 6), but there are other instances known in which the barrier to remating and reinsemination is provided by male accessory secretions of the kind that are not normally incorporated into a spermatophore.

1.6 Common Features of Spermatophore, Copulatory Plug, and Sphragis

Mating or copulatory plugs are a characteristic feature of reproductive function in a large variety of animals, including many vertebrates. In some ways, the mechanism of their formation is similar to that of spermatophore production. In both instances the underlying cause is the action of male accessory secretions resulting in gelation, coagulation, and solidification of parts of the ejaculate. As set forth in detail elsewhere (Mann and Lutwak-Mann 1981), gelation and coagulation of mammalian semen are brought about by interactions, often enzymatic in nature, between proteins secreted by individual male accessory glands such as the prostate, seminal vesicle, bulbo-urethral (Cowper's) gland and coagulating glands. In some mammals, man included, semen coagulates immediately upon ejaculation, but later it liquefies spontaneously. Liquefaction of human semen usually occurs within 20 min of ejaculation. In the rhesus monkey, on the other hand, liquefaction is delayed for much longer, and as long as the semen remains in a coagulated state, about two thirds of the entire ejaculate, including a high proportion of spermatozoa, are caught up in the coagulum. In rats and mice, coagulation of the ejaculated male accessory secretions leads to the formation of structurally distinct plugs which effectively seal the vagina (bouchon vaginal). Similar plugs are formed during copulation in some Insectivora (mole, hedgehog), Chiroptera (bats), and Marsupialia (opossum).

Presumably the main function of copulatory plugs is to prevent spermatozoa from being lost by "backflow" of semen from the vagina, but there are other indications of a much wider role of the plugs. One distinct possibility is that in common with spermatophores, plugs can put up in females an effective barrier to remating and reinsemination, mostly of mechanical nature, but in some instances combined with anti-aphrodisiac, that is, sex-repellent properties. A well-known example of sexual discrimination of this kind, is the sex-repellent activity displayed by mating plugs of natricine snakes (Devine 1975). Snakes often gather in courting groups comprising one female and several males, and the effectiveness of the plug is displayed in that the unsuccessful males of such an aggregate cease courting and disperse once their successful rival has achieved intromission; the female will not mate until the plug has been expelled. Male acanthocephalan worms possess special cement glands associated with the vas deferens, which secrete material that seals the vagina after copulation (Abele and Gilchrist 1977). Of special significance are the plug-like devices encountered amongst insects, especially Diptera, such as mosquitoes (Giglioli and Mason 1966), Hymenoptera, such as the honeybee, *Apis mellifera,* in which the coagulable protein is contributed by the mucous glands of the drone (Laidlaw 1944; Taber 1977), and certain spermatophore-carrying Lepidoptera.

In some Lepidoptera, the so-called sphragis or spermatophragma (Petersen 1929) provides an alternative plugging device with properties similar to those of copulatory plugs on one hand, and to spermatophores on the other. It, too, consists of coagulated material derived from male accessory secretions, which though not actually encapsulating spermatozoa, serves at least two distinct purposes, namely, to prevent losses of inseminated spermatozoa by leakage or backflow from the female genital tract, and to put up in the female an effective barrier against remating and reinsemination. The formation of sphragis takes place after the spermatozoa have been transferred to the female within the spermatophore. The butterfly *Euphydryas editha* provides a typical example. Following the transfer of a spermatophore, the external genital opening of the female becomes sealed by male accessory secretions, and so long as the seal remains intact it prevents the insertion of a second spermatophore (Labine 1964). The literature on the subject of mating plugs and sphragis in insects is extensive (reviewed by Parker 1970). In some insects, as for instance the cockchafer *Melolontha melolontha* (Landa 1960), the African migratory locust, *Locusta migratoria migratoides* (Gregory 1965 b) and two other shorthorned grasshoppers, *Euthystira brachyptera* and *Chrysochraon dispar* (Renner and Kremer 1980), spermatophores themselves act as copulatory plugs, providing at the same time an efficient barrier to remating.

In view of what has been said above, one must surely feel very reluctant to endorse the view voiced by those entomologists who still regard the mating plugs in insects as a primitive device and persist in the opinion that the next step in the evolutionary sequence might be the elimination of mating plugs and the development of methods of direct transfer of semen into the spermathecal duct opening (reviewed by Gerber 1970). Like so many other evolutionary concepts relating to the origin and function of spermatophores, this one unnecessarily clouds discussions on comparative merits of reproductive devices in so-called lower and higher animals. Above all, it ignores the basic fact that mating plugs of one sort or another are as much a feature of reproduction in insects as in rats or mice.

1.7 Conflicting Views on the Origin and Purpose of Spermatophores

Another pronouncement of doubtful merit, but often encountered in entomological literature, is that, in common with mating plugs, spermatophores constitute a primitive character in phylogenetic development, which appeared during the transition of animals from aquatic to terrestrial life in creatures which for some unknown reason failed to acquire proper copulatory organs prior to choosing freedom from the sea, copulation having been evolved considerably later than the spermatophore, and the use of liquid semen constituting an *advance* over the primitive and uneconomical method of sperm transfer by means of spermatophores (reviewed by Alexander 1964; Gerber 1970; Parker 1978).

To begin with, spermatophores and copulation can coexist, among not only terrestrial, but also aquatic animals. As a matter of fact, few terrestrial creatures can be said to have achieved the same degree of complexity and sophistication in patterns of copulation and function of spermatophores as some of the aquatic animals, amongst cephalopods particularly (Chap. 3). In order to deposit the spermatozoa in the female genital tract, the copulating male giant octopus of the North Pacific has to haul the large volume of semen (10–20 ml) over a distance of 1 m, from the male-oriented to the female-oriented end of the spermatophore. A feat of this kind could not have been accomplished by transporting semen in a free liquid form. Clearly, under these circumstances, the method of sperm transfer by means of a spermatophore constitutes an *advance* over the use of liquid semen. Equally relevant is the fact that not only within a single phylum or class of animals, but amongst members of the same order, there are some species that depend on spermatophores and others that use liquid semen. Furthermore, amongst species that are generally regarded as being closely related to each other, one finds some that have copulatory organs and others that have none. Arachnida provide in this respect some illuminating examples (Chap. 8).

There is in addition some evidence, though admittedly not extensive, that within a single species, a switchover from spermatophores to free semen is possible under certain circumstances. The oriental fruit moth, *Grapholitha molesta,* provides an interesting example. Normally, the spermatozoa of this insect reach the female in the form of a spermatophore, but insemination can also be effected by free spermatozoa (George and Howard 1968). During the first mating, the male moth produces a large spermatophore, but subsequent matings result in much smaller and, eventually, in no spermatophores. Yet regardless of whether the spermatozoa have reached the female as packed or naked, the numbers of eggs and progeny resulting from the matings are apparently the same.

Yet another frequently voiced opinion is that spermatophores have evolved coincidentally with the ability of the males to introduce spermatozoa directly into the female, thereby disposing of the risk of sperm losses that would have occurred during external fertilization. That argument, too, is difficult to accept for several reasons. First, spermatophores are frequently deposited by a male on a substratum that the female has to locate herself. Next, there are situations, for example one encountered in hermaphroditic rhynchobdellid leeches, where a spermatophore is initially attached externally to the surface of the body, and the spermatozoa are then injected subcutaneously so that having migrated through coelomic cavities, some of them (by no means all) can reach the ovary (Chap. 4).

Finally, as regards the almost universally held view that external fertilization in general is not only more primitive, but at the same time less economical than internal fertilization, it must be said that this notion, too, rests at present on a somewhat shaky experimental basis. As a matter of fact, we still know little about the actual numbers of spermatozoa that must reach the site of fertilization before male and female gametes can fuse together. Amongst vertebrates with internal fertilization, even mammals have received scanty attention in this respect (cf. Mann and Lutwak-Mann 1981). As regards invertebrates, of special significance are observations made on some Chelicerata. Of the two classes, Merostomata and Arachnida, the latter includes a number of animals that reproduce by means of spermatophores and internal fertilization (Chap. 8). Amongst Merostomata reproducing by external fertilization, the horseshoe crab, *Limulus polyphemus* (Xiphosura) has been studied with special attention to spermiogenesis and the number of spermatozoa needed to ensure successful fertilization (Brown and Humphreys 1971; Fahrenbach 1973). Between 10^5 and 10^6 spermatozoa must actually become attached to one egg, and their acrosomal filaments have to penetrate the investment, if that egg is to be fertilized successfully. At first glance, this would seem to be a very high sperm number indeed, particularly when related to events in mammals. But it has to be remembered that to ensure the fertilization of just one or a few mammalian eggs, a very large number of spermatozoa must reach the vagina and uterus after ejaculation; a rabbit ejaculate, for example, contains on the average about 10^8 spermatozoa. Furthermore, we know that a fair proportion of the inseminated spermatozoa must pass from the uterus into the oviduct even though finally only one spermatozoon is needed to fertilize an egg.

All in all, the currently available evidence concerning the origin and purpose of spermatophores makes one very reluctant to endorse the concept of "phylogenetic development," as vaguely expressed in the present form. Much more convincing, at least for the moment, is the alternative view that the use of spermatophores by aquatic and terrestrial animals alike constitutes a device "more directly related to features of habitat occupied than to phylogeny of the group concerned" (Clark 1981).

Chapter 2

Platyhelminthes, Aschelminthes, and Phoronida

2.1 Early Observations on the Attachment of Spermatophores to the Skin of Turbellaria

Arnold Lang (1884), an early authority of Platyhelminthes, especially of the turbellarians, was the first to apprehend clearly that the peculiar small white threads attached to bodies of *Leptoplana* and certain other Polycladida inhabiting the Gulf of Naples are not some kind of a mysterious parasite, but spermatophores in the form of thread-like packages of spermatozoa, wrapped in tough membranes. He also deserves the credit for pointing out that in Polycladida, the so-called male copulatory organ is not necessarily inserted into the female genital pore, but may serve the purpose of attaching the spermatophores to the surface of the partner's body. The fixed spermatophore, having penetrated the skin and musculature, then releases the spermatozoa so as to enable them to invade the body.

2.2 Mechanisms of Sperm Transfer in Turbellaria and Monogenea

Though many years have passed since Lang's observations, the question of the extent to which the mechanisms of "hypodermic" and "vaginal" entry of spermatozoa operate in turbellarians is still under active investigation (Ax 1969; Ax and Apelt 1969; Ax and Borkott 1969). While in *Macrostomum romanicum* (Macrostomida) the cork-screw-shaped spermatozoa are pumped during copulation straight into the vagina, in *Archaphanostoma agile* (Acoela), the copulating partners exchange their spermatozoa by hypodermal injection into the "bursal tissue," whole copulation lasting no longer than 3 s.

As regards sperm transfer mechanisms in platyhelminthes, Monogenea, next to Turbellaria, have been receiving special attention in recent years. All known monogeneans are hermaphroditic, and most of them have a "penis". Among the relatively few species in which copulation has been actually observed, only some reproduce by exchange of spermatophores, while others transfer their spermatozoa either by intromission of a penis into a vagina or by hypodermic impregnation.

Two studies of monogenean skin parasites of fish may serve as examples of trends in research on spermatophores in these platyhelminthes. The first concerns *Entobdella soleae*, a skin parasite of the common sole (Kearn 1970). The jelly-like matrix of the spermatophore is manufactured in a tubular organ called penis sac which communicates with the ejaculatory tube opening a short distance from the end of the penis, while the spermatozoa are contained in the vesicula seminalis, from where they pass into the ejaculatory tube and are injected into the matrix at the time when the latter is extruded into the surrounding seawater. Extrusion of the spermatophore and the injection of spermatozoa into the matrix are almost instantaneous, and a complete spermatophore consisting of the semi-transparent jelly and a central core of spermatozoa can be either of elongated or spherical shape, the sphere measuring in diameter from 250 to 400 μm. On the outside of the spermatophore there is a thin transparent layer of a few microns in thickness. The extruded spermatophores are first attached to the ventral surface of the body in the region of the vaginal opening and from there they are probably sucked into the vagina after the mating has taken place.

The second study deals with *Acanthocotyle greeni* and *Acanthocotyle lobianchi*, skin parasites of the ray (Macdonald and Llewellyn 1980). In *A. greeni*, the 12–18 testes, roughly spherical in shape, lie in two rows along the midline of the body. Each testis opens by a short duct into the vas deferens which runs longitudinally between the two rows of testes. Anteriorly the vas deferens swells to form a vesicula seminalis. The apparatus for manufacturing the matrix of the spermatophore consists of a pair of spermatophore matrix reservoirs, called also prostatic vesicles, which join with each other and with the most distal region of the vesicula seminalis to lead into a very short common duct opening to the outside. The actual process of spermatophore extrusion has been observed in *A. lobianchi*. In this species, the expelled spermatozoa, in the form of an ovoid sperm cluster, become mixed with the matrix and thus form a spermatophore which, like that of *Entobdella soleae,* is surrounded by a thin layer and not by any sclerotized capsule of the kind that one encounters in spermatophores of certain other Monogenea, such as *E. diadema.*

Hypodermic impregnation in Monogenea, as described in the gill parasite *Gastrocotyle trachuri* (Llewellyn 1983), is believed to be effected by suction pressure in a muscular chamber of the penis, which brings into contact with each other the tube of the penis and almost any part of the tegument of a co-copulant that is penetrated. This gastrocotylid, which is readily obtainable from the gills of *Trachurus trachurus,* has no vagina, and the general structure of the penis suggests that it is capable of grasping and perforating a small area of the tegument. The muscular contraction of the chamber could easily bring about a pinching of the skin, perhaps accompanied by perforation, and this, together with the discovery of sperm masses remote from the male system and from the uterus, strongly indicates that in *G. trachuri* hypodermic impregnation provides the mechanism for the transfer of spermatozoa.

2.3 Hypodermic Impregnation in Rotifera and Other Aschelminthes

About the same time as Lang (1884) was exploring the reproductive function of Turbellaria, several investigators were engaged in similar research on Rotifera. Plate (1885), in his doctoral dissertation on the *Natural History of Rotatoria,* came out strongly in support of the thesis that just as in some turbellarians, the penis of the copulating *Hydatina* (*Epiphanes,* Ploima) is not inserted into the cloaca of the female, but instead pierces her body wall at any arbitrary place, thereby enabling the spermatozoa to invade her body. The same author, in another of his papers (Plate 1887), provided also a fairly detailed description of the spermatophores in the vasa deferentia and ductus ejaculatorius of *Paraseison* (Seisonidae). Working with *Asplanchna* (Ploima), Hudson (1883) reported:

"I attempted on several occasions to see the union of sexes, but without success. The male would play round a female, and thrusts the penis backwards and forwards, but I never observed any copulation. Once I found the male adhering by the tip of the penis to the female; but it was on the outside of the centre of the ventral surface, and not at the opening of the oviduct."

Nonetheless, Hudson must have had serious doubts about the actual capability of male rotifers to inject spermatozoa subcutaneously, since in his presidential address to the Royal Microscopic Society in 1891, he dismissed Plate's assertion "that the penis bores through the body wall, anywhere, and ejects the spermatozoa and the rod-like bodies which accompany them into the body cavity", as a "strange theory" upon which it is "not necessary to comment further". Instead, he preferred to leave wide open the vexed question "How did they" (the spermatozoa) "get there"?

The same year, Whitman (1891), reporting on his discovery of "Spermatophores as a means of hypodermic impregnation" in leeches, responded to Hudson's question as follows:

"Plate's observations answer this question, and it will not do to dismiss them as ,a strange theory', since the same strange thing is reported from so many different sources."

In later years, the occurrence of spermatophores and hypodermic impregnation has been amply confirmed by observations on other aschelminthes, including Gastrotricha (Ax 1969; Teuchert 1968). Two mechanisms of sperm transfer have been observed in macrodasyoid gastrotrichs. Direct sperm transfer occurs in some, while others depend on dermal copulation.

An interesting recent contribution, providing yet another example of an unusual pattern of spermatophore transfer, concerns the interstitial marine nematode *Prorhynchonema warwicki* (Gourbault and Renaud-Mornant 1982, 1983). In this nematode, marked by the display of peculiar alterations in male and female copulatory apparatus, the spermatophores are transferred by the male spicular system to female genital apertures distinct from the vulva, and are kept there hanging externally.

2.4 Spermatophores of *Phoronis vancouverensis* and *Phoronopsis harmeri,* and the Role of Lophophoral Organs in Phoronida

Within the phylum Phoronida, individual species have frequently been noted to possess special glandular structures lining the lophophoral concavity (space formed by indentation of the tentacular ring into which open the anus and

16

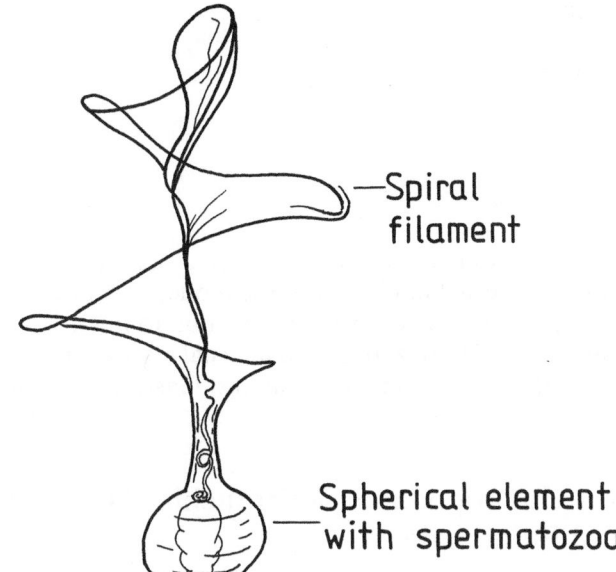

—Spiral
filament

Spherical element
with spermatozoa

Fig. 3. Spermatophore of
Phoronopsis harmeri. The two
main portions shown
diagrammatically are the
spherical element containing the
spermatozoa, and the spiral
filament functioning as a sail
(× 70). (Redrawn from
Zimmer 1967)

nephridiopores). It took a long time, however, after Dyster (1858) first described
the lophophoral glands, before the chambers enclosed by these organs were
shown to serve as sperm receptacles (Brooks and Cowles 1905). The manner in
which the spermatozoa are stored remained unexplored until Zimmer (1967) discovered that spermatophores are present in two phoronids, namely, in the hermaphroditic *Phoronis vancouverensis* and the functional males of *Phoronopsis
harmeri,* and that the lophophoral organs function as "accessory spermatophoral
organs" in spermatophore elaboration.

In both species, coelomic spermatozoa accumulate within the nephridia and
from there they pass into the accessory spermatophoral organs where the sperm
mass is moulded into a characteristic shape and encased before the spermatophore is released into the ambient water. The spermatophore of *Phoronis vancouverensis,* ovoid in shape and about 130 µm in length, has only one bounding
membrane, decorated on the surface with numerous stalked spherules; the spermatozoa are aligned along the long axis of the ovoid space and when examined
in a freshly deposited spermatophore, they show synchronized waves of movement passing down their length. The spermatophore of *Phoronopsis harmeri*
(Fig. 3), about 1 mm in total length, is composed of two portions – a spherical
part with a mean diameter of 230 µm, containing the packed spermatozoa and
bounded by an outer and inner membrane, and a long spiral membrane of two
or three turns, called sail because it is believed to make possible the sailing of the
spermatophore in water. Both the outer membrane and sail stain metachromati-
cally with Toluidine Blue O and are probably made up of the same acid mucopolysaccharide material. The structure of phoronid spermatozoa has been studied
by Franzén (1956, 1970), and more recently, by Franzén and Ahlfors (1980).

17

Chapter 3

Mollusca

The structural complexity and wide distribution of terrestrial and aquatic molluscs is reflected in highly diverse modes of reproduction: from primitive broadcast spawning and external fertilization to involved patterns of courtship behaviour, internal fertilization, and internal embryonic development. Spermatophores are frequently encountered amongst Gastropoda and Cephalopoda.

3.1 Gastropoda: Prosobranchia, Opisthobranchia, and Pulmonata

Gastropods, of whatever kind, have but a single gonad. In most prosobranchs the two sexes are separate, but opisthobranchs and pulmonates are hermaphroditic. All three divisions include species which discharge semen in free form, that is as sperm suspension in seminal plasma, and others in which, prior to ejaculation, the spermatozoa are either aggregated into loosely assembled naked conglomerates (spermatozeugmata) or encapsulated into spermatophores. In many instances, the spermatophores are of firm consistency, their bodies having hardened properly prior to extrusion, but in some polygyrid snails, for example, the freshly discharged sperm mass is of soft appearance and acquires the final shape and consistency later; the name paraspermatophore has been bestowed on the latter kind (Webb 1977). Characteristic of many gastropod families is also the occurrence of two distinct types of spermatozoa, the smaller "normal" (eupyrene) endowed with fertilizing ability, and the larger "abnormal" with chromatin either deficient (oligopyrene) or absent altogether (apyrene). [For details see Fretter and Graham (1964), Roosen-Runge (1977), and Webber (1977).]

3.1.1 Special Features of Reproduction in Archaeogastropods, Mesogastropods, and Other Prosobranchs

Of the three major orders of prosobranchs, Archaeogastropoda, Mesogastropoda, and Neogastropoda, the archaeogastropods are generally held to be the most primitive group of snails. Many of them depend on external fertilization as the principal mode of reproduction. A notable exception are the Neritidae. In neritids, the reproductive pattern is, in fact, more like that in mesogastropods than in other archaeogastropods, so much so that some authors view them as a separate order: Neritacea (Morton and Yonge 1964). Unlike most archaeogastro-

18

pods, which have only one type of spermatozoa, the neritids, in common with mesogastropods and neogastropods, frequently exhibit the phenomenon of sperm dimorphism. The ultrastructure of the eupyrene spermatozoa in *Neritina communis* has been the subject of a detailed study (Kohnert and Storch 1984).

There are some prosobranchs which produce spermatozeugmata, and others in which the spermatozoa are encapsulated into spermatophores. In both instances, sperm dimorphism has been encountered, but the most extreme forms of "abnormal" (oligopyrene or apyrene) sperm cells are found in the mesogastropods, especially those that produce the very large "giant spermatozeugmata." A giant spermatozeugma can be occasionally as much as 1 mm in diameter, and in addition capable of active swimming movements, using its "undulating plate" for this purpose. In spite of much research, the function of the "abnormal" cells still remains to be clarified. While some authors are of the opinion that their prime function within a spermatozeugma is to help with the transport of the eupyrene spermatozoa to the receptaculum seminis, others favour the view that the main role of the apyrene or oligopyrene cells is to provide nutrients for the eupyrene spermatozoa (Battaglia 1953; Bulnheim 1962 a, b; Fretter 1953; Fretter and Graham 1962; Hanson et al. 1952).

Neritina (Andrews 1936, 1937) and *Fissurella* (Medem 1945) are typical examples of animals which produce spermatophores. Those of *Neritina reclivata,* which Andrews (1936) studied in Florida, are 2.4–3 cm long and 1 mm wide; the length of the female's body is only 2 cm. The spermatophore of this snail looks like a "white turgid cylinder bent in a loop and with both attenuated ends stuck together" (Fig. 4). The two end portions which enclose the spermatozoa are of unequal width, the wider one displaying in addition some "spicules, unique in being chitinoid secretions", which are supposed to provide an anchoring device that keeps the tips of the spermatophore in position while the spermatozoa are being discharged and moved to the receptaculum seminis. *Nerita birmanica* (from Malayan mangrove swamps) is another archaeogastropod in which spermatophores have been described, and the effect of season on their numbers determined (Berry et al. 1973). Yet another snail, in which the formation of spermatophores has been investigated more recently, is *Tanzaniella souverbiana* (originally named *Neritina souverbiana*) from the coast of the Indian Ocean (Lupu 1979). In this snail, too, distinct spines, analogous to Andrews' "spicules" in *Neritina reclivata,* have been described, and their supposed function stated to be that of anchoring the spermatophore firmly in the female genital pore, thereby preventing loss during transfer from the male.

Amongst mesogastropods, the occurrence of spermatophores is particularly characteristic of the superfamily Cerithiacea, which comprises numerous herbivorous snails with high-spired shells and aphallic males, and includes *Gonobiasis* (Jewell 1931; Dazo 1965), *Modulus* (Houbrick 1980) and *Gourmya* (Houbrick 1981). *Modulus modulus,* which Houbrick (1980) studied in Florida, is a snail with average shell dimensions 8×9 mm, and a body which measures 9 mm in its natural coiled state, but 15 mm when uncoiled. Males become ripe in mid-winter and produce spermatophores from January until May; females begin spawning in the spring. The spermatophores are immobile, thinly walled, crescent-shaped, about 3 mm long, and they contain equal numbers of eupyrene (36 µm long) and

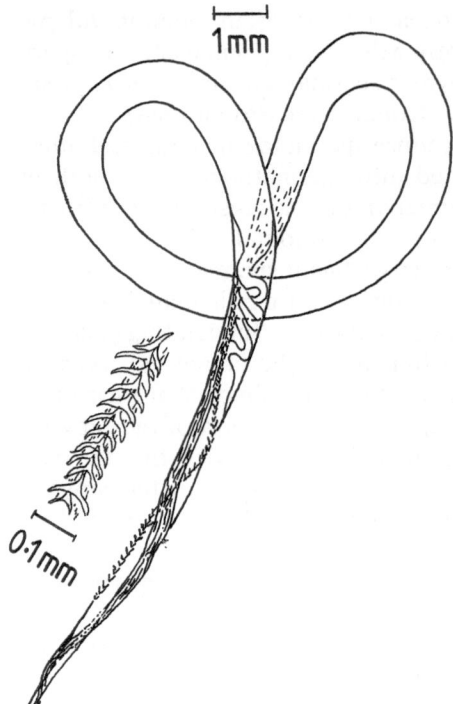

1mm

0.1mm

Fig. 4. Spermatophore of *Neritina reclivata* (Gastropoda, Prosobranchia, Diagrammatically shown are the whole spermatophore, freshly removed from the female, and a portion with a row of triods. (Redrawn from Andrews 1936)

apyrene (48 µm long) spermatozoa. The wall of the spermatophore is about 5 µm thick and composed of three layers: an outer fibrous layer, a thick middle layer filled with globules, and a thin inner layer. Of the two ends of the crescent-shaped spermatophore, one is rounded and pointed, and the other flattened and bifurcate. The spermatophores leave the mantle cavity of males by the exhalant siphon and presumably enter the female's mantle cavity via her inhalant siphon, since males lack any kind of intromittent organ. In the female's mantle cavity, the spermatophores are moved first by ciliary currents into the open ciliated gutter and then into the spermatophore chambers (spermatophore receptacle), which they enter with the pointed end. An individual spermatophore fills the entire length of the spermatophore receptacle, the bifurcated end sticking out. Normally, the chamber holds one spermatophore only, but occasionally two may be found. The spermatozoa leave the spermatophore by emerging from the lower branch of the bifurcate end, where there appears to be a special tube-like exit. From there they move into another ciliated groove and finally into the receptaculum seminis. The discharged spermatophore remains in the spermatophore chamber and there it undergoes gradual disintegration and dissolution.

Gourmya gourmyi, the cerithiid studied by Houbrick (1981), lives on subtidal coral reefs of island groups around New Caledonia. It is probably the sole survivor of a long lineage that can be traced back to the Eocene of the Paris Basin. The pale yellow spermatophores extruded by the aphallic males are lodged with their rounded ends in the proximal part of the spermatophore chamber (receptacle), thereby causing it to bulge. The spermatophore chamber, which may contain as many as two spermatophores, opens into the lumen of the oviduct. Along the edge of the spermatophore chamber lies a slit within which is located the receptaculum seminis. From here the spermatozoa move into the lumen of the oviduct to fertilize the ova.

3.1.2 Sporadic Occurrence of Spermatophores in Cephalaspidean and Acochlidiacean Opisthobranchs

Except for some of the shell-free slugs (Acochlidiacea), in which two sexes exist, Opisthobranchia are hermaphrodites. In many of them the ejaculated semen is mutually exchanged in the course of a reciprocal copulation, when the penis of each mate is inserted into the female gonopore (or common genital aperture) of the other, but some practice unilateral copulation (reviewed by Beeman 1977). Relatively few opisthobranchs bundle their spermatozoa into spermatophores. Among these are some cephalaspideans, notably the genera *Haminoea* (Perrier and Fischer 1914) and *Runcina* (Ghiselin 1963), and the acochlidiaceans *Microhedyle* (Hertling 1930; Swedmark 1968; Westheide and Wawra 1974) and *Ganitus* (Marcus 1953). In some cephalaspideans, a so-called prostate, which forms part of the vas deferens, has the function of producing spermatophores, the prostatic duct serving also as the ejaculatory duct (Ghiselin 1966). In nudibranchs (Acoela), Schmekel (1971) described two types of prostatic secretion: fine floccular material and osmiophilic granules; her work also indicates that the prostate cells are rich in Golgi complexes, mitochondria and endoplasmic reticulum. Apart from the intragenital mode of reproduction, some opisthobranchs depend on an extragenital route for the transfer of spermatozoa. Hypodermic impregnation through the body wall into a sperm chamber of the haemocoel has been described in several Sacoglossa (Gascoigne 1956; Purchon 1968; Reid 1964).

Cutaneous fertilization without copulation is characteristic of Acochlidiacea; only a few of these slugs are known to copulate. In a manner resembling the events in some Platyhelminthes and Annelida (Chaps. 2 and 4), the slugs attach their spermatophores at various places of the animal to be impregnated. A frequent site of attachment is the visceral sac, but deposition on the head–foot complex, most commonly in its middle, has also been observed (Swedmark 1968). One or more spermatophores is attached to each animal. Discharge of spermatozoa begins within a few days and gradually the whole of the spermatophore is emptied, making it possible for the spermatozoa to penetrate through the skin of the receiving animal. In histological sections, the changes that can be seen in the contact area between the base of the spermatophore and the epidermis include disappearance of nuclei in the epidermal cells, most probably resulting from cell autolysis which had been caused by the penetrating spermatozoa.

Microhedyle cryptophthalma, an animal thus named by Westheide and Wawra (1974) because its small eyes are difficult to spot, provides a typical example of an acochlidiacean with two separate sexes, in which the accumulation of spermatozoa under the epidermis has been reported. The body of this slug, which inhabits the sandy surf beaches of the Mediterranean Sea, is 0.7–1.6 mm long. The spermatozoa are transferred in spindle-shaped, 80–187 µm long and 17–30 µm wide spermatophores; each contains 200–300 spermatozoa. Attachment occurs mostly in the lateral parts of the visceral sac. Some recipients were seen with their empty capsules still attached some time after the spermatozoa had been discharged.

3.1.3 Love-Darts and Spermatophores in Pulmonates

Basommatophora and Stylommatophora are hermaphrodites in which male and female gametes are produced by a common gonad. From the ovotestis the mature gametes first pass to the little hermaphroditic duct, of which one section, commonly designated as seminal vesicle, provides the storage place for spermatozoa. The little hermaphroditic duct opens into the much larger glandular duct at a point where it is joined by the duct leading from the "albumen gland." Copulation is the usual mode of reproduction, the "penis" commonly matching the "vagina" in structure and length, but some pulmonates practise cutaneous fertilization (Berry 1977; Fretter and Graham 1964).

An early indication that in pulmonate gastropods the "penis" can provide a site for sperm encapsulation into a spermatophore came from Simroth (1900). His observations were followed by an extensive study of Meisenheimer (1907), who explored the sequence of the complex and prolonged copulatory events in *Helix pomatia*. His study has shown that the inserted spermatophore does not only fill the vagina but extends some distance towards the receptaculum seminis. It also included a detailed description of the snail's spermatophore and its four main component parts: head portion, neck portion, semen container, and the long filamentous tail. In Fig. 5 are shown side by side a spermatophore as seen by

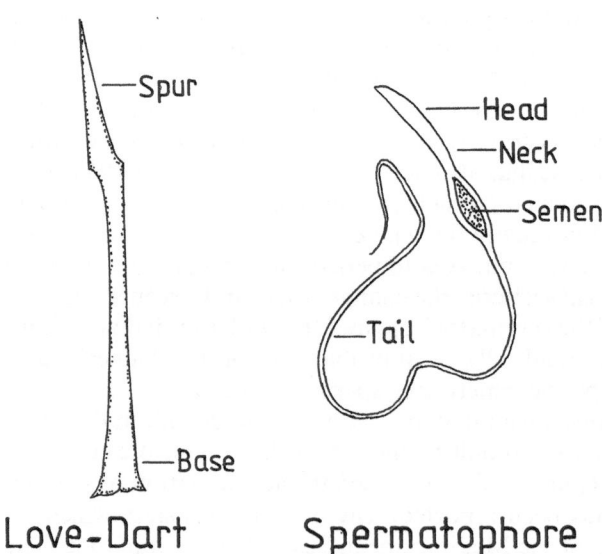

Fig. 5. Love-dart and spermatophore of pulmonate snail. *Left* a love-dart (*telum amoris, Liebespfeil*), about 8 mm long, of *Zonitoides arborens;* based on a drawing by Ihering (1892); *right* the spermatophore of *Helix pomatia*, $\frac{4}{5}$ of natural size. (Based on a drawing by Meisenheimer 1907)

Meisenheimer (1907) in *Helix,* and next to it a so-called love-dart, as described by Ihering (1892) in *Zonitoides arborens* (like *Helix,* a stylommatophore). The love-dart (*telum amoris, Liebespfeil*) a stiletto-like, finely ridged calcareous shaft (in *Helix* about 8 mm long), is a body that snails extrude, not during copulation itself, however, but some time before, as part of their pre-copulatory love-game (Liebesspiel). Love-darts are supposed to act as releasers of courtship behaviour; they do not carry spermatozoa. Unlike the capsule of a spermatophore, which is secreted in the tubular outgrowth of the penis, love-darts are formed in the caecumlike portion of the vagina, the dart-sac (Meisenheimer 1921).

The "penis" of pulmonates, an organ involved in both copulation and spermatophore production, can be of variable size and shape, and highly composite nature. In *Siphonaria* (Basommatophora), the structural component most directly involved in sperm encapsulation is the epiphallus gland, whose epithelium generally resembles that of prostatic tissue and voids a secretion containing both acidophil and basophil granules. During copulation, the freshly assembled spermatophore is extruded by the penis into the bursa copulatrix. The size and shape of spermatophore varies in different siphonariid species, but in general, it is of vermiform appearance. In *Siphonaria obliquata,* the spermatophores are hooked, and in some other species such as *S. cochleariformis* and *S. kurracheensis* they are equipped with spines. Most are less than 1 cm long, but that of *S. gigas* is nearly 2 cm long and bears an angular crook at one end (Berry 1977; Hubendick 1947; Marcus and Marcus 1960). Spines are also characteristic of spermatophores in other pulmonates. In *Milax gracilis* (Stylommatophora), the long, pale-brown spermatophore is spirally twisted, like a cork-screw, at one end, and smooth at the other end, but the rest of the spermatophore is covered mostly by branched spinose processes bent towards the smooth end (Phillips and Watson 1930).

The number of spermatophores, coiled and held in a brown gelatinous matrix, is 3–5 in the bursa copulatrix of *Siphonaria australis,* and 1–4 in that of *S. propria;* in a 9.9 mm long *S. propria,* the spermatophore is 4.5 mm long and consists of a blunt cylindrical head (50–60 μm in diameter) tapering to a thin, thread-like, transparent tail (Jenkins 1983).

Apart from those already mentioned, the mode of transfer, structure, and fate of spermatophores have been studied in a good many other pulmonates, including some urocyclid slugs, and bulimulid and polygyrid snails (Breure and Eskens 1977; Lanzieri et al. 1972; Van Goethem 1977; Webb 1974, 1977). Arionidae and Clausiliidae are two other families of pulmonates in which the occurrence of spermatophores has been known for some time, but more information about them has become available in the course of more recent studies (Lupu 1974, 1980). *Helix pomatia,* the snail favoured by pioneers in research on reproduction in gastropods, still continues to be the preferred subject of current studies on the structure and function of spermatophores. An interesting contribution to this area of research is the observation made by Hryniewiecka-Szyfter and Redziniak (1976) that the bursa copulatrix of the snail does not normally show any acid phosphatase or β-glucuronidase activity, but both enzymes appear in the epithelium of the bursa some time after copulation. Acid phosphatase activity appears 8 h after copulation and its activity diminishes from the 6th day onwards, while β-glucuronidase can first be demonstrated 3 days after copulation but thereafter it persists for another 11 days. Both enzymes were detected in the spermatophores inside the bursa. The same study also provided further evidence that the spermatophores of gastropods, just as those of cephalopods, are largely composed of secretory materials containing mucopolysaccharides and mucoproteins. The functional significance of the presence of lysosomal enzymes in spermatophores is not yet clear, but from the above-mentioned findings concerning acid phosphatase and β-glucuronidase, and from other observations on the changes associated with the discharge of spermatozoa and ultimate breakdown of the spermatophore's capsule in *Helix pomatia* (Lind 1973), it would appear that the activation of lysosomal enzymes can form an important part of the mechanism underlying the function of spermatophores after deposition in the bursa copulatrix.

3.2 Cephalopoda: Main Features of Male Reproductive Function in Nautiloidea, Decapoda, Vampyromorpha, and Octopoda

Tetrabranchia (Nautiloidea) and Dibranchia (Decapoda, Vampyromorpha, and Octopoda) are sea inhabitants reproducing by sexual means. Sperm transfer depends most frequently on spermatophores, which are as a rule totally assembled within the male reproductive tract and extruded by the male only during copulation. The release of spermatozoa occurs when following the spermatophoric reaction, the spermatophore has finally ruptured. Fertilization takes place some time later, that is, after the liberated spermatozoa have established contact with the eggs which passed down the oviduct.

Various aspects of cephalopodan testicular and ovarian function, its dependence on hormones and environmental factors, the process of fertilization itself, and the different stages of embryonic development, have been reviewed by Arnold (1971), Arnold and Williams-Arnold (1977), and Wells and Wells (1977). A great deal of information about spermiogenesis and spermatozoa of several cephalopods is contained in the publications by Franzén (1967), Longo and Anderson (1970), Maxwell (1974, 1975), and Fields and Thompson (1976).

3.2.1 The Male Genital System of Cephalopoda

There is only one testis in Tetrabranchia and Dibranchia. On leaving it, the spermatozoa enter first the vas deferens and subsequently a complex tubular system of other accessory glandular structures where they become engulfed and encapsulated by the secretions. From there, the spermatophores pass for storage into the spermatophoric sac.

The gross morphology of the male genital system varies among genera and species, but in many cephalopods the main anatomical features are essentially similar. As a representative example can serve the male reproductive tract of the

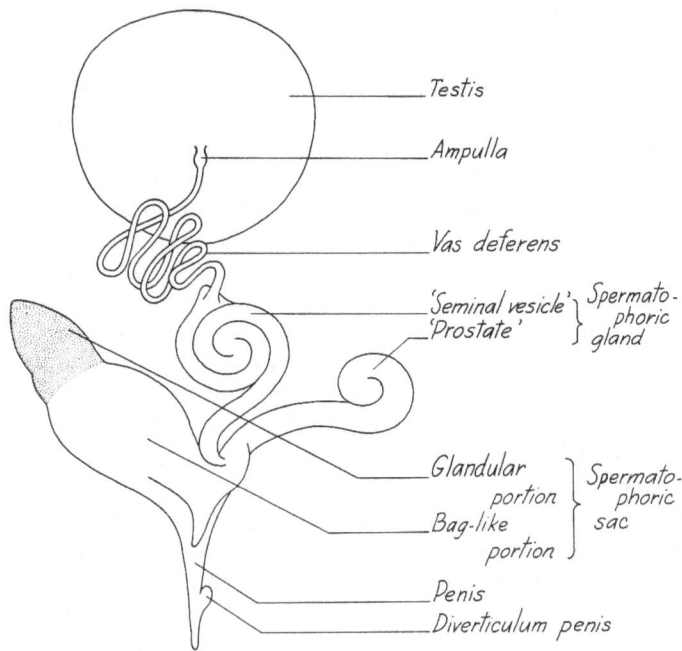

Fig. 6. Diagrammatic outline of the male reproductive tract (unravelled) of *Octopus vulgaris*. (Mann 1963)

common octopus, *Octopus vulgaris,* first described by Cuvier (1817) and later extensively explored by others (Belonoschkin 1929 a, b; Brock 1879, 1882; Mann 1963; Marchand 1907, 1913). Its diagrammatic outline is presented in Fig. 6.

Upon leaving the testis the spermatozoa of octopus enter the vas deferens proximale through the slightly enlarged portion known as the ampulla. The vas is a highly convoluted tube, in its gross appearance resembling the mammalian epididymis. Through it, the spermatozoa pass tightly squeezed together into a "sperm-rope." Emerging from the vas deferens proximale, they enter the "spermatophoric gland," a composite accessory organ in which the spermatozoa are enfolded by the secretions and encased into the tube-shaped spermatophore. The portion of the spermatophoric gland that the spermatozoa enter first, is the helically wound, tube-shaped Cuvier's vesicula seminalis ("seminal vesicle") in which the two chief component-parts of the spermatophore, the "sperm-container" and "ejaculatory apparatus" (Chap. 1) assume their characteristic shape. Next, the still somewhat loosely assembled spermatophore passes into the second portion of the spermatophoric gland, a blindly ending tube corresponding to Cuvier's prostata or Marchand's *Rangierdrüse.* The latter name originated from the concept that, like a shunting train which, having entered a platform, then leaves it in reverse, the spermatophore enters the prostate with its sperm-containing portion pointing forward, but after reaching the blind end of this organ, goes into reverse, that is, with the ejaculatory apparatus placed frontally. In this formation, the spermatophores, now completely assembled, enter a short tubular vas deferens distale (Brock's vas efferens), and having traversed it, they pass into the spermatophoric sac. This is the organ in which the numerous spermatophores, arranged in matchbox-like fashion, are commonly held. From there they are discharged into a short "penis" and then, at the time of mating, removed by the male with the help of his third right arm for transfer to the female.

On closer examination, the capacious spermatophoric sac of *Octopus vulgaris* can be seen to consist of two distinct portions, one adjoining the vas deferens distale, shaped like a pouch or bag, and serving as a reservoir for the spermatophores, and the other, located at the tip of the sac, and glandular in texture. The function of this gland remains, as yet, to be explored, but my own study (Mann 1963) leads me to believe that it involves the secretion of a substance which by eliciting the contractions of the spermatophoric sac makes the discharge of the spermatophores possible. Upon analysing the glandular portions of spermatophoric sacs taken from males caught in the Gulf of Mexico during February and March, I found in them a high concentration of 5-hydroxytryptamine (serotonin), a well-known muscle-stimulating substance. The demonstration of a large amount of 5-hydroxytryptamine (12–169 mg/100 g of glandular tissue) in the spermatophoric sac of octopus recalls an earlier finding made on the clasper-siphon sac of another aquatic animal, the spiny dogfish *Squalus acanthias*. This organ, too, has been shown to secrete large amounts of 5-hydroxytryptamine (Mann 1960), and in a subsequent study, this substance was found to provide a powerful stimulus for the long-lasting contractions of the fish's uterus, presumably facilitating thereby the ascent of the ejaculated spermatozoa (Mann and Prosser 1963).

Apart from contraction-inducing agents, the formation and discharge of spermatophores depends probably on other stimuli. Ciliary action of the epithelia lining the vas deferens and the tubular ducts of male accessory glands in octopus is likely to be of pivotal importance in this respect. In turn, these glandular tubes are well known to be fairly richly innervated and, as Young (1967) rightly pointed out, it is highly probable that the ciliary action required for the manufacture of the spermatophores is at least partly under nervous control.

3.2.2 Modes of Spermatophore Transfer

Of all the spermatophores found in animals, those of the cephalopodan molluscs bear the unique distinction of having been discovered first, and of possessing the most beautiful and highly organized structure. Small wonder that for several centuries they have served continuously as a highly successful model on which many investigators have based their observations and descriptions of spermatophores in other invertebrate animals.

A brief description of one cephalopodan spermatophore, that of the common squid, has been already presented in Chap. 1, and from the more detailed accounts which follow later (this chapter), it will become obvious that, in spite of the enormous variability in size and shape, in many cephalopodans the spermatophores are constructed in a similar manner, their "body" enclosing within its various membranes and tunics, two essential units: the "sperm mass" and the "ejaculatory apparatus", linked to each other by the "cement body" (Fig. 2). This rule, however, is not without exceptions, as we shall see later when dealing with the spermatophore of *Nautilus,* which is of coiled, and not rod-like appearance.

Whereas in some cephalopods, such as octopus, the spermatophores are inserted during copulation into the oviduct directly, in others, as for example, the common squid, they can initially be placed by the male elsewhere inside the female. In the common squid, the site of deposition is primarily dependent on the position which the male and female adopt during copulation, that is, side-by-side or head-to-head. In the side-by-side position, which is more common in copulating squids, the spermatophores are placed by the male inside the female's mantle cavity near the opening of the oviduct. In the head-to-head position, on the other hand, they are mostly deposited by the male in a special seminal receptacle located below the female's mouth (reviewed by Arnold 1962, 1971; Arnold and Williams-Arnold 1977). A pouch-like receptacle on the female's head, and not her mantle, is the common place of spermatophore deposition in *Vampyroteuthis* (vampire squid, Vampyromorpha) (Pickford 1946, 1949).

3.2.3 Role of the Hectocotylus

With a few exceptions (*Vampyroteuthis*), copulation occurs in most cephalopods at a distance, that is, with the male contacting the female with one arm which carries the hectocotylus at its distal end. In *Loligo,* the hectocotylized arm, though present, is sometimes hard to distinguish from the other nine arms, but in *Octopus,* the third right arm which bears the suckerless hectocotylus is modified in such a characteristic manner that it can be easily recognized as being different from the other seven arms, the more so as one side of it is converted into a special groove (to hold the spermatophore). In many octopods the structure of the hectocotylus is so characteristic as to provide the observer with a marker to identify the species to which the male belongs, but in this connection one has to remember that even within the same species there can be considerable variation in the form of the hectocotylus (Robson 1929, 1932). Another fact of general significance is that the hectocotylus does not regress after castration. Moreover, if cut off, it regenerates, also in castrated males. It will grow even in a male that had been castrated prior to reaching puberty, that is, before the hormonal system controlling male sexual maturity became functional (Wells 1964; Wells and Wells 1977).

It seems that the adaptation of one special "tentacle" for copulatory function may have been known as far back as three centuries BC, at the time when Aristotle was writing the following in his *Historia animalium* (transl. by D'Arcy Wentworth Thompson):

> "Molluscs such as the octopus, the sepia, and the calamary have sexual intercourse all in the same way, that is to say, they unite at the mouth, by interlacing of their tentacles.
> Some assert that the male has a kind of penis in one of the tentacles, the one in which are two largest suckers; and they further assert that the organ is tendinous in character, growing attached right up to the middle of the tentacle, and that the latter enables it to enter the nostril or funnel of the female."

The extraordinary ease with which in some caphalopods a portion of the hectocotylized arm can detach itself and yet remain contractile and capable of swimming has been a source of endless confusion to those who have tried to establish its identity. Cuvier (1829), whose otherwise brilliant and pioneering efforts contributed so much to advances in molluscan reproductive biology, mistook it for a hitherto unknown parasitic worm («un ver parasite de nouveau genre»), which he named *Hectocotylus,* even though he must have been greatly surprised that his newly discovered parasite of the octopus, *Hectocotylus octopodis,* should bear close resemblance to the host animal itself: «Voilà le corps du Poulpe qui a pour parasite un ver tellement semblable à un bras de Poulpe». In the same paper he also bestowed a similar identity on a second "worm", parasitizing on *Argonauta* (Octopoda), which he accordingly named *Hectocotylus argonautae.* Another brilliant investigator, Kölliker (1845), came to a different, but equally erroneous conclusion, insisting that *H. tremoctopodis* and *H. argonautae* are not some "Epizootic Worms", but whole males of the two

"genera on which they are found; inasmuch as they have – a. The same *spermatozoa;* b. Contractile pigment-cells; c. Similarly formed and similarly organized suckers; d. The same remarkable arrangement of the muscular fibres – the *Hectocotylae* in the muscular envelope of the body, the *Cephalopoda* in their arms".

These and other mistaken ideas, including an opinion that the hectocotylus of *Argonauta* is no less than the spermatophore itself, persisted for a number of years, that is, until Vérany and Vogt (1852) and Müller (1853) proved conclusively that the hectocotylus forms part of the modified arm that distinguishes a male from a female. While conceding defeat, Kölliker (1853) had this to say in a *Postscript* to Müller's paper:

"It thus turns out that though in general, I was right in ascribing the hectocotyls to the cephalopods, they are not whole animals, but obviously, only very peculiar parts thereof, in their great independence of organization and appearances in life closely reminiscent of the independent animals".

Even Müller (1853), however, has only partly grasped the exact role of the hectocotylus. He remained convinced that "the hectocotyls are designed to detach themselves from the rest of the body before taking refuge in the females". It was Steenstrup (1856) who showed conclusively that the detachment of a hectocotylus occurs rarely, and that while copulating, most male cephalopods preserve the hectocotylized arm in an intact form. As "hectocotylisation" he defined the type of extensive structural modification that affects the third right arm of *Octopus* and *Eledone* (Octopoda), and either the first left arm (*Rossia, Sepiola*) or the fourth left arm (*Sepia, Sepioteuthis, Loligo, Loliolus*) in Decapoda.

For some time thereafter, the hectocotylus continued to be designated as penis, but in the end most investigators agreed that the name of penis should be reserved for the short terminal organ through which the spermatophore leaves the male reproductive tract, before it is seized upon by the hectocotylized arm.

3.3 Nautiloidea: *Nautilus pompilius* and *Nautilus macromphalus*

Considering their lengthy fossil record, it is not altogether surprising that the nautiloids (Tetrabranchia) should differ from Dibranchia in gross morphology and function of the male reproductive organs and spermatophores (reviewed by Haven 1977a).

Nautilus pompilius. Though in possession of a single, oval testis, similar to that of Dibranchia, the pearly nautilus features male accessory structures of a relatively simple nature. The testis of *Nautilus pompilius* which, according to Griffin (1902), is 41 mm long, 36 mm wide, and 24 mm thick, opens freely into the coelom, but close to the aperture of the vas deferens, of which one portion, glandular in character, and commonly called seminal vesicle, provides the main site for the formation of spermatophores. Here the spermatophore is already loosely coiled, but the coils become much tighter and more regular after the spermatophore has passed from the seminal vesicle into the spermatophoric sac. Inside the sac runs a septum over the rim of which the spermatophore is draped in a U-shaped manner.

Anteriorly the spermatophoric sac extends into a tubular "penis," about 5 mm long, the tip of which is protruding into the mantle cavity. When taken from the lumen of the penis, the spermatophore does not appear to be firmly encapsulated, but upon extrusion, at the place of storage above the male buccal cone, it is found to be surrounded by a spherical capsule of chitinous material about 13 mm in diameter. An empty capsule is left behind over the buccal cone at the time when the unencapsulated spermatophore is transferred to the female and becomes attached below her ventral cirri (Haven 1977a, b; Willey 1902).

In addition to the internal organs of reproduction, there are several secondary sexual organs in the male. Two of them, both located in the oral region, are the spadix (on one side of the buccal mass) and the antispadix (on the opposite side) (Griffin 1902). Most probably both these organs, each composed of four modified tentacles, participate in some manner, first in encapsulation of the spermatophore, and next in shedding of the capsule.

Nautilus macromphalus. The growth cycle and copulation in *Nautilus macromphalus* have been studied by Martin et al. (1978) in New Caledonia. In the winter months of June to September, the captive animals were observed to copulate even while being transported in the surging water of plastic containers, and their copulation was also frequently witnessed in the aquarium tank at Noumea. Cephalopods generally are believed to die after they have ceased reproducing, but this has not been assumed hitherto to happen in nautiloids. However, Martin et al. (1978) found that the females did not recover their weight after the last period of egg-laying and often died spontaneously. Males, too, have been observed to cease feeding, and to die spontaneously.

As in *N. pompilius,* the spermatophore of *N. macromphalus* is of a characteristically coiled appearance (Fig. 7).

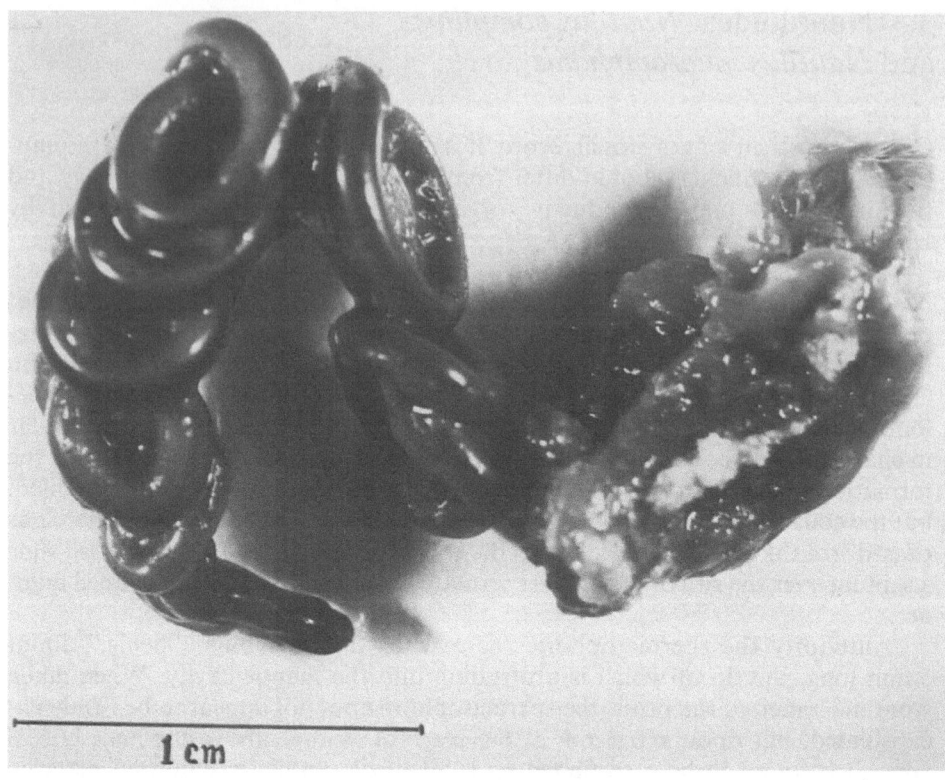

Fig. 7. Spermatophore of *Nautilus macromphalus.* (Courtesy of Dr. Arthur Martin)

3.4 Decapoda: *Loligo pealii, Rossia macrosoma, Rossia pacifica, Sepiola rondeletii,* and *Sepia officinalis*

A number of distinguished investigators, including Brock (1879) and Racovitza (1894 b, c) in the last quarter of the XIXth century, and Blancquaert (1925), Drew (1911, 1919 a), Marchand (1907, 1913), and Williams (1910) early in the present century, deserve the credit for laying down sound foundations for the present-day knowledge of reproductive processes in Decapoda. The common Atlantic (American) squid *Loligo pealii,* especially, provided over many years much experimental material for research on the formation, structure, and transfer of decapodan spermatophores.

As mentioned earlier (Chap. 3.2), the male squid transfers the spermatophores to the female during copulation in either a side-by-side or a head-to-head position, and deposits them either near the oviduct or in the buccal pouch; during the mating season, the buccal pouches of females are invariably filled with spermatophores. Other decapods differ from the common squid in patterns of sexual behaviour and copulation, but in all of them, the males transfer the spermatophores

30

to the females directly. An account of the different variants has been given by Arnold and Williams-Arnold (1977); apart from *L. pealii*, it includes *L. vulgaris, L. opalescens, Sepia officinalis, Sepioteuthis sepioidea, Euprymna scolopes, Idiosepius pygmaeus paradoxus,* and *Todarodes sloani pacificus.* Verco and Cotton (1931) made a special study of the spermatophore and spermatophoric reaction in the common South Australian squid, *Sepioteuthis australis.* Fields (1965) has given special attention to *L. opalescens.* Spermatophores have also been closely observed in *L. reynaudii* (Badenhorst 1974) and *Ilex illecerebrosus* (Durward et al. 1980). In most squids that have been investigated, the spermatophores are quite small, and even in the giant squid, *Architeuthis physeteris,* they are less than 10 cm long.

Loligo pealii. Gross morphology of the male genital system in the Calamary, first explored by Needham (1745), was later described in great detail by Williams (1910) in his important treatise on *The anatomy of the common squid, Loligo pealii Lesueur.* Like that of the common octopus (Chap. 3.2), the male genital system of the common squid includes a single testis and vas deferens; a complex, compartmentalized spermatophoric gland where the spermatozoa become encapsulated by secretions into spermatophores; a spacious spermatophoric (Needham's) sac, providing storage place for the spermatophores; and, connected to the sac by a duct, the muscular "penis," from which the spermatophores are taken by the male during copulation for transfer to the female.

Needham (1745) who discovered the "milt-vessels" in the Calamary, was also the first to describe in detail their structure (Chap. 1), but more information about the formation, structure, and ejaculation of the common squid's spermatophores became available later, chiefly as a result of an extensive study by Drew (1911, 1919a).

In contrast with morphological studies, there have been relatively few attempts to determine either the chemical composition or the metabolism of decapodan spermatophores. In her study of spermatophores in two decapods, *Loligo* and *Sepia,* and two octopods, *Eledone* and *Octopus,* Hamon (1939a, b) used only qualitative colour tests to demonstrate the presence of polysaccharides and certain protein-bound amino acids (phenylanine, tyrosine, cysteine, arginine, and histidine). Nixon (1966), using a microbiological assay method, determined inositol (free and combined) in the spermatophores of *Sepia officinalis;* the concentrations of total meso-inositol (mg/100 g) ranged from 2 to 5.4, as compared with 53.1 and 62.7 in *Octopus vulgaris,* and 56.7 in *O. macropus.* In our investigation of the spermatophores of *Loligo pealii* (Austin et al. 1964), we employed histochemical methods for studying the distribution of mucopolysaccharides and mucoproteins, and both manometric and biochemical methods for defining the extent to which respiration and glycolysis operate in the spermatozoa. Included in our study were observations on the fine structure of squid spermatophores and a photographic record of the sequence of events which accompany the spermatophoric reaction; the photographs were taken in quick succession at a shutter speed of 1/500 s. The main results of this study are described below.

A diagrammatic representation of the squid spermatophore has been already given in Fig. 2 (Chap. 1), to illustrate the relationships between the cap thread and

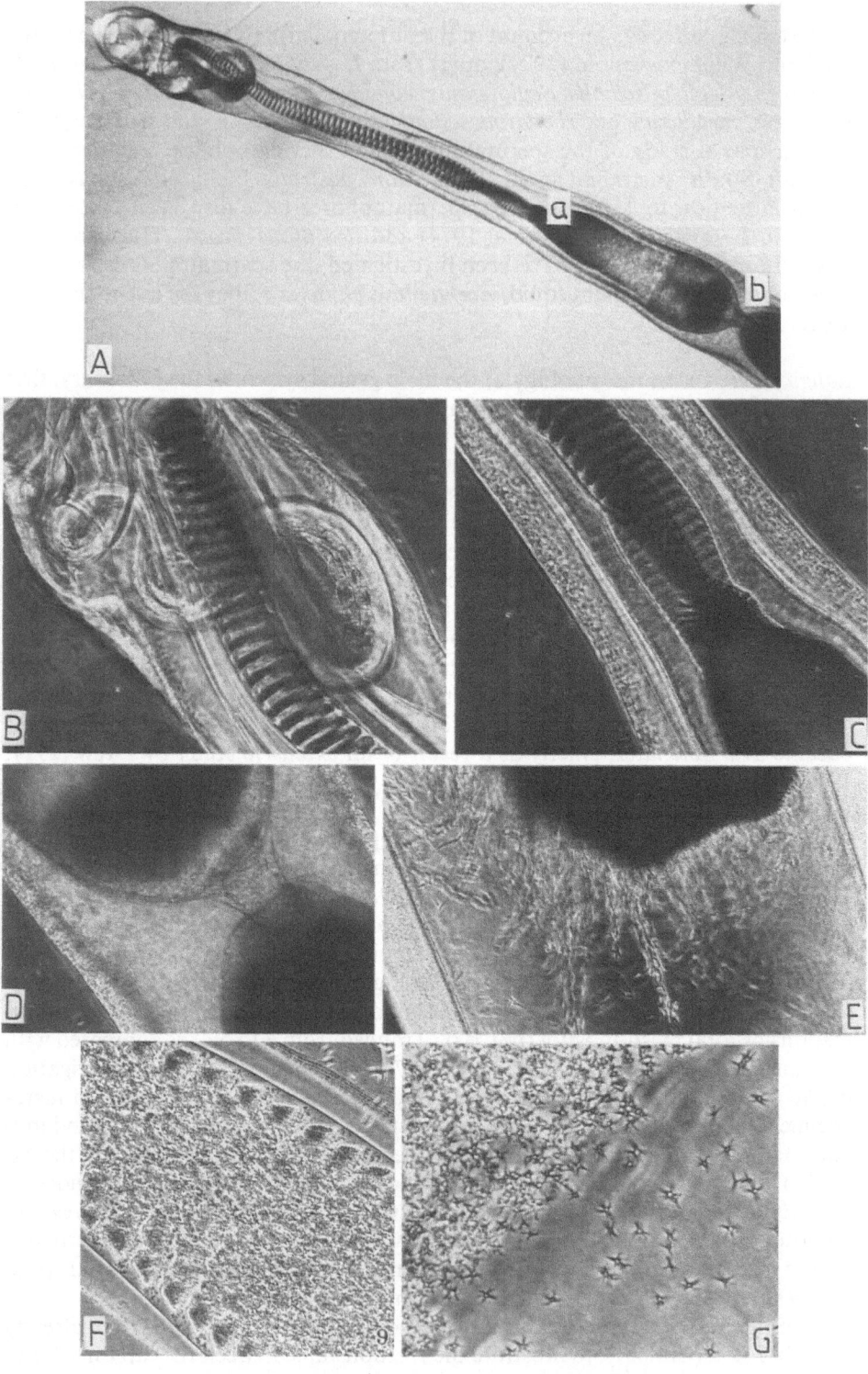

32

the main body in which the spermatophoric semen and ejaculatory apparatus are enclosed.

In a mature squid, 20–30 cm long, the spermatophoric sac holds about 100 1-cm-long spermatophores; 3/4 of that length is taken up by semen and the rest mainly by the ejaculatory apparatus. In the semen inside each spermatophore there are 7–10 million densely packed spermatozoa, which means that the total sperm content of the spermatophores enclosed within the spermatophoric sac of a single male must represent at least 700 million spermatozoa, an impressive number if compared with the total sperm content of a human ejaculate which hardly ever exceeds 300 million spermatozoa.

Structural details of several distinct regions of the spermatophore are illustrated in Fig. 8. The two chief portions of the ejaculatory apparatus, that is, the ejaculatory tube in the front and the bulbous cement body behind it, can be seen in Fig. 8 A; the anterior pole of the cement body is continuous with the ejaculatory tube, and the posterior pole projects into the sperm mass. The long straight portion of the ejaculatory tube carries a distinctive spiral filament (Fig. 8 A and C), which extends into the coiled position at the anterior end of the spermatophore. On closer examination, the spiral filament can be seen to consist of particulate matter embedded in a gelatinous matrix (Fig. 8 F), and at a still higher magnification, these particles present a remarkable stellate, crystalline formation (Fig. 8 G). Much of the space behind the cement body is occupied by the sperm mass, except the pocket near the posterior end of the spermatophor (Fig. 8 E), which is filled mainly with the viscous spermatophorie plasma and contains only a few projecting streamers of spermatozoa.

Of the three tunics, only the outer one surrounds the body of the spermatophore throughout; the arrangement of the two other tunics differs between regions of the spermatophore and is particularly complex around the ejaculatory apparatus (Fig. 9).

Spermatophores removed with the aid of fine forceps from the spermatophoric sac of a recently killed squid are prone to undergo within seconds an explosive spermatophoric reaction when transferred to seawater or sometimes even in mid air. The aerial reaction may well be due to an unintentional slight traction on the cap thread. Certainly, an intentional tug on the thread causes an immediate break in the outer tunic, and triggers the reaction. As shown by the photographic record of the spermatophoric reaction in vitro (Fig. 10), the rupture of the outer tunic is followed by the extrusion of the ejaculatory apparatus, that is, a rapid eversion of both the cement body and the sperm mass, through the lumen of the spiral filament. As a result, the spiral filament is found outside the tube which by

---

Fig. 8 A–G. Parts of the spermatophore in the squid, *Loligo pealii* (for the diagram of the whole spermatophore, see Fig. 2). *A* Ejaculatory apparatus; (*a*) junction between the ejaculatory tube (on the left) and the bulbous cement body (on the right); (*b*) junction between the cement body and the anterior end-portion of the sperm mass. *B* and *C* Finer details of the ejaculatory apparatus and the region of the juncture with the cement body. *D* Junction of cement body (*left*) and anterior end of the sperm mass. *E* The pocket at the posterior end of the spermatophore, filled with spermatophoric plasma into which are projecting thin streamers of spermatozoa (in Fig. 2 this region is marked *c*). *F* Portion of the ejaculatory tube. *G* Detail from the edge of the region shown in *F*. [Magnifications: *A* ×45 (direct illumination); *B–F* × 185 (phase contrast); *G* ×938 (phase contrast)]. (Austin et al. 1964)

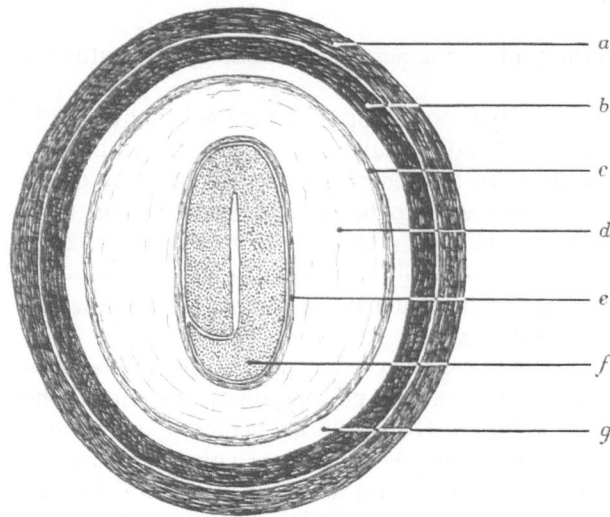

Fig. 9. Diagram of the appearance in cross-section of the ejaculatory apparatus of the squid spermatophore. *a* Outer tunic; *b* middle tunic; *c* a layer representing Drew's (1919 a) inner tunic and outer membrane combined; *d* middle membrane; *e* inner membrane; *f* spiral filament; *g* fluid space. (Austin et al. 1964)

now is fully extended and about five times wider than originally (Fig. 10 E). After all the contents had been evacuated, the tube contracts (Fig. 10 F), and in the end, the empty spermatophore can be seen to carry the casing of the former cement body as the terminal portion of the tube (Fig. 10 G).

In vivo, that is, after a bunch of spermatophores had been withdrawn from the penis and placed in the female, the spermatophoric reaction probably follows the same course as in vitro, but what causes the outer tunic to rupture is not clear. It could be that the cap thread is pulled mechanically. On the other hand, the extrusion of the sperm mass might be triggered in some other way; in the aforementioned squid *Ilex illecerebrosus,* the jelly secreted by the nidamental glands is believed to provide the trigger for sperm release from implanted spermatophores (Durward et al. 1980).

Unlike the spermatozoa recovered from a fresh spermatophore, those that had been set free by the spermatophoric reaction in vitro (Fig. 10 H) are intensely motile. A squid spermatozoon is about 50 µm long (that is, of roughly the same length as a human spermatozoon). It has a slim, slightly curved head narrowing towards the anterior extremity onto which is attached the acrosome; the midpiece appended to the head is thin and long. Respiration and glycolysis were determined by using sperm suspensions diluted with seawater so as to give a final concentration of $0.25–0.5 \times 10^9$ spermatozoa/ml (Austin et al. 1964). Two main findings emerged from the metabolic study, (1) squid spermatozoa are endowed with marked respiratory and glycolytic activity which is related to their motility, and (2) the viscous spermatophoric plasma (obtained by centrifugation of whole spermatophoric semen) contains a factor capable of suppressing both sperm motility and metabolism. At 20°–25 °C, squid spermatozoa consume oxygen at a rate of $10–20 \,\mu l \; O_2/10^8$ sperm/1 h, which is similar to that of mammalian semen but as one can see from Fig. 11, the oxygen uptake follows a linear course for a short time only, unless glucose, fructose or pyruvate are added. Both hexoses are metabolized by squid spermatozoa to lactic acid at a rate of $0.05–0.4$ mg/10^9 sperm/

34

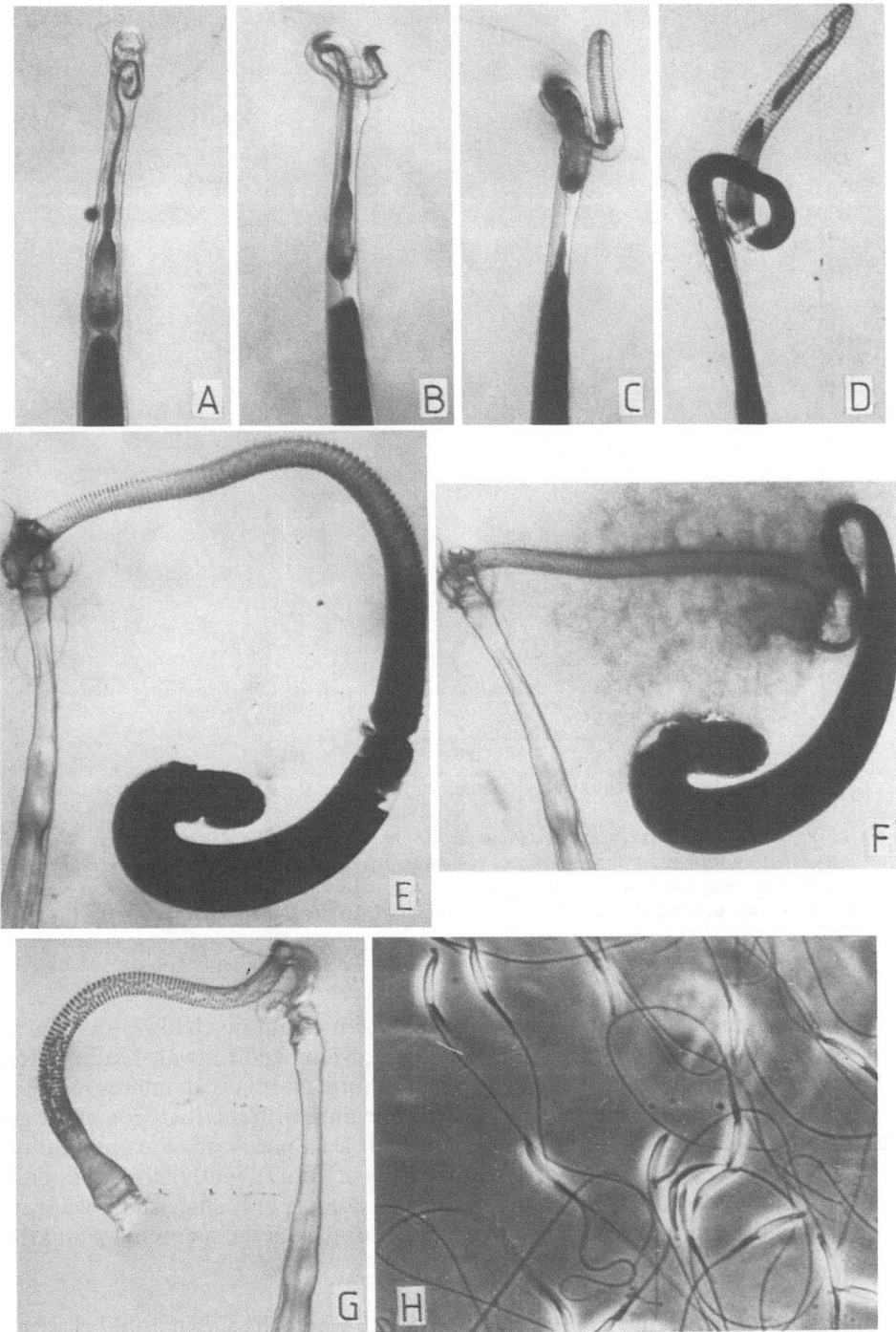

Fig. 10 A–H. The spermatophoric (ejaculatory) reaction and the spermatozoa of the squid *Loligo pealii.*
A–D The rapid initial part of the spermatophoric reaction; the photographs were taken at quick succession, at a shutter speed of 1/500 s. *E* The spiral filament outside of the ejaculatory tube. *F* The contracted tube at the end of the spermatophoric reaction, with contents outside. *G* Anterior part of the empty spermatophore. *H* Spermatozoa which have been ejected during the spermatophoric reaction. [Magnifications: *A–G* × 19 (direct illumination). *H* × 1500 (phase contrast)]. (Austin et al. 1964)

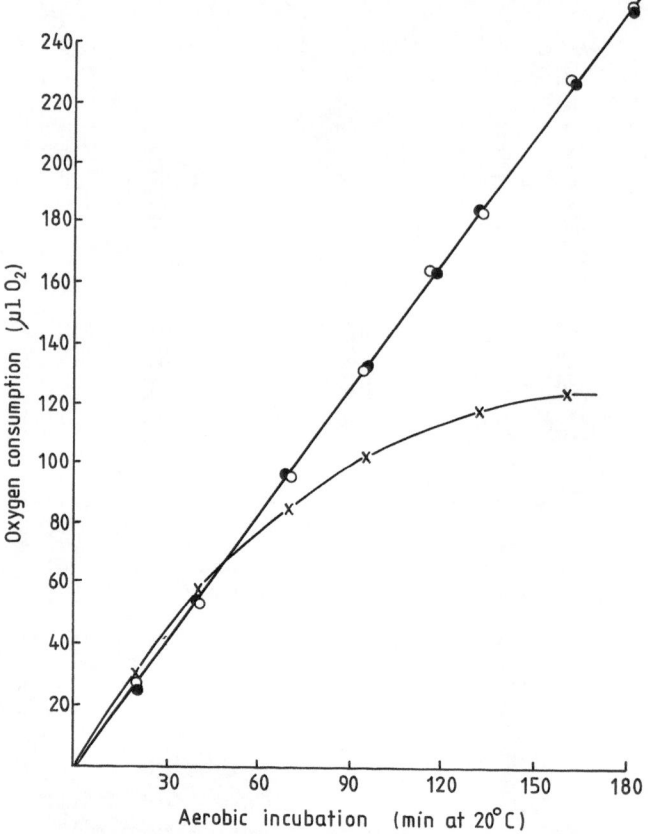

Fig. 11. Respiration of spermatozoa of *Loligo pealii.* The oxygen uptake was measured in Warburg manometers filled with air, in the presence of potassium hydroxide, at 20 °C. Each manometer vessel contained 2 ml suspension of 0.8×10^9 spermatozoa in seawater. The additions were as follows: ×—×, none; ●—●, 5 mg glucose; ○—○, 5 mg sodium pyruvate. (Austin et al. 1964)

1 h, which is similar to that in boar or stallion semen, but much lower than in either bovine or human semen (Mann 1948, 1964; Mann and Lutwak-Mann 1981). The inhibition of sperm motility by spermatophoric plasma is an interesting phenomenon to which we shall return later when dealing with spermatozoa of the giant octopus (Chap. 3.6). It may well account, at least partly, for the observation that inside the spermatophore the spermatozoa are only slightly motile, if at all: in intact spermatophores placed under the microscope, only sluggish movements are seen in the streamers of spermatozoa projecting from the sperm mass into the posterior region of the spermatophore.

Rossia macrosoma. Unlike in the common squid, where the spermatophoric reaction is completed within seconds, in other decapods it takes much longer. Furthermore, in some of them, the spermatozoa are not released immediately after

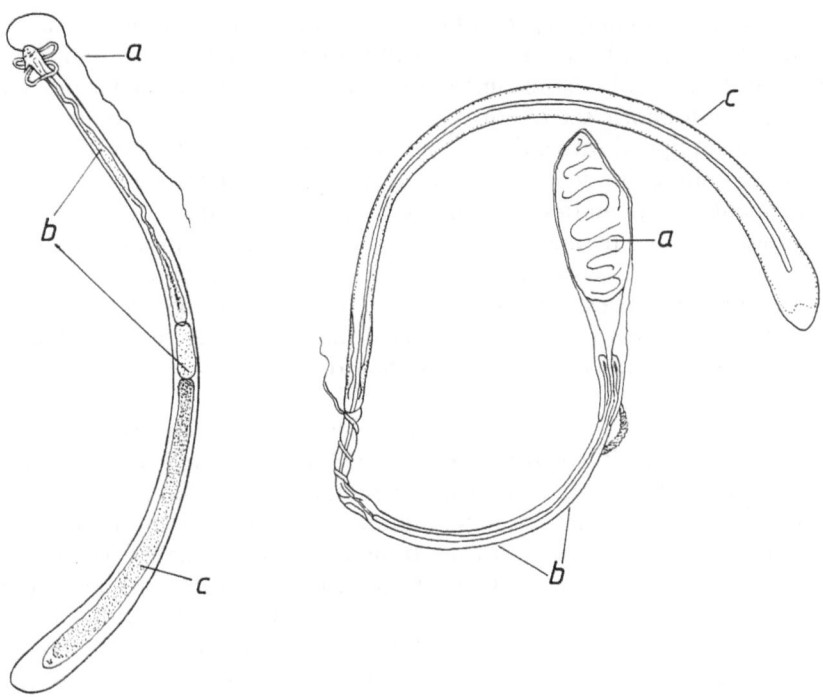

Fig. 12. Spermatophore of *Rossia macrosoma. Left* the fresh spermatophore before the onset of sper-matophoric reaction; *a* cap thread; *b* ejaculatory apparatus; *c* the primary sperm container; *right* spermatophore at the end of the spermatophoric reaction; *a* the secondary sperm container ("sper-matophoric bladder"); *b* ejaculatory apparatus; *c* the everted capsule. (Based on Racovitza 1894b)

the evagination of the spermatophore has taken place, but the everted portion is transformed first into a bulbous, bladder-like structure which does not rupture until much later.

Racovitza (1894 b), the first to describe the formation of this type of structure in the spermatophore of *Rossia macrosoma,* named it reservoire spermatique de second ordre, so as to differentiate from the primary sperm container in which the sperm mass is enclosed within the body of an intact spermatophore. In Fig. 12 are depicted side-by-side two spermatophores of *R. macrosoma,* one freshly re-moved from the spermatophoric sac, and the other after it has reached the stage of the terminal enlargement during the spermatophoric reaction. Similar bulbous structures have later been described in spermatophores of other cephalopods: by Marchand (1907, 1913) in *Eledone moschata* and *Octopus macropus,* as *sekun-däre Spermabehälter;* by Fort (1937) in *Eledone cirrhosa* as *spermatange;* by Mann et al. (1966,1970) in the giant octopus of the North Pacific, *Octopus dofleini mar-tini,* as *spermatophoric bladder* (on account of the distinct bladder-like appearance and elasticity: Chap. 3.6).

The publications of Racovitza (1894 a–c), based on his extensive studies at Banyuls and Roscoff, contain a wealth of information, not merely about spermatophores but on many aspects of cephalopodan reproduction, such as copulatory behaviour, role of hectocotylus in sperm transfer, and fertilization. According to him, at the peak of the breeding season (in Banyuls: late August and September), the spermatophoric sac of *Rossia* contains about 60 spermatophores (each 2 cm long and equipped with a thread and three tunics), and it takes 90 s for a spermatophore undergoing spermatophoric reaction to reach the stage of the bulbous enlargement, but thereafter little change takes place until 2–3 h later, when the bulb ruptures and the spermatozoa begin to ooze out. When looking through the wall of an oviduct in dissected females, he could actually spot there some of the white bulbous bodies. They were embeddied in the region of the oviductal opening, and connected to open tubes (derived from internal tunics of the spermatophores) providing an escape route for the spermatozoa.

Rossia pacifica. Spermatophores have also been observed in other kinds of *Rossia,* including *R. australis* (Cotton 1938) and *R. pacifica* (Fields and Thompson 1976). In the latter, the ridged area in the wall near the opening of the oviduct provides the normal site of implantation for the spermatophoric bladders after the spermatophores had first been transferred by the male's hectocotylized (left dorsal) arm to the pallial cavity of the female. The intact spermatophore of a mature male is a slender, gently curved cylinder, 9–10 mm long, with rounded ends. A complex ejaculatory apparatus occupies its anterior (oral) part. The cylindrical sperm mass in the posterior part consists of spermatozoa embedded within a mucilaginous matrix which is enclosed by a delicate membrane. The spermatozoa have a complex and highly organized structure, including an acrosome containing membrane-bound vacuoles, an elongated nucleus, mitochondria lying in a separate, spur-like appendage, and a tail. They contain glycogen mainly in the mitochondrial spur near the junction with the nucleus.

Sepiola rondeletii. *Sepiola,* whose reproductive pattern in some ways resembles that of *Rossia,* is another decapodan included by Racoviza (1894 b, c) in his studies. As a matter of fact, the occurrence of spermatophores in "La Sépiole de Rondelet" was already known to Duvernoy (1853), and the reproductive organs of this animal (the male is much smaller than the female) have been the subject of an early investigation by Brock (1879). Copulation in *S. rondeletii* lasts about 8 min, during which time the spermatophores are transferred to the female by the hectocotylized arm. The spermatophoric reaction has been described by Marchand (1913), whose publication also includes two figures illustrating diagrammatically special features of the spermatophore's structure. Weill's (1927) publication contains much information about the structure and function of spermatophores in *Sepiola atlantica.*

Sepia officinalis. Cuttlefish, *Sepia,* the first of all animals in which the spermatophores were discovered and described by Swammerdam in the *Biblia Naturae* (Chap. 1), has long been a favoured object of experimental studies. Among the early investigators, Denys-Montfort (1802) deserves the credit for carrying out the first detailed microscopic investigation of the «machine ou fibrille spermatique de la sèche mâle». *Sepia officinalis* was one of the five cephalopods used by

Milne-Edwards (1842) in his collaborative study with Peters (Chap. 1), in which the name "spermatophore" was first introduced. It was also one of the five species used by Brock (1879) in the extensive anatomical study of male reproductive organs in Cephalopoda, which included the observation that even at the onset of male puberty, the spermatophoric sac of cuttlefish is already filled with spermatophores, but unlike in a fully mature male, these spermatophores are still empty. Once more, one is reminded of an analogous situation encountered in mammals, namely, that the onset of secretory activity of male accessory organs precedes considerably in time the production of spermatozoa in the testis (Mann 1964, 1967; Mann and Lutwak-Mann 1981).

Blancquaert's (1925) extensive study of the development of spermatophores in cephalopod decapodes was primarily based on cuttlefish. Tinbergen (1939) described in detail the copulatory behaviour in *S. officinalis* as being marked by an extraordinarily aggressive attitude of the male towards the female; eventually the male grips the female by the head and passes the spermatophores into her buccal pouch with the hectocotylus. More information about spermatophores in *Sepia* and several other decapods is contained in the review by Arnold and Williams-Arnold (1977).

3.5 Small Octopoda: *Octopus vulgaris, Octopus hummelincki, Octopus bimaculatus,* Other Small Octopuses, *Eledone moschata,* and *Eledone cirrhosa*

Among the smallest octopods is the colourful ringed octopus *Octopus maculosus* found along the Australian beaches, which hardly ever exceeds 10 cm in total body length. No-one seems to have looked at its spermatophores, not surprisingly perhaps in view of this animal's reputation as one of the deadliest creatures in the world, said to be carrying enough poison to kill 10 men. By contrast, a great deal is known about spermatophores in many other octopods, both small, such as *Octopus vulgaris,* and large, such as *Octopus dofleini.* In most of them the male and female look alike, but there are exceptions, the male *Argonauta* for instance is considerably smaller than the female. Another distinguishing feature of *Argonauta* and certain other dimorphic pelagic octopods (*Tremoctopus, Ocythöe*) is the behaviour of hectocotylus. After the third right or third left arm of the male has been charged with sperm, it breaks off to remain in the mantle cavity of the female, a fact that at one time led to erroneous ideas that hectocotylus is a parasite, an independent organ, or a whole male (Chap. 3.2).

Reproduction in Octopoda, including the properties of their spermatophores, has been the subject of a review by Wells and Wells (1977). What follows below is meant to provide a few specific examples of structure and function of spermatophores in the small octopods. The giant octopus will be dealt with separately later.

Octopus vulgaris. The main features of the male reproductive tract in the common octopus have been outlined already (Chap. 3.2). Spermatophores appear in the spermatophoric sacs of males early in life, usually within a few months of hatching, at the time when the animals weigh 200–300 g. This is in contrast to females, which mature sexually after reaching the relatively large size of 1500 g (Wells and Wells 1969, 1977). Off the northwest coast of Africa there are two spawning seasons of common octopus, May–June and September; most of the males larger than 50 cm in total length have spermatophores, but amongst females only those that are larger than 70 cm in total length are mature (Hatanaka 1979).

Light is one of the factors that can advance the onset of male puberty. In the Mediterranean Sea, under natural conditions, males weighing less than 150 g never display signs of sexual maturity, but under laboratory conditions, young males from 70 g onwards, illuminated 16 h daily, show clear signs of gonadal development, and similarly treated males of over 150 g have a larger number of spermatophores than individuals illuminated only 8 h daily (Laubier-Bonichon and Mangold 1975). If the optic glands are removed from a mature male, the testis declines in weight and spermatophore production ceases altogether after 5–6 weeks. On the other hand, enlargement of the optic glands, brought about by cutting the nerve supply to the glands, causes the testis to enlarge (Wells and Wells 1972 a, b). Normally, a male weighing 500–700 g carries about 50 spermatophores in the Needham's sac, but after the optic glands have enlarged, the count is liable to rise to 100 or more within 3–4 weeks. Unlike optic gland removal, castration by removal of the testis does not appear to have a dramatic effect on the state of the male tract, nor does it produce any significant change in male copulatory behaviour, but of course, as a result of orchidectomy, the supply of spermatozoa for the spermatophores ceases immediately.

The spermatophores of the common octopus are of the same gross-morphological appearance as those of the common squid, but they are larger (Philippi 1839), more variable in size, and require more time to complete the spermatophoric reaction. Interesting in this connection are the measurements made by Drew (1919 b) in *Octopus americana* (sold by Jamaican fishermen as sea-cats). According to this author, as many as 79 spermatophores can be present in the spermatophoric sac of a single male, but the spermatophores "differ in size with the size of the individual from which they are taken. Large ones measure as much as 50 mm in length, small ones may not be more than two-thirds as long"; and as regards the spermatophoric reaction, "the whole process is usually complete in less than 10 s in the squid and may occupy from 1½ to 3 min in *Octopus*".

Copulation and transfer of spermatophores in the common octopus, extensively studied already by Racoviza (1894 a), has ever since attracted the attention of careful and patient observers. It may take a whole hour or even longer before copulation is completed, and it seems that during that period successive spermatophores can be passed to the female (Wells and Wells 1972 b), but according to Wodinsky (1973) only a few spermatophores are actually transferred to the female during each single copulation.

Unlike Kollmann (1876), who insisted that copulation in the common octopus constitutes nothing less than a desperate struggle for life and death (grimmiger Kampf auf Leben und Tod), Racovitza's opinion was that no such struggle takes place («Il n'y a pas de lutte, à proprement parler»), on the contrary, the male remains quiet while moving the tip of his hectocotylized third right arm into the mantle cavity of the female, and inserting spermatophores into the canal which runs along that arm. Much, it seems, depends on the attitude of the female. According to Orelli (1962), the male will act violently if the female refuses his approaches, but "if the female remains quiet, normal copulation will take place, as described by Racovitza". The more recent account of sexual displays and mating in *Octopus vulgaris*, and also *O. cyanea*, by Wells and Wells (1972 b), is essentially in agreement with Racovitza's description. According to Wells (1983), who recorded the heartbeat of a copulating male octopus, "the male's heart missed a beat when the female was dumped into his tank, but he lost no time in grabbing her and inserting his hectocotylus into her mantle cavity", and furthermore, "his heart also missed a beat at each ejaculation. But there is no sign of a racing pulse; octopuses take these things very calmy". At the moment of inserting the spermatophore into the hectocotylus, the male inflates the mantle cavity, places the tip of his funnel over the edge of the interbranchial web between the third and fourth right arms, and after a short pause withdraws the funnel from the web. The withdrawal is accompanied by an "explosive exhalation" from the mantle cavity through the funnel, and at the same time a wave of contractions passes along the edge of the third arm where the canal is located, triggering a series of "pumping" movements. The penis is now brought forward through the funnel to expel the spermatophore which can then be moved by peristaltic contractions to the canal along the arm and to the tip of the hectocotylus. Finally, the spermatophore is placed by the tip of the hectocotylus in the oviduct. There the spermatophore ruptures and the spermatozoa are carried by peristaltic contractions of the oviduct in the direction of the spermathecae in the oviducal glands. In the spermatheca, the spermatozoa become attached to the wall by means of their acrosomes and there they can remain for months if necessary, that is until the female reaches full maturity. When the mature eggs have been freed from their follicular sheath they descend along the oviduct; the spermatozoa must be released from the spermatheca into the cavity of the oviducal gland for fertilization to occur (Froesch and Marthy 1975).

Octopus hummelincki. The appearance of spermatophores early in life of the male is by no means a phenomenon restricted to the common octopus. Male *Octopus hummelincki* will copulate and pass spermatophores which already contain spermatozoa, at the time when its body weighs as little as 20 g. Later in life the male will grow to about 50–60% of the female's maximum weight (which can be as much as 300 g). After achieving sexual maturity, as expressed in the presence of spermatophores, the male lives for another 7–9 months. Ageing males begin to eat less, gradually lose weight, copulate erratically, and die. In the females, removal of both optic glands after spawning results in cessation of broodiness, resumption of feeding, increased growth, and greatly expanded life-span. Like females, the males without optic glands surpass the species-typical maximum body weight of normal males, and they also live longer (Wodinsky 1977).

Octopus bimaculatus. The anatomy and histology of the reproductive system in *O. bimaculoides* have been described by Peterson (1959), whose study included also observations on spermatophores. More recently, the spermatophores and spermatozoa of *O. bimaculatus* have been investigated in detail by Longo and Anderson (1970).

The spermatophores excised from the Needham's sac are sausage-shaped, about 1 cm long and 0.5 mm wide, and they have a body composed of two parts: the sperm mass and the ejaculatory apparatus. The wall consists of three regions: an outer discontinuous layer of loose fibrillar material, a middle compact region composed of densely packed and parallel-oriented fibrils, and an inner layer composed of a flocculent substance. The spermatozoa are embedded within a flocculent material forming a highly coiled cord which is bounded by a fibrous layer. Glycogen particles fill the intracellular space between the plasma membrane and the outer fibres of the sperm tail.

Other Small Octopuses. Apart from those mentioned above, spermatophores have been described in other octopuses of the smaller kind. Marchand (1913) in his study of spermatophores listed 17 cephalopods, among them a number of octopods, including *Octopus macropus.* Wells and Wells (1972b), along with *Octopus vulgaris,* studied in detail the sequence of events of spermatophore transfer in *O. cyanea.*

Eledone moschata and Eledone cirrhosa. The main anatomical features of the male genital system in *Eledone* (the "lesser octopus") have been known since the time of Brock's (1879) study of reproductive organs in cephalopods. Meyer (1911) described the spermatophores of *E. moschata,* but under the name of *Polypus (Octopus) vulgaris.* Among the South-African cephalopods studied by Voss (1962) was *E. thysanophora,* a species that can be distinguished from *E. moschata* by the presence of special "teeth" on the inner wall of the spermatophore tube. As mentioned earlier (this chapter), *Eledone moschata* was the octopod that Marchand (1913) used to demonstrate that, as in *Rossia macrosoma* (Fig. 12), the spermatophore expands during the terminal stage of the spermatophoric reaction into the bulbous bladder-like "secondary sperm container". Fort (1937, 1941), in his subsequent studies which extended over both *Eledone moschata* and *Eledone cirrhosa,* referred to the same kind of bulbous enlargement formed by the evaginated spermatophore, as *spermatange.* In both eledones spermatophores freshly removed from the spermatophoric sac are tube-like in shape. That of *E. cirrhosa* has been described by Fort (1937) as consisting of a 6–7-cm-long tapering tube which encloses within its envelope (l'étui) the transparent ejaculatory apparatus in the anterior part, and the white helically wound sperm mass (le boudin spermatique) in the longer and wider posterior portion. He also showed that after copulation some spermatanges actually reach the ovary. Fort's observations were confirmed and extended by Orelli (1962).

Copulation in *Eledone* takes half an hour or occasionally even longer. The male approaches the female from behind and embraces her with all his arms so that only her head remains visible. He then inserts his third right arm into the opening on the right side of the female's mantle cavity. According to Orelli (1962), in *E. moschata* two spermatophores are passed on to the females every minute, so that, presumably, a total of 60 must reach the female during half an hour. In the ovaries of mated females, he could spot up to six bulbous structures, in size and pear-like shape very similar to those that would have been formed by spermatophores during a spermatophoric reaction in vitro. After the spermatanges have ruptured, he could also see the liberated spermatozoa passing in the form of a "white cloud" into the sperm receptable proper, their transport being probably assisted by the peristaltic movements of the ovary.

3.6 The Giant Octopus of the North Pacific, *Octopus dofleini martini*

A re-examination of the taxonomic status of several types of "giant octopus" led Grace Pickford (1964) to the conclusion that at least two, *Octopus honkongensis* Hoyle (Hoyle 1885, 1886) and *Octopus dofleini* (named after F. Doflein, by Wülker 1910), are in fact independent species, and furthermore, that the second of the two should be recognized as existing in the form of three distinct subspecies: *O. d. dofleini* (Wülker, emend. Sasaki) from Japan, *O. d. apollyon* (Berry) from the Pacific Far North, and *O. d. martini* (new subspecies) from the North Pacific along the coast of Washington State. By naming the third subspecies after Arthur Martin of the University of Washington, she fittingly recognized the outstanding contributions of this investigator to physiology of Cephalopoda in general, that of the North Pacific Giant Octopus in particular.

The occurrence of "enormous" spermatophores in the larger kinds of octopus has been casually commented upon by several investigators. Robson (1929, 1932) in his *Monograph of the Recent Cephalopoda in the Collection of the British Museum* refers to *Benthoctopus ergasticus* as an outstanding example of a cephalopod in which for "quite inexplicable" reasons, "the spermatophore is single and of enormous size." When, on request by Pickford (1964), a specimen of *Octopus honkongensis* at the British Museum was subjected to a reexamination, "remains of two spermatophores were recovered from the penis, and another less severely damaged fragment was taken from Needham's sac", but all of these were not only damaged but relatively short.

From a specimen of *Octopus dofleini* (Wülker) from Japan, Sasaki (1929) recovered 12 spermatophores. Two were lodged in the penis and diverticulum and ten in Needham's organ, all "more or less damaged and opaque from alcohol preservation". Of those from Needham's organ the anterior pieces

> "averaged 598 mm in length, and a single sperm reservoir recovered from this organ in three pieces measured 472 mm; the total length of the spermatophore is therefore about 1072 mm. The better preserved of the two penis spermatophores was about the same length; the anterior piece measures 540 mm and the sperm reservoir 590 mm, totalling 1130 mm".

3.6.1 The Animal

Octopus dofleini martini is a large octopus, averaging about 15 kg in body weight at sexual maturity, which "has proved to be a docile animal, easily anesthetized and with vascular and renal systems superbly available for preparative surgery"

(Martin 1983). In our collaborative study with Drs. Arthur Martin and John Thiersch at the Department of Zoology, University of Washington, Seattle, extending over a period of 16 successive years, the material consisted of 53 animals, most of them mature males, weighing 10–51 kg. The average number of spermatophores recoverable from a spermatophoric sac was seven, and in addition, at least one was invariably present in the terminal spermatophoric duct, but in one male, which suffered from occlusion of the terminal portion of the reproductive tract, we found 32 spermatophores inside the greatly extended spermatophoric sac, all of them about 1 m long (Mann et al. 1970).

Most of the animals were trapped in the Puget Sound–San Juan Archipelago area. The males could easily be distinguished from the females by the appearance of the third right hectocotylized arm, differing from the third left arm by having a smooth, sucker-free tip and a groove, the spermatophore channel, on one side. On most occasions, the animals were brought to Seattle by sea-plane or car, usually in large plastic bags filled with seawater, and on arrival in the laboratory they were immediately transferred to tanks filled with aerated seawater at 8°–10 °C. Such animals survived well for several months. On a few occasions, the animals were transported in bags without water. Provided that they were kept cool, they sustained the 1–2-h journey well, even though deprived of air.

The ability of the giant octopus to withstand oxygen deficiency is remarkable, much in contrast to the behaviour of other cephalopod molluscs (Martin 1961). It has been known for some time that the heart of octopus is extraordinarily resistant to anoxia, and that heart-tissue slices are capable of reverting to a normal rate of respiration after anaerobic incubation for 48 h, during which period they continued to consume their own glycogen (Pritchard et al. 1963). We found that spermatozoa released from the spermatophores and suspended in buffered seawater are capable of remaining motile for several days under strictly anaerobic conditions, consuming all the time their own glycogen and converting it to $D(-)$lactic acid (Mann et al. 1974, 1977; Martin et al. 1976).

3.6.2 Male Gonad and Accessory Organs

In their general appearance the male genital organs of *O. dofleini martini* resemble those described by Winkler and Ashley (1954) in the animal presumed to be *O. apollyon*. Though much bigger and differing in shape from those of *O. vulgaris* (Chap. 3.2), they also conform to the general pattern characteristic of the common octopus (Mann et al. 1966, 1970).

The whole male reproductive tract, except the "terminal organ" or "penis", lies enclosed in a membranous sac, the "genital bag", which occupies a large portion of the posterior region of the mantle cavity.

A genital bag (extirpated from a 23.6-kg male, and weighing 1051 g) is shown in Fig. 13 (top left). By stripping off the external capsule of the bag, it is possible to free the rest of the "terminal spermatophoric duct" and expose at the same time a second, much thinner membrane of the bag, within which lie the testis and male accessory organs. From Fig. 13 (top right) it can be seen that the terminal spermatophoric duct consists of two branches: the longer, loop-shaped "diverticulum"

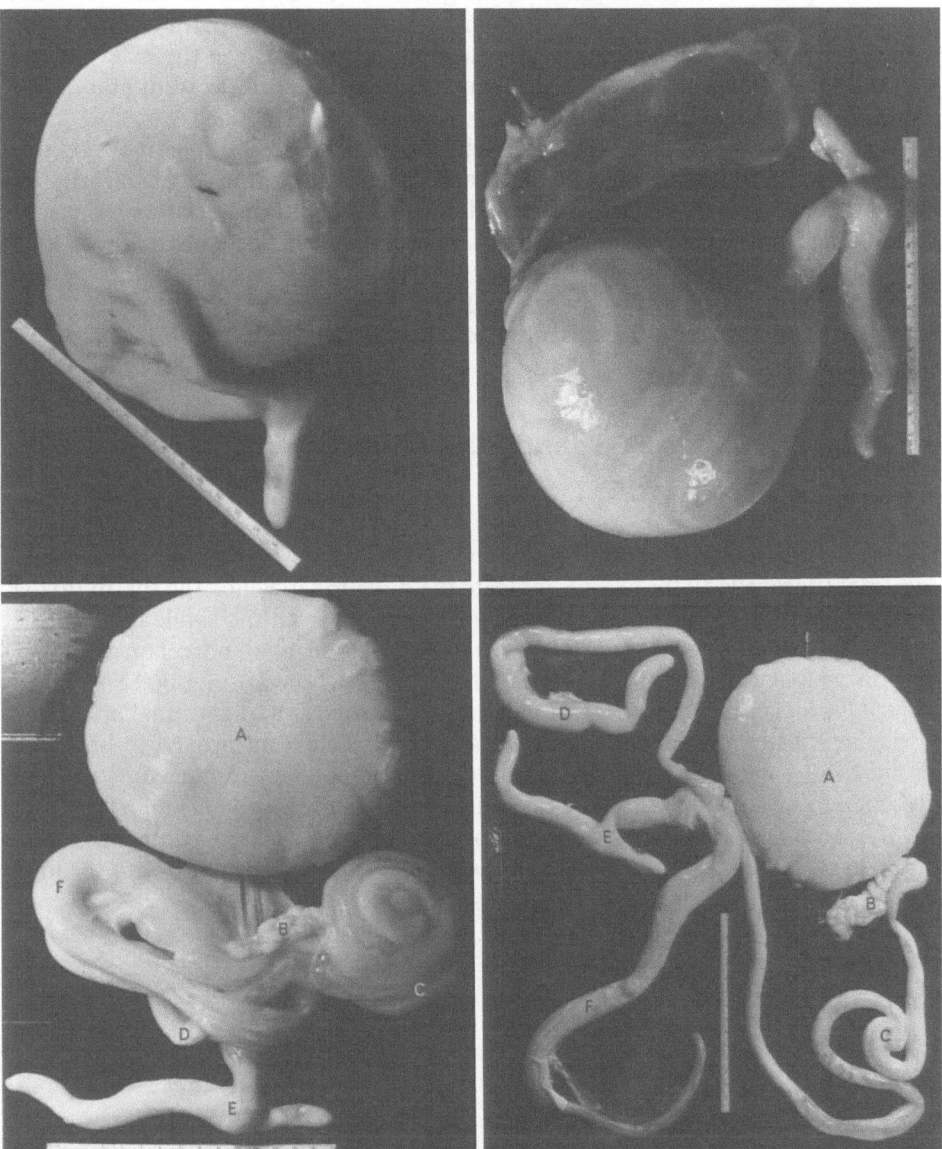

Fig. 13. Male reproductive tract of *Octopus dofleini martini* (scale in cm). *Top left* the genital tract in toto, with the genital bag intact. The diverticulum of the terminal spermatophoric duct can be seen on the left, closely adjacent to the genital bag. The terminal organ ("penis") hangs free and points downwards; its orifice is on the left side; *top right* after stripping the external wall of the genital bag and exposing the second, thinner wall. Visible through the latter are some spermatophores; the one on the extreme left (coiled and broad) is inside the seminal vesicle, those on the right are in the spermatophoric sac. On the extreme right is the terminal spermatophoric duct with its diverticulum pointing downwards, and the terminal organ ("penis") upwards; *bottom left* after stripping the second wall of the genital bag; (A) testis; (B) vas deferens proximale; (C) spermatophoric gland system I or seminal vesicle; (D) spermatophoric gland system II or prostate (most of it underneath the spermatophoric sac); (E) terminal spermatophoric duct (diverticulum on the left, "penis" on the right side); (F) spermatophoric (Needham's) sac; *bottom right* the male reproductive tract completely unravelled; the different parts marked as before (A–F); the spermatophoric sac has been cut open at the distal end to expose one of the spermatophores. (Mann et al. 1970)

occupies a position closely adjacent to the genital bag, while the shorter "terminal organ" or "penis", marked by a side opening through which spermatophores are normally extruded, hangs freely from the genital bag. Between the external and internal membrane of the genital bag one can usually find a certain amount of a clear, non-viscous "genital bag fluid".

By stripping off next the internal membrane of the genital bag, it is possible to obtain a clear view of the testis itself and adjacent to it, the accessory organs, tubular in shape, convoluted and coiled. Their convolutions and coils must be straightened out before one can view all the unravelled parts of the male tract (Fig. 13, bottom right), that is, the vas deferens proximale, spermatophoric gland system I ("seminal vesicle"), spermatophoric gland system II ("prostate"), and the spermatophoric sac ("Needham's sac") connected to the terminal spermatophoric duct.

3.6.3 Formation and Maturation of the Spermatophore in the Male Reproductive Tract

Figure 14 shŏws the interrelationship between various parts of the male reproductive tract (unravelled); the arrows indicate the direction in which the spermatozoa and spermatophores traverse the length of the tract, from the single testis to the terminal spermatophoric duct. The testis itself is a large, oval, porcelain-white organ, in weight occasionally exceeding 1 kg (the one shown in Fig. 13 weighed 587 g). Upon gentle homogenization it yields a suspension of germinal and other testicular cells in a fluid, the "testicular plasma". This fluid proved to be a rich source of D(−)lactate dehydrogenase, the enzyme necessary for the conversion of glycogen to D(−)lactic acid (Martin et al. 1976; Mann et al. 1977).

Upon leaving the testis the spermatozoa as yet not encased into a spermatophore pass first through the membranous, funnel-shaped "ampulla" into the long

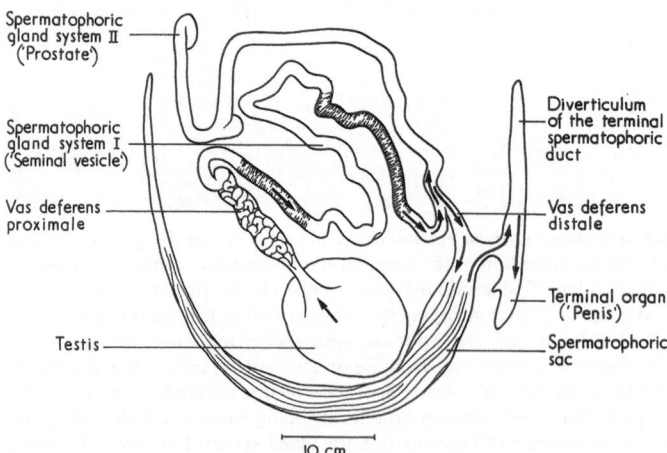

Fig. 14. Diagrammatic outline of the male reproductive tract (unravelled) of *Octopus dofleini martini;* *arrows* indicate the direction in which the spermatophore traverses the tract. (Mann et al. 1970)

46

convoluted "vas deferens proximale" which in gross appearance resembles the mammalian epididymis (Fig. 15 A). On viewing it in a freshly dissected state, one is struck by the occurrence of waves of contractions running from the ampulla onwards. By homogenizing the vas, and subsequent centrifugation, it is possible to collect some fluid, the "vas deferens plasma". Emerging from the vas, the spermatozoa, by now tightly packed together into a "sperm rope", enter the spermatophoric gland system I, tubular in shape, about 1 m long, and marked by a tall secretory epithelium (Fig. 15 B). Here the sperm rope undergoes coiling, the coils become engulfed by glandular secretion, and a loosely fitting membrane is formed around the coils. In addition, a gel column is produced here, which having become attached at one end to the coiled sperm rope, constitutes the core of the future ejaculatory apparatus.

From the spermatophoric gland system I, the spermatophore, as yet grossly immature, but already carrying the gel column at its posterior end, enters the spermatophoric gland system II, also tubular in shape. Here the coils of the sperm rope become more tight and regular, the gel column of the ejaculatory apparatus more rigid, and further additions are formed to the membrane. At his point, the spermatophore goes into reverse, that is, it leaves the spermatophoric gland system II with the ejaculatory apparatus pointing forward. In this formation, the spermatophore enters the next segment of the reproductive tract, the short tubular "vas deferens distale", and having traversed that tube, it passes into the spacious bag-like spermatophoric sac. This sac, when completely unravelled, measures about 1 m in length, and the mature spermatophores which are housed within it are lying parallel to each other, their ejaculatory apparatus pointing towards the blind (distal) end of the sac. When the external wall of the apical portion of the spermatophoric sac is cut carefully (so as to prevent damage to the spermatophores), a small amount of fluid oozes out: the "spermatophoric sac fluid". The spermatophores leave the spermatophoric sac singly and enter the terminal spermatophoric duct with the portions containing the sperm rope.

The formation and maturation of the spermatophore is accompanied by structural and biochemical changes (Mann et al. 1981 b). Among the latter is a transformation of the spermatophoric plasma, that is, of the fluid surrounding the sperm rope from a watery salt solution containing little organic material into a remarkably viscous fluid with low NaCl concentration but a high dry weight (nearly 25 g/100 ml). Table 1 lists the results of Na, K, Cl, Ca, and Mg in the blood, genital-tract fluids and spermatophoric plasma. The most striking difference in the electrolytes pattern of the spermatophoric plasma, between the immature and mature spermatophores, is in the concentration of sodium chloride, which declines precipitously during the final stage of maturation, and is replaced by organic substances, including glycoproteins, glycopeptides, phosphoglycopeptide, glycerylphosphorylcholine, and carnitine (Brooks et al. 1974 b, 1981; Mann 1975; Mann et al. 1970, 1973, 1981 b; Martin et al. 1973). Similar events are associated with the maturation of spermatozoa in the mammalian epididymis, and with the transformation of the dilute testicular plasma, which is rich in salt but poor in organic matter, into the more concentrated and viscous epididymal plasma, wherein the electrolytes have been largely replaced by organic substances (including glycoproteins, glycopeptides, glycerylphosphorylcholine, and carnitine; for review see Mann and Lutwak-Mann 1981).

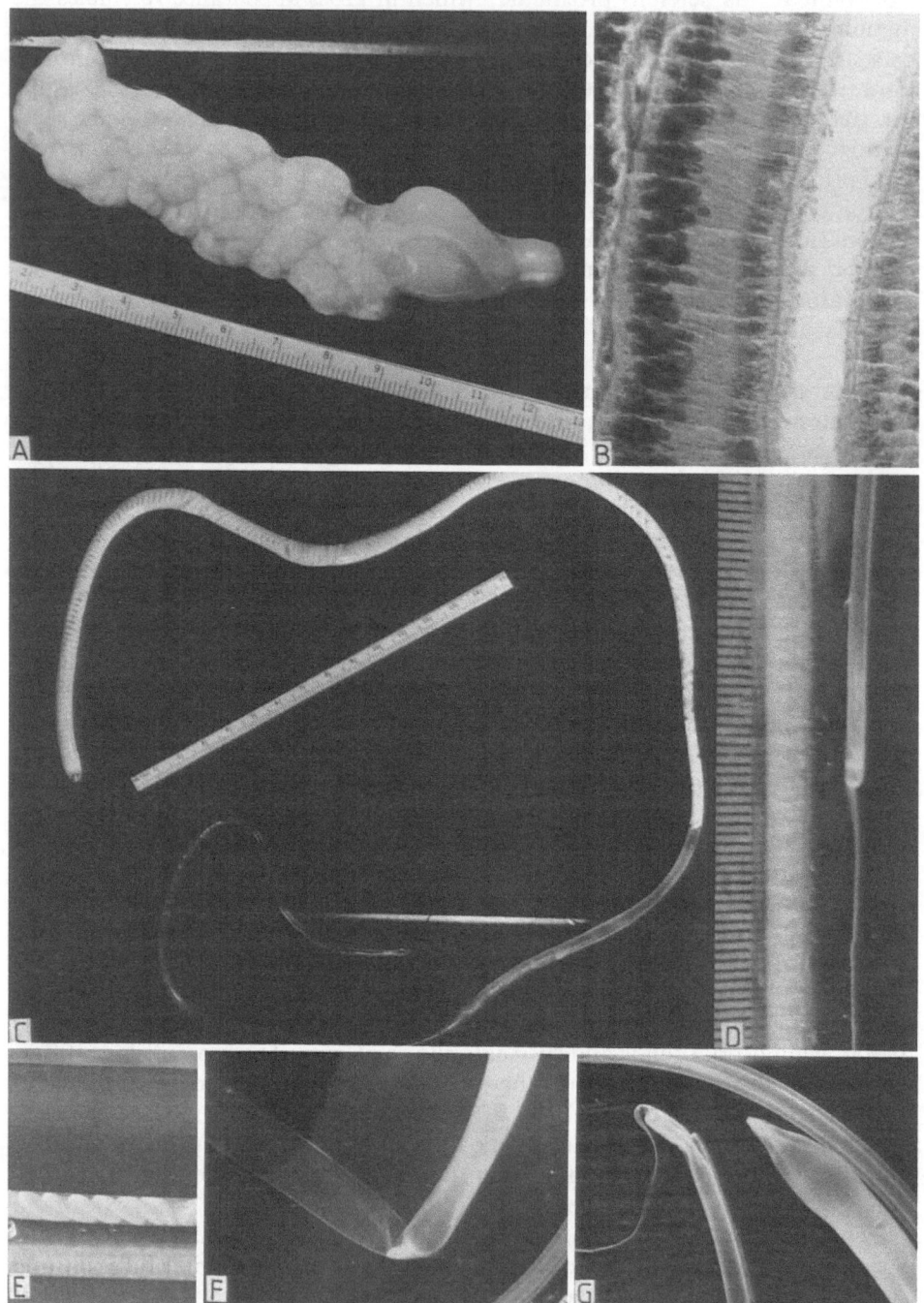

48

Table 1. *Octopus dofleini martini*. Content of sodium, potassium, chloride, calcium and magnesium (in mEq l^{-1}) in the blood, reproductive-tract fluids and spermatophoric plasma of the same octopus. (Mann et al. 1981b)

	Na	K	Cl	Ca	Mg
Blood	392	11	443	10	46
Genital-bag fluid	458	12	472	11	46
Testicular plasma	375	40	462	8	40
Vas deferens plasma	420	18	436	8	40
Spermatophoric sac fluid	474	13	420	8	26
Spermatophoric plasma from:					
spermatophore in seminal vesicle, located close to vas deferens	386	11	470	8	31
spermatophore in seminal vesicle, located close to prostate	380	12	465	8	31
prostatic spermatophore	350	11	450	9	33
mature spermatophore in spermatophoric sac	277	12	260	6	15

3.6.4 The Mature Spermatophore

To extricate the mature spermatophores from the spermatophoric sac, it is necessary to dissect the animal. From the terminal spermatophoric duct, however, where either one or sometimes two mature spermatophores are lodged, it is possible to obtain the spermatophore situated nearest to the orifice of the freely hanging terminal organ, by pulling it out by hand, from the live male. Quite often, to do this one can dispense with anaesthesia, especially in animals that had been used to being handled in the laboratory. Should, however, the male fail to respond to such a manipulation, general anaesthesia has to be induced, best by slowly adding ethanol to the seawater in the tub, until the concentration reaches about 2%. Within 15 min the animal becomes sufficiently relaxed and the spermatophore, occasionally already slightly protruding, can then be pulled out without any difficulty. As soon as a spermatophore had been recovered, the animal must be returned to the tank with fresh seawater, so as to recover from the effects of ethanol. Usually it is possible to repeat the removal of a further spermatophore next day, or occasionally even on the same day. As a matter of fact, it has been noticed that

Fig. 15 A–G. Parts of the male reproductive tract (*A* and *B*), and the mature spermatophore (*C–G*) of *Octopus dofleini martini*. *A* Vas deferens proximale with the bulbous gland (*right*) through which the sperm rope (seen in the centre of the gland) enters the spermatophoric gland system I (not shown). *B* Secretory epithelium of the spermatophoric gland system I (seminal vesicle); [× 200 (haematoxylin stained)]. *C* A mature spermatophore freshly removed from the spermatophoric sac; the thick portion of the body contains the coiled sperm rope; the thinner portion contains the ejaculatory apparatus; the cap is at its end; the cement body is located between the sperm rope and the ejaculatory apparatus. *D* The cap and the filamentous cap-thread. *E* A portion of the sperm rope inside the spermatophore. *F* A spermatophore from which the sperm rope has been removed to expose the transparent outer tunic and the portion of the ejaculatory apparatus with the cement body. *G* Cap and cap-thread (*left*), and the cement body (*right*). (Mann et al. 1970)

in a normally copulating male octopus the second spermatophore moves into a position close to the orifice of the penis within less than an hour after the extrusion of the first spermatophore.

A freshly procured mature spermatophore is shown in Fig. 15 C; this particular spermatophore weighed 24.7 g, and its body was 1.08 m long, and 0.7 cm wide (at the thicker end). The spermatophore's cylindrical body, all of it enclosed within the tough and highly elastic outer tunic, consists of two main parts, each roughly 0.5 m in length, but one much thinner than the other. The thinner half of the spermatophore contains the ejaculatory apparatus, presenting as a golden-yellow-coloured hyaline rod of rubbery consistency. At the point of contact with the sperm rope, the ejaculatory apparatus forms a bulbous enlargement, that is, the cement body (Fig. 15 F and G). In this region there is also present a small amount of a syrupy, amber-coloured and highly viscous "cement liquid". At the thin, distal end of the ejaculatory apparatus, the outer tunic of the spermatophore forms a round cap to which is attached a fine filamentous appendage, the cap thread. This thread extends in two directions. Proximally it runs attached to the outer covering of the spermatophore, and distally it continues as a free filament about 0.5 m long (Fig. 15 D and G).

In the thicker half of the spermatophore, that is, the one that during copulation would leave the "penis" first, is located the spermatophoric semen consisting of the sperm rope and 5–10 ml of spermatophoric plasma. After piercing the spermatophore and removing the semen, one can gain the best view of the transparent outer tunic (Fig. 15 F). The tubular, spirally wound sperm rope enclosing the tightly packed mass of spermatozoa (seen in Fig. 15 C and E) resembles a mammalian epididymal tubule in general appearance. When completely unwound, the sperm rope measures about 4 m. On eight occasions a sperm count was made in the contents released from sperm ropes. The lowest count was 3.8×10^8, the highest 4.6×10^{10} spermatozoa per single spermatophore; a count of 10^{10} would be roughly equivalent to the total sperm count of 30 human ejaculates.

The spermatozoa obtained by fragmenting and dispersing the sperm rope in seawater or isotonic salt solution are only feebly motile. The total length of an individual spermatozoon is 0.5 mm (roughly ten times the length of a human spermatozoon), most of it derived from the tail. The sperm head, slim, and oblong, is composed of a highly condensed nucleus and a cork-screw-shaped acrosome. At first sluggish and non-progressive, the movements of spermatozoa gradually increase on standing, provided that the sperm suspension is kept all the time at temperatures not exceeding 8°–10 °C. Motility can be enhanced further by the addition of cysteine and also some other amino acids. Spermatophoric plasma produces an opposite effect. Its addition slows down considerably the movements of spermatozoa (Mann et al. 1970).

3.6.5 Osmoregulation as the Driving Force for the Spermatophoric Reaction in Vitro

Below are described briefly the observations which led us to conclude that osmoregulation constitutes the main driving force for the spermatophoric reaction, and that the spermatophore acts as an "osmotic cell", its outer tunic performing the role of a semipermeable membrane (Hanson et al. 1973; Mann et al. 1970).

For observations on the mechanism of the spermatophoric reaction in vitro, most spermatophores were extricated manually from the male reproductive tract and transferred into a trough or tray filled with seawater. Under these conditions, it took usually about 15 min for the spermatophoric reaction to begin, and up to 2 h, occasionally even a little longer, to be completed. By compressing the cap or pulling the cap thread very gently, it has been possible on many occasions to advance the initial (but not the later) stages of the reaction.

The first signs that the reaction is about to start come from a change evident at the proximal, that is, the thicker end-portion of the spermatophore which swells owing to the influx of seawater and consequent dilution of the spermatophoric plasma. At the same time the sperm rope begins to uncoil and slowly moves forward, pushing before it the ejaculatory apparatus. The next change to take place affects the opposite, that it the distal end of the spermatophore. Within 20–30 min of placing the spermatophore in seawater, a break occurs at the point where the cap is joined on to the outer tunic (this point is shown in Fig. 15 D). This causes the extrusion of the ejaculatory apparatus and leaves the cap remnants behind. Sometimes, when no attempt was made to accelerate the onset of the spermatophoric reaction by pulling the cap thread, the cap refused to rupture at all, and under these circumstances the pressure inside the spermatophore continued to rise until reaching 400 mm H_2O or so, when instead of the cap, the outer tunic itself ruptures and the whole sperm rope consequently erupts, thereby bringing the spermatophoric reaction to an abrupt end.

Normally, after the cap has burst, it takes another 30 min or so for the next phase of the spermatophoric reaction to occur. During this period, the extrusion of the ejaculatory apparatus continues, the spermatophoric plasma undergoes further dilution due to the influx of seawater and the spermatophore gradually increases in volume and length, so that ultimately it measures about 1.5 m (not counting the cap thread). Consequently, the ejaculatory apparatus is pushed still further and at the same time it begins to shorten under pressure from the steadily advancing sperm rope and diluted spermatophoric plasma, forming a characteristic spiral (Fig. 16 A). As more seawater enters through the outer tunic, the sperm rope continues to push further and the spiral of the ejaculatory apparatus becomes increasingly thicker and shorter (Fig. 16 B).

At this stage there is usually a pause in the spermatophoric reaction, lasting 30 min or more, during which time the spermatophore elongates only slowly, and the ejaculatory apparatus undergoes a gradual, slowly progressing evagination. Then, quite suddenly, the process of evagination gains momentum and the sperm rope rapidly pushes forward, covering a distance of several centimetres within a few seconds. Consequently it fills quickly the whole distal end of the spermatophore which balloons out into the egg-shaped "spermatophoric bladder"; the "bladder", shown in Fig. 16 C and D, measured roughly 8 × 4 cm. As a result of the influx of seawater during the spermatophoric reaction, the spermatophoric plasma undergoes an at least fivefold dilution, so that when completely formed, the spermatophoric bladder usually has a volume of 40–50 ml, and occasionally is even larger.

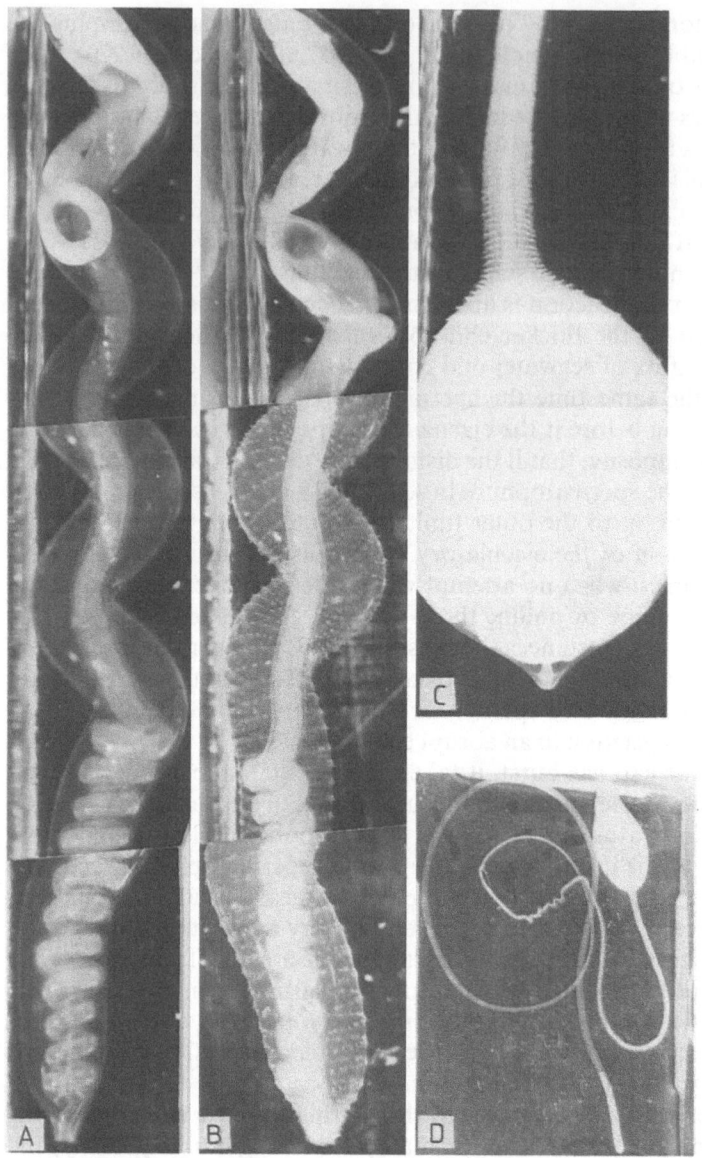

Fig. 16 A–D. Spermatophoric (ejaculatory) reaction in *Octopus dofleini martini. A* The stage after the extrusion of the ejaculatory apparatus, now undergoing the formation of a spiral; also shown is the sperm rope (× 1.5). *B* Same spermatophore, stage later, when the spiral became thicker and shorter; the sperm rope has wound its way around one part of the ejaculatory apparatus. *C* Same spermatophore, final stage: the spermatophoric bladder with the sperm rope inside, and the evaginated ejaculatory apparatus. *D* The whole spermatophore at the final stage of the spermatophoric reaction. (Mann et al. 1970)

Evidently, the elasticity of the outer tunic and the pressure arising from the influx of seawater transmitted by the diluted spermatophoric plasma constitute the two most important factors in the mechanism that drives the spermatophoric reaction all the time, but most forcibly during evagination of the spermatophore and the formation of the spermatophoric bladder. To record quantitatively the pressure changes inside the spermatophore, we used indwelling catheters (25 cm ong polythene tubes with 0.3 mm inside diameter), attached to a pressure transducer by means of a hypodermic needle on a three-way stopcock, and connected to a recorder; the experimental set-up used for this purpose is shown in Fig. 17 (Hanson et al. 1973).

When freshly pulled out from the male, the spermatophore is already in a state of high turgidity, corresponding to a transmural pressure of 143 ± 31 cm water (mean from 15 experiments), and as soon as a spermatophore had been placed in seawater, the pressure and turgidity begin to increase further owing to the influx of seawater. A complete record of a typical experiment, in which pressure changes have been followed throughout the whole period of the spermatophoric reaction, is presented in Fig. 18. As can be seen, in this particular experiment the pressure initially recorded inside the proximal end-portion of the spermatophore was 155 cm H_2O, but as soon as contact with seawater was permitted, it began to rise steeply to reach a maximum of 355 cm H_2O at the moment when the cap ruptured, that is, immediately prior to the extrusion of the ejaculatory apparatus. The rate at which pressure rises inside the spermatophore in vitro during the time interval between the onset of the spermatophoric reaction and the moment of cap rupture has been determined in a number of experiments. It varied around a mean value of 8 cm H_2O/min. The average pressure at which the cap bursts was 330 cm H_2O.

As soon as the ejaculatory apparatus begins to be extruded, the transmural pressure falls precipitously; in the experiment illustrated in Fig. 18, the pressure fell from 355 cm down to 60 cm H_2O. From that moment onwards and throughout the period of the next hour or more, the seawater that enters the spermatophore causes primarily eversion of the ejaculatory apparatus rather than a further increase of pressure. In the experiment shown in Fig. 18 the pressure inside the spermatophore was 80 cm at the moment of spermatophoric-bladder formation, as compared with 60 cm an hour previously, and it fell to 40 cm after the formation of the bladder was completed. The mean pressure value at the end of the spermatophoric reaction (six experiments) was 36 ± 13 cm H_2O in the proximal portion of the spermatophore, but only 9 ± 4 cm H_2O, that is four times lower, inside the spermatophoric bladder. This difference in transmural pressure between the two compartments of the same spermatophore could not have been established without the development of an effective physical barrier between them. A barrier of this kind must have been formed as the result of a constriction and subsequent closure of one segment of the outer tunic, namely near the site formerly occupied by the cap. It seems highly probable that the pressure gradient, though not very large ($36 - 9 = 27$ cm H_2O), constitutes an important factor in driving the sperm rope towards the distal end of the spermatophore.

Yet in spite of all the changes that accompany the spermatophoric reaction, and the manifold increase in the volume of spermatophoric plasma resulting from

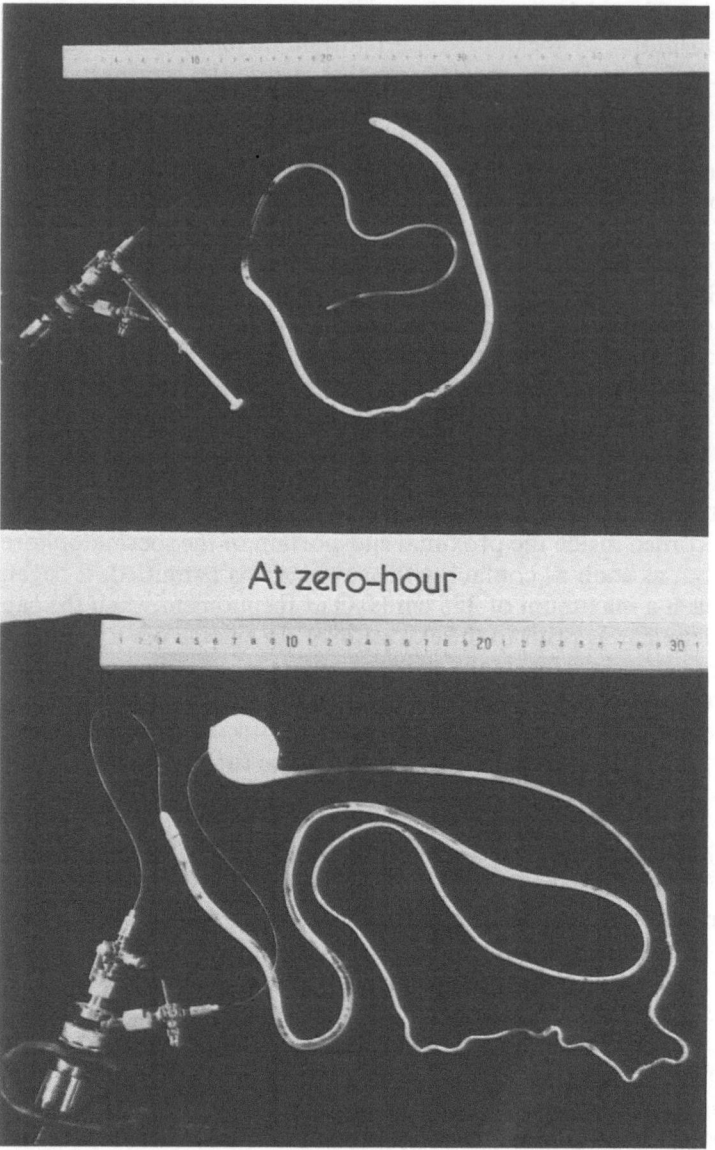

At zero-hour

2 h later

Fig. 17. Pressure-recording inside the spermatophore of *Octopus dofleini martini,* during the spermatophoric reaction in vitro. *Top* before placing the spermatophore in seawater, a catheter, connected with the pressure transducer, has been inserted into the proximal (sperm-rope containing) end of the spermatophore; *bottom* same spermatophore after a 2-h period in seawater, at the end of the spermatophoric reaction, when the spermatophoric bladder has been formed; a second catheter has been inserted at the distal end of the spermatophore, that is, into the spermatophoric bladder. Appropriate adjustment of two stopcocks permitted measuring pressure at either end of the spermatophore with the same pressure transducer. In order to perform the experiment, syringes filled with seawater had to be attached to the sidearm of each stopcock, but to simplify the figure, the syringes have been removed. (Hanson et al. 1973)

54

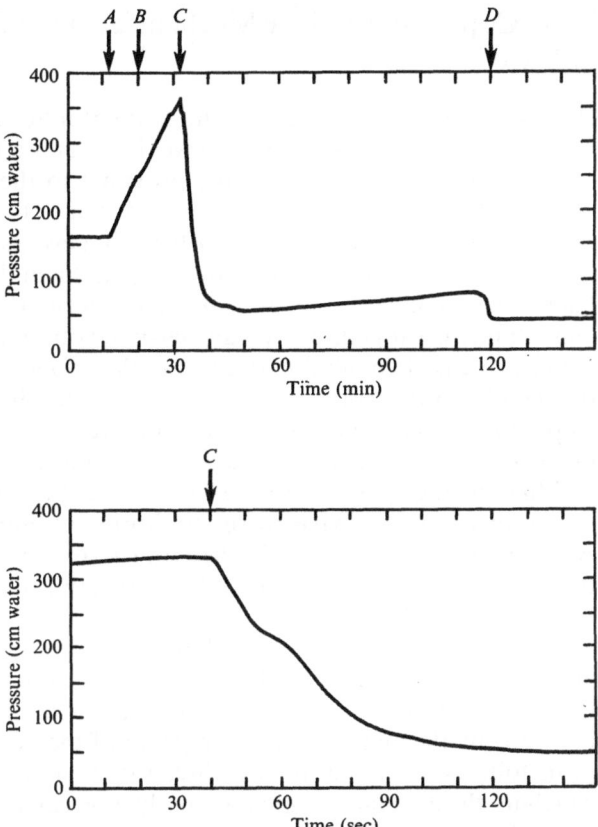

Fig. 18. Changes in transmural pressure (cm water) during the spermatophoric reaction in vitro, at 10 °C, recorded with the apparatus shown in Fig. 17. The time scale is in minutes in the upper part of Fig. 18, and in seconds in the lower part of this figure. *Arrows* indicate the following: (*A*) after inserting the catheter into the proximal end of the spermatophore, but before placing the spermatophore in seawater; (*B*) cap-thread has just been pulled free (when pressure reached 250 cm); (*C*) cap has just burst (when pressure reached 355 cm); (*D*) evagination of the ejaculatory apparatus has been completed with the resulting formation of the spermatophoric bladder. (Hanson et al. 1973)

the influx of seawater, the osmolality of the fluid milieu inside the spermatophore remains practically unaltered throughout the whole period of the spermatophoric reaction, but all the time it is definitely higher than in the surrounding seawater. Freezing point determinations have shown that the difference between the internal and external milieu of the spermatophore is 32 mOsm/kg water at the peak of intraspermatophoric pressure, that is, just prior to the extrusion of the ejaculatory apparatus, and 28 mOsm/kg water after the pressure has fallen to its lowest level, that is, at the time of spermatophoric bladder formation. A relationship of this kind clearly indicates that the mechanism underlying the spermatophoric reaction is one of osomregulation, that is, governed by the Donnan distribution law. In line with this concept are also observations on the chemical changes in the composition of spermatophoric plasma that accompany the spermatophoric reaction; the results of these observations will be described later in this chapter (3.7).

55

3.6.6 Copulation and the Mechanism of the Spermatophoric Reaction in Vitro

Ostensibly, the main purpose of the spermatophoric reaction, as revealed by the observations in vitro described above, is to haul a large mass of spermatozoa a distance of about 1.5 m, from the proximal to the distal end of the spermatophore, and thereby to create a sperm reservoir in the form of the spermatophoric bladder. There is every reason to believe that events analogous to those observed in vitro take place also during copulation which in the giant octopus of the North Pacific takes about as long as a spermatophoric reaction in vitro. By placing the female and male together in a seawater-filled observation tank, one can follow the whole process of copulation right from the moment when the male has mounted the female, to the time when he withdraws, leaving her with either one or two empty tubes of spermatophores hanging freely from the body cavity over a distance of 0.5 m. On several occasions we found it also possible to interrupt coitus at different stages, and remove for observation spermatophores that have undergone only a partial in vivo spermatophoric reaction (shown in the Frontispiece). A spermatophore of this kind, when transferred to a tray with seawater, then continues to undergo the spermatophoric reaction and in due course, produces a spermatophoric bladder.

The function of the male copulatory organ is performed by the hectocotylized arm (and not by the penis). The proximal portion of the spermatophore's body, that is, the one containing the sperm rope, is the first to leave the penis and to be seized by the male. The distal portion of the body with the ejaculatory apparatus follows next, and the cap thread is the last to emerge from the penis. By watching the copulating animals closely, one can actually observe soon after the start of copulation how the cap thread, attached at one end to the siphon, is gradually pulled away from the hectocotylized arm. The traction that thus develops is probably sufficiently strong to cause the cap to burst and to set in motion the next stage of the spermatophoric reaction in vivo, that is, the extrusion of the ejaculatory apparatus. As pointed out previously, the pressure rise during the early phase of a spermatophoric reaction in vitro from 140 cm H_2O initially, to 330 cm at the moment when the cap bursts occurs at an average rate of 8 cm H_2O/min, and requires therefore $(330-140)/8 = 24$ min. This means that under similar conditions in vivo, the copulating male would have no more than 24 min to pull out the entire spermatophore from the penis, and to manoeuvre the ejaculatory apparatus by means of the hectocotylized arm into a position near the external orifice of the female reproductive tract, thus ensuring that the cap does not burst before the distal part of the spermatophore's body can reach the female's genital orifice. Unless the male is able to accomplish this feat within the 24-min interval, the pressure within the spermatophore would have to rise beyond the 330 cm H_2O-upper limit, soon making it impossible for the outer tunic to resist a further influx of seawater, and causing it to rupture and void its contents into the seawater. As a matter of fact, our observations of copulating animals have led us to conclude that the cap thread is pulled away from the hectocotylized arm and that the cap bursts within much less than 24 min, in most instances it seems within 1 min of the cap thread's emerging from the penis. Obviously, the risk of a spermatophore rupturing and emptying its contents prematurely is much smaller in vivo than in vitro.

The next stages of the spermatophoric reaction in vivo, that is, the events associated with the evagination of the ejaculatory apparatus and the formation of the spermatophoric bladder, also appear to be the same as during a spermatophoric reaction proceeding in a tray filled with seawater. In vivo, however, the terminal portion of the ejaculatory apparatus is guided by the hectocotyle into the oviduct before it balloons out into the spermatophoric bladder. The exact moment at which that balloon bursts and releases the spermatozoa inside the female reproductive tract could not be determined, but on the few occasions when the females were dissected next day after copulation, that is, after they started laying eggs, their oviducts were filled with large masses of free, motile spermatozoa, particularly in the lumen of the oviducal gland.

3.7 Biochemistry of the Giant Octopus Spermatophore

3.7.1 Outer Tunic and Its Permeability to Chemicals

The outer tunic, which forms a capsule around the whole body of the spermatophore, is tough and elastic. When stripped off and washed, it appears as a transparent tube, elastic to such an extent that one can stretch it manually to nearly twice the original length without causing a rupture. When ground in cold acetone and converted into acetone-dried powder, it contains 7.7% ash, 10.5% orcinol-reactive carbohydrate, 13.5% bound aminosugar, and 11.75% nitrogen, a large portion of the latter present in the form of protein which is rich in proline, lysine, aspartic acid, and threonine. The results of amino acid analyses in the acetone powder hydrolysed in $6N$-HCl for 72 h are shown in Table 2, side by side with corresponding values for hydrolysates of the ethanol-insoluble protein of spermatophoric semen and cement liquid (Mann et al. 1970).

No less striking than the high elasticity and resistance to pressure is the ability of the outer tunic to act as an efficient dialysis membrane. When filled with 0.1 M solutions of either sodium chloride, ammonium sulphate, glucose, fructose or sucrose, and suspended in water, the outer tunic tube permits a free exchange of all these substances, so that within less than 24 h the internal and external concentrations become identical. Inulin on the other hand, dialyses poorly, cytochrome c in traces only, and albumin not at all. Of exceptional interest is the behaviour of the outer tunic towards small-molecular constituents of spermatophoric plasma itself. When outer tunics stripped off spermatophores, and pieces of Cellophane tubing of roughly equal diameter and length, were filled with 2 ml samples of spermatophoric plasma (separated by centrifugation of spermatophoric semen) and suspended in 50 ml seawater, the volume of fluid inside the two dialysis sacs increased in both instances from 2 to 12 ml, and at the same time, out of a total of 29.6 mg acid-hydrolysable aminosugar, initially present in 2 ml spermatophoric plasma, 11.7 mg was found to have passed through the outer tunic, and

Table 2. *Octopus dofleini martini*. Amino acids in acid hydrolysates from the outer tunic of spermatophore, the semen (sperm rope and spermatophoric plasma) and the cement liquid. Prior to hydrolysis in 6N-HCl for 72 h, the membranous material of the outer tunic was converted into aceton powder, and ethanol-soluble material removed from the semen and the cement liquid. The results are expressed in terms of µmol/mg dry weight. (Mann et al. 1970)

	Outer tunic	Semen	Cement liquid
Aspartic acid	0.9040	0.477	0.645
Threonine	0.7035	0.367	0.360
Serine	0.5025	0.258	0.318
Glutamic acid	0.5875	0.582	0.548
Proline	0.3963	0.363	0.272
Glycine	0.5510	0.378	0.307
Alanine	0.4155	0.385	0.336
Valine	0.5010	0.322	0.303
½ Cystine	0.1894	0.069	0.209
Methionine	0.1041	0.047	0.100
Isoleucine	0.5205	0.207	0.264
Leucine	0.4547	0.181	0.306
Tyrosine	0.1603	0.075	0.167
Phenylalanine	0.3532	0.100	0.197
Lysine	0.9545	0.343	0.477
Histidine	0.2025	0.266	0.229
Arginine	0.3845	0.197	0.181
Total amino acids	7.8850	4.617	5.219

12.8 mg through Cellophane tubing, into the external seawater milieu (Mann et al. 1970).

An opposite example of permeability is provided by a series of analyses demonstrating that the outer tunic does not present a barrier to the entry of foreign substances from the external milieu. Glucose and inorganic phosphate, for example, when added to seawater, pass readily into spermatophoric plasma during a spermatophoric reaction in vitro (Mann et al. 1981 b). Evidently, like the wall of the male reproductive tract in mammals, which forms no barrier to the entry of drugs and other toxic chemicals into the semen, the outer tunic which forms the external wall of the spermatophore can be also penetrated by foreign substances. These, having entered the spermatophoric plasma from the seawater, may well be capable of affecting the spermatozoa adversely (Mann and Lutwak-Mann 1982, 1983).

3.7.2 The Gel Rod of the Ejaculatory Apparatus

By placing a ligature around the spermatophore where the sperm rope meets the cement body, and subsequently puncturing the two halves of the body, it is possible to collect separately the spermatophoric semen, that is sperm rope plus spermatophoric plasma, from the proximal half, and the ejaculatory apparatus from the distal half. At the same time one can withdraw by means of a fine syringe at least a few drops, occasionally as much as 1 ml, of the cement liquid located near the cement body, readily distinguishable from the colourless spermatophoric plasma by its amber colour.

The gel rod of the ejaculatory apparatus, washed free of the cement liquid, has a remarkably high dry weight. The average value recorded in material obtained from five mature spermatophores was 33.3 g/100 g, and of that dry weight 32.7 g/100 g was ethanol-insoluble, and distinguished by a high content (11.5%) of bound orcinol-reactive carbohydrate. Very little of that carbohydrate is derived from glycogen (the highest value was 0.13%), and only 12 mg carbohydrate/100 g dry weight was present in a free, that is, ethanol-soluble form (Mann et al. 1970).

3.7.3 Cement Liquid and the Role of Glycosidases

The highly viscous cement liquid has a dry weight about the same as the ejaculatory apparatus, and most of it derived from ethanol-insoluble material rich in protein, bound carbohydrate and aminosugar (Table 3). The amino acid composition of the protein is shown in Table 2; among the 17 amino acids listed, the two most highly concentrated are glutamic acid and aspartic acid; equally characteristic is the unusually high content of lysine.

A most unusual feature of the cement liquid is the presence of several highly active glycosidases. This provides a particularly instructive example of the strikingly close biochemical similarity between two sperm storage organs, structurally differing as widely as the mammalian epididymis (Conchie and Mann 1957) and the cephalopod spermatophore (Mann et al. 1973). The two most active glycosidases within the octopus spermatophore are α-mannosidase and β-N-acetylglucosaminidase. Both are more concentrated in the cement liquid than in either the ejaculatory apparatus, sperm rope or spermatophoric plasma (Fig. 19). In view of the importance of the area in which the cement liquid is located, namely, the junction between the sperm rope and cement body, it would appear that the two highly active glycosidases play a cardinal role in the spermatophoric reaction, most likely by facilitating a degradation of those structural mucoproteins that might otherwise have constituted a mechanical barrier to advance of the sperm rope and evagination of the ejaculatory apparatus. A possibility of this kind is supported by the finding that in the course of the spermatophoric reaction some of the large-molecular mucoprotein material within the spermatophore is split into smaller fragments which are diffusable and pass through the outer tunic into the surrounding seawater (Hanson et al. 1973).

3.7.4 Main Physicochemical Properties of Spermatophoric Semen

The dry weight, nitrogen, phosphorus, carbohydrate, and amino sugar contents of whole semen, removed by puncture from the spermatophore's body, are listed in Table 3. The amino acid composition of protein is shown in Table 2 (Mann et al. 1970). On viewing Table 3, one is struck first of all by the high dry weight of semen, 35.05 g/100 ml, most of it contributed by material insoluble in 66%

Table 3. *Octopus dofleini martini*. Dry weight, nitrogen, phosphorus, carbohydrate and amino sugar in semen and cement liquid of a fresh spermatophore (Mann et al. 1970). Semen (14 ml) containing both the sperm rope and spermatophoric plasma, and the cement liquid (1 ml) were treated with 2 vol. ethanol, and centrifuged. The ethanol-extractable material was concentrated in vacuo to near-dryness. The ethanol-precipitable material was washed with ethanol and dried in vacuo

	EtOH-soluble		EtOH-insoluble		Total
	mg/100 ml	% of dry wt	mg/100 ml	% of dry wt	mg/100 ml
SEMEN					
Dry weight	2250		32800		35050
Nitrogen (Kjeldahl)	167.7	7.4	3542	10.8	3709.7
N × 6.25	1048	46.1	22137	67.5	23185
Phosphorus					
inorganic	2.6	0.11	26	0.08	28.6
acid-soluble			131	0.4	
total	18.8	0.83	243	0.74	261.9
Carbohydrate					
directly reducing	189	8.3			
orcinol-reactive	200	8.8	3674	11.2	3874
soluble in 0.1 N-NaOH and reducing after					
HCl-hydrolysis			3575	10.9	
Glycogen	0	0	2247	6.85	2247
Amino sugar (total)	630	27.7	1525	4.65	2155
Sialic acid (total)	2	0.09	20	0.06	22
CEMENT LIQUID					
Dry weight	746		32630		33376
Nitrogen (Kjeldahl)	46	6.2	3782	11.6	3826
N × 6.25	287		26637		23924
Phosphorus					
inorganic	0		0		0
acid-soluble			37.3	0.114	
total	3	0.40	71	0.217	74
Carbohydrate,					
directly reducing	23	3.1			
orcinol-reactive	25.8	3.5	2160	6.61	2185.8
soluble in 0.1 N-NaOH and reducing after					
HCl-hydrolysis			1780	5.45	
Glycogen					
Amino sugar (total)	11.3	1.51	1640	5.02	1651.3
Sialic acid (total)			0		

ethanol. Other interesting features are the high contents of organically bound phosphorus, amino sugar and carbohydrate, the latter mostly in ethanol-insoluble form, and derived partly from glycogen (6.85% of dry weight) and partly from polysaccharide material containing a large amount of amino sugar and several sugars (galactose in particular), but hardly any sialic acid.

The analyses referred to above were initially carried out on whole semen, but subsequently they were repeated on spermatozoa and spermatophoric plasma

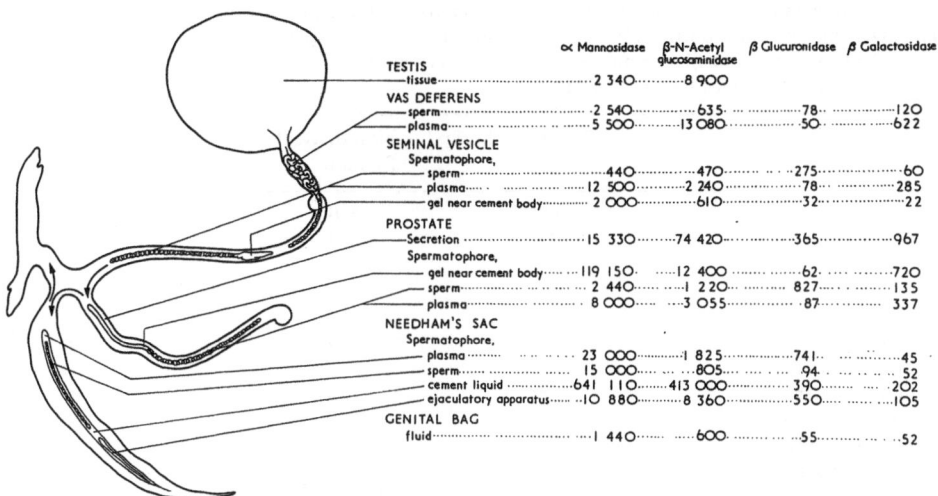

Fig. 19. Distribution of glycosidases in the male reproductive tract and spermatophores of the giant octopus. Results are expressed as enzyme units/g, 1 unit being the amount of enzyme required for the liberation in 1 h, of (1) 1 μg nitrophenol from *p*-nitrophenyl α-D-mannoside, *p*-nitrophenyl *N*-acetyl-β-D-glucosaminide and *o*-nitrophenyl β-D-galactopyranoside, resp., and (2) 1 μg phenolphthalein from phenolphthalein β-glucuronic acid. (Further details: Mann et al. 1973)

separated from each other by centrifugation. Considerable differences have been encountered in respect of many parameters during analyses of individual samples of spermatophoric plasma, much less so, however, of spermatozoa.

3.7.5 Proteins, Peptides, Galactophosphopeptide, and Certain Other Major Constituents of Spermatophoric Plasma

The spermatophoric plasma separated from the sperm rope by centrifugation, represents a clear colourless and highly viscous fluid with most unusual physico-chemical properties, but variable in quantitative terms (Mann et al. 1970): dry weight 24–35%, as compared with 3% of the seawater from Puget Sound; density 1.080–1.087; viscosity (cP) 10.5–16.9. Osmolality, when measured in samples of spermatophoric plasma from two spermatophores, corresponded to a freezing point depression of -1.488 °C in one, and -1.254 °C in the other. If the osmolality of these two samples had been due exclusively to sodium chloride, one would have expected to find in them 2.573% and 2.221% NaCl, respectively. In fact, however, the NaCl contents determined chemically were less than half, namely 1.22% and 1.02% respectively.

The high dry weight content of spermatophoric plasma is due chiefly to organic material, 87–94% of it in ethanol-insoluble form; of about 1.5 g total carbohydrate present in 100 ml spermatophoric plasma, roughly 1.4 g is ethanol-insoluble; similarly out of a total of 1.28 g amino sugar/100 ml, present in plasma,

Table 4. Octopus dofleini martini. Composition of spermatophoric plasma, and of its non-dialysable portion (Mann et al. 1970). All results are expressed as concentration per 100 ml spermatophoric plasma. The amino acids were determined after 24 hydrolysis in 6 *N*-HCl

	Spermatophoric plasma	Non-dialysable fraction
Dry weight (mg)	24 690	15 360
Nitrogen (mg)	2 470	2 080
Protein (mg)	16 115	11 650
Carbohydrate (mg)		
reducing	47	0
orcinol-reactive	1 477	735
Amino sugar (mg)		
directly determined	44	22
after acid hydrolysis	1 380	673
Amino acids (mmol)		
aspartic acid	14.522	12.059
threonine	13.530	11.305
serine	12.077	9.731
glutamic acid	15.745	14.250
proline	21.605	10.251
glycine	11.143	9.883
alanine	18.744	13.431
valine	13.201	11.087
$\frac{1}{2}$ cystine	16.456	13.656
isoleucine	2.200	1.718
leucine	2.147	2.195
tyrosine	1.448	1.192
phenylalanine	1.267	1.019
lysine	14.824	8.844
histidine	12.525	6.839
arginine	1.402	1.048
Total amino acids	172.836	128.508

1.26 g is in a bound, ethanol-insoluble form. The large-molecular character of the organic material becomes equally evident when note is taken of the ratio between dialysable and non-dialysable fractions of spermatophoric plasma, and in particular, of the fact that out of the total nitrogen content of 2470 mg N/100 ml spermatophoric plasma, 2080 mg is non-dialysable. Nonetheless, as can be seen from Table 4, by no means all of the bound carbohydrate, amino sugar, and amino acids present in the spermatophoric plasma is in non-dialysable form. A particularly interesting feature of the dialysable portion is the presence of peptides. On hydrolysis in 0.5 *N*-HCl for 5 h, these peptides yield a number of amino acids, the two most prominent being aspartic acid and serine which together account for 71.9% of the total amino acid content.

Another most unusual feature of the peptide fraction is the presence of organically bound but acid-soluble phosphate. This is derived from a phosphoglycopeptide. Electrophoretic and chromatographic purification of it, followed by chemical and enzymic hydrolysis, yielded D-galactose phosphate as a degradation product; the galactose-acid peptide moieties of the peptide are held together by the phosphoryl linkages (Brooks et al. 1981). Much more of the galactophosphopeptide is found in mature spermatophores than in those undergoing development in the spermatophoric gland systems I or II. It is probably first secreted by the epithelia of the male reproductive tract and only later concentrated during the final stages of spermatophore maturation. Another possibility is that it constitutes a natural derivative of some of the large-molecular weight glycoproteins which abound in the spermatophoric plasma.

Roughly 90% of the characteristically high total content of acid-soluble organically bound phosphate in the spermatophoric plasma of mature spermatophores (2.57–4.53 mg P/ml) is accounted for by the phosphoglycopeptide. The remainder is largely in the form of glycerylphosphorylcholine.

3.7.6 Distribution of Glycerylphosphorylcholine and Carnitine Between Spermatophoric Plasma and Spermatozoa

Glycerylphosphorylcholine and carnitine are well known constituents of mammalian epididymal semen (Brooks et al. 1974a; Dawson et al. 1957; Marquis and Fritz 1965). Both are equally characteristic of spermatophoric semen (Brooks et al. 1974b).

As in epididymal semen, most of glycerylphosphorylcholine in spermatophoric semen occurs extracellularly. Carnitine on the other hand, again in close analogy to epididymal semen (of the bull at any rate), is largely concentrated within the spermatozoa, and relatively little of it is present in the spermatophoric plasma. The distribution of both substances in the spermatophoric semen of three octopuses (A, B, and C) is shown in Table 5, from which it can also be seen that carnitine is much more concentrated in the spermatozoa of the sperm rope than those obtained from the vas deferens. This indicates that at least some carnitine

Table 5. *Octopus dofleini martini*. Distribution of carnitine, glycerylphosphorylcholine and acid-soluble phosphorus in semen. (Brooks et al. 1974b)

Material	Carnitine	Glyceryl-phosphoryl-choline	Acid-soluble phosphorus
Sperm-rope (µmol/g wet tissue), octopus B	5.35	1.89	50
Spermatozoa separated from sperm-rope fluid (µmol/g wet tissue), octopus C	14.6	–	56
Spermatozoa separated from vas deferens plasma (µmol/g wet tissue), octopus C	1.53	–	23
Spermatophoric plasma (µmol/ml)			
octopus A	0.80	5.3	111
octopus B	0.98	9.8	83
Cement liquid (µmol/ml), octopus B	1.15	0.26	22
Sperm-rope fluid (µmol/ml), octopus C	0.58	–	140
Vas deferens plasma (µmol/ml), octopus C	1.2	–	23

must have been acquired by the spermatozoa during their passage through the male reproductive tract after they have left the vas deferens.

3.7.7 Glycogen and Glykogenolysis in Spermatozoa

After the two parts of semen were separately subjected to Pflüger's method of isolation, the glycogen was found to be confined mostly to spermatozoa; in the sperm-containing portion, there was 3070 mg glycogen/100 ml semen, but the spermatophoric plasma yielded merely 210 mg glycogen/100 ml semen (Mann et al. 1970). On further investigation involving histochemical methods, coupled with the use of amylase, the bulk of sperm-glycogen was shown to be concentrated around the heads of spermatozoa, very close to their surface. Same investigation revealed the presence in sonicated spermatozoa, of three highly active enzymes: phosphorylase, phosphoglucomutase, and glucose-6-phosphate isomerase (Martin et al. 1970). The presence of glycogen in octopus spermatozoa, coupled with the occurrence of phosphorylase and phosphoglucomutase, the two enzymes required for the initial steps of glycogenolysis (phosphorylation of glycogen to glucose-1-phosphate, followed by formation of glucose-6-phosphate), led us to conclude that glycogenolysis may represent the pathway of carbohydrate metabolism in octopus spermatozoa, this in direct contrast to mammalian spermatozoa which depend on free glycolysable sugars, mainly fructose, as substrates for glycolysis and neither contain nor metabolize glycogen (Mann 1964).

Subsequent investigations (Mann et al. 1974, 1977; Martin et al. 1976) fully justified this supposition, by demonstrating that when spermatozoa obtained from the sperm rope of a freshly dissected spermatophore are suspended in seawater and incubated anaerobically at or below 10 °C (a temperature close to the normal habitat of the giant octopus in the Puget Sound), they maintain vigorous

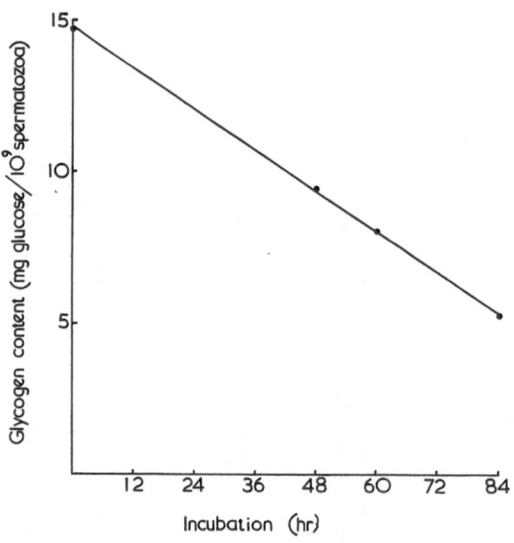

Fig. 20. Decrease in the glycogen content of octopus spermatozoa upon incubation at 8 °C. (Mann et al. 1977)

motility for several days and at the same time produce lactic acid. As Fig. 20 shows, the spermatozoa utilize glycogen at a characteristically constant rate throughout the whole 84-h period of anaerobiosis, bringing the glycogen content down from 14.7 mg/10^9 fresh spermatozoa to 5.2 mg/10^9 incubated spermatozoa. The constancy of the rate at which glycogen is broken down by octopus spermatozoa is the more remarkable since their motility remains at a high level only for up to 48 h, but thereafter it declines, becoming distinctly poorer after 60 h, and almost extinct at the end of the 84-h incubation period. This behaviour of octopus spermatozoa towards utilizable carbohydrate is again in direct contrast to mammalian semen in which fructolysis proceeds at a constant rate only for as long as motility is high, but thereafter declines and comes to a complete halt at the time when motility ceases (Mann 1946, 1948, 1964). But most remarkable of all is the fact that while mammalian spermatozoa produce L(+)lactic acid during glycolysis, the product of glycogenolysis in octopus spermatozoa is D(−)lactic acid (formula in Fig. 21) (Mann et al. 1974; Martin et al. 1976).

3.7.8 D(−)Lactic Acid and D(−)Lactate Dehydrogenase

The identity of lactic acid formed during glycogenolysis was established by four methods, (1) colorimétric, by means of the hydroxydiphenyl reagent, (2) iodometric, by distilling the acetaldehyde formed by oxidation with potassium permanganate and determining the acetaldehyde bound to sodium sulphite by titration with iodine, (3) gas-liquid chromatographic separation of the methyl derivatives of non-volatile organic acids extracted with ether from the incubation mixture, and (4) enzymatic. The latter method, based on the use of nicotinamide adenine nucleotide (NAD), proved to be of decisive value for the following reason: When, at first, an attempt was made to confirm the chemical evidence (methods 1–3) by using L(+)lactate dehydrogenase (from rabbit muscle) to reduce NAD in the presence of the lactic acid derived from octopus sperm, no NADH was formed. This alerted us to the possibility that the lactic acid which is formed by the spermatozoa during glycogenolysis is not L(+)lactic acid, but the enantiomorph form, namely, D(−)lactic acid, a supposition that proved fully justified when, instead of L(+)lactate dehydrogenase, a D(−)lactate dehydrogenase preparation (from *Lactobacillus leichmannii*) was used to reduce NAD to NADH.

Following the identification of D(−)lactic acid as a metabolite of octopus spermatozoa, we turned our attention to the possibility that these spermatozoa may also contain D(−)lactate dehydrogenase and thereby differ from mammalian spermatozoa which have only L(+)lactate dehydrogenase. By extracting the enzyme from octopus spermatozoa and testis, we obtained a preparation of a highly active D(−)lactate dehydrogenase, that is, a NAD-linked D(−)lactate oxidoreductase, highly specific towards D(−)lactic acid, but completely inactive towards L(+)lactic acid. On electrophoretic examination, the purified enzyme could be clearly distinguished from the D(−)enzyme from bacteria and that from the skeletal-musculature of the horseshoe crab, *Limulus polyphemus* (Fig. 21). No appreciable reduction of NAD in the presence of D(−)lactic acid could be observed in spermatozoa from man, bull, ram, and cock; all these contained the

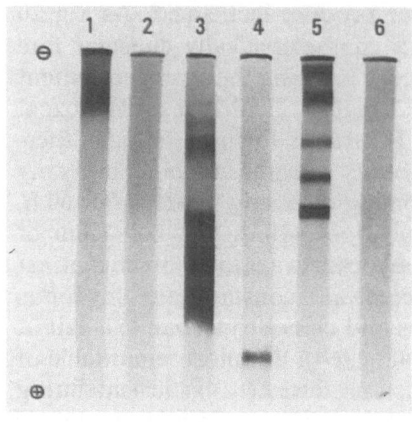

COOH
|
H—C—OH
|
CH₃

D(−)Lactic acid

COOH
|
HO—C—H
|
CH₃

L(+)Lactic acid

Fig. 21. Polyacrylamide gel disk-electrophoretogram of lactate dehydrogenase from various sources. *1* and *2* enzyme prepared from the testis of *Octopus dofleini martini*, plus D(−)lactate and L(+)lactate, resp.; this enzyme can be seen to be active against D (−) lactate only, *3* enzyme prepared from the muscles of *Limulus polyphemus*, plus D(−)lactate; like the octopus-testis enzyme, the *Limulus*-enzyme is also a D(−) lactate dehydrogenase, but not identical with the testicular dehydrogenase; *4* enzyme from *Lactobacilus leichmannii*, plus D(−)lactate; it is a D(−)lactate dehydrogenase, but not identical with either the testicular or the *Limulus* dehydrogenase; *5* and *6* enzyme prepared from rabbit skeletal muscle, plus L(+)lactate and D(−)lactate, resp.; this enzyme, unlike the octopus-testis enzyme, is a L(+)lactate dehydrogenase. (Martin et al. 1976). Below the electrophoretogram are shown the formulae of the two isomers of lactic acid

L(+) enzyme only. On the other hand, a homogenate prepared from the testis of a squid (*Loligo opalescens*) contained the D(−)lactate dehydrogenase (Martin et al. 1976).

3.7.9 Relation of Glycogen, Arginine Phosphate, and Adenosine Polyphosphates to Sperm Motility

Even though octopus spermatozoa continue to utilize glycogen anaerobically at a constant rate for 84 h at 8 °C (Fig. 20), the D(−)lactic acid production ceases much earlier, namely, at about 48 h, coinciding with the time when motility is lost. As can be seen from Fig. 22, the amount of D(−)lactic acid accumulated at the end of the first 48 h, corresponded nearly stoichiometrically to the amount of utilized glycogen, but later the largely immotile spermatozoa continued to break down glycogen without producing any lactic acid. Evidently under these conditions they metabolize glycogen to substances other than lactic acid while no longer able to sustain their motility. So far only one of these substances, namely glucose, has been identified; presumably it is formed as a product of the diastatic action of the spermatozoa (Mann et al. 1977).

The possibility that the levels of energy-rich phosphorus compounds, such as adenine or guanidine derivatives rather than glycogen alone, may be directly related to sperm motility was explored in a separate study (Brooks et al. 1971). This brought to light the occurrence in octopus spermatozoa of arginine phosphate (ArgP) and two adenine nucleotides, namely, ATP and ADP. Creatine phosphate, on the other hand, could not be demonstrated, yet another feature diamet-

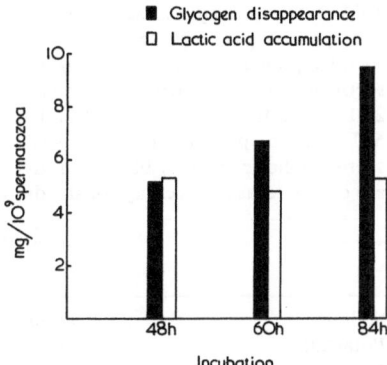

Fig. 22. Relation between disappearance of glycogen and the formation of D(−)lactic acid in octopus spermatozoa; ■ glycogen disappeared; □ lactic acid accumulated. (Mann et al. 1977)

Table 6. Octopus dofleini martini. Changes in adenosine triphosphate (ATP), adenosine diphosphate (ADP) and arginine phosphate (Arg P) during incubation of sperm suspensions. (Brooks et al. 1971)

Results are expressed as nmol/10^9 spermatozoa

Incubation	Motility	Arg P	ATP	ADP
0 min	+ +	1670	470	30
15 min at 3 °C	+ +	1160	420	20
3 days at 3 °C	+ +	660	540	20
5 days at 3 °C	+	0	460	30

rically opposed to those of mammalian semen. Changes in ArgP, ATP, and ADP concentrations during incubation of octopus sperm suspensions are shown in Table 6. In this experiment, the spermatozoa were incubated at 3 °C, a temperature at which their motility does not begin to decline until day 5.

In addition to the three phosphorus compounds, the glycogen content was determined in the spermatozoa at the start and end of the 5 days' experiment, but the difference was small. On the other hand, the decline in ArgP at 3 °C was very marked, in contrast to ATP and ADP, which remained fairly constant. Arginine phosphate is well known to share with ATP the presence of energy-rich phosphate bonds, and presumably both phosphorus compounds play a role in octopus sperm energetics, but ArgP appears to be utilized preferentially. In this respect, it may be relevant to recall some early observations which we made on contracting frog muscles deprived of glycogenolytic activity by iodoacetate poisoning (Mozolowski et al. 1931). These showed that when in the absence of glycogenolysis, muscle phosphagen (creatine phosphate) and ATP are the only source of energy available to the muscle, the phosphagen is used up first, and the breakdown of ATP follows only later, when its resynthesis at the expense of the phosphagen is no longer feasible. As regards the levels of ATP and ADP in octopus spermatozoa, both are much lower than that of ArgP. Nonetheless, both ATP and ADP probably fulfil in glycogenolysis of these spermatozoa the same pivotal role of phosphate-transferring coenzymes as in mammalian and echinoderm sperm (Mann 1945, 1964; Rothschild and Mann 1950).

Table 7. *Octopus dofleini martini.* Constitutents of the ash in spermatozoa and spermatophoric plasma (Martin et al. 1973). The contents of two spermatophores yielded on centrifugation 9.8 g wet weight of closely packed sperm and 14.2 g of spermatophoric plasma. The dry weight of sperm was 20.1%, and that of plasma 24.9%. The weight of the ash (dry ashing at 500 °C) from sperm represented 1.96% of the dry matter, and that from spermatophoric plasma 7.06% of the dry weight. The results of ash analysis are expressed as a percentage of the dry ash

	Spermatozoa	Spermatophoric plasma
Sodium	11.00	30.00
Potassium	4.90	3.50
Magnesium	0.07	0.66
Calcium	0.22	0.91
Zinc	7.85	0.22
Copper	0.001	0.001
Iron	0.002	0.005

3.7.10 The High Zinc Content of Octopus Spermatozoa

Another characteristic feature of octopus spermatozoa is the extraordinarily high content of zinc as shown by the analysis of ashed material (Martin et al. 1973). It can be seen from Table 7 that the zinc content of the ash from spermatozoa is 7.85%, as compared with 0.22% in the spermatophoric plasma. Calculated on a fresh weight basis, these values correspond to a concentration of 154 mg Zn/ 100 g wet weight in sperm and 4 mg Zn/100 g wet weight in spermatophoric plasma. Human semen, usually given as an example of mammalian semen with a particularly high content of zinc, has only 10 mg Zn/100 g wet weight, the concentration being roughly the same in spermatozoa and spermatophoric plasma.

The zinc which occurs in octopus spermatozoa is mostly in a bound form as a large-molecular complex, from which it can be displaced by ethylenediaminetetra-acetate (EDTA). Such treatment, however, results in complete loss of sperm motility.

3.7.11 Biochemical Changes Associated with the Spermatophoric Reaction

The fivefold (or more) increase in the volume of intraspermatophoric fluid, caused by the influx of seawater during the spermatophoric reaction, is accompanied by a number of changes in the composition of spermatophoric plasma, the pattern of electrolytes especially. Figure 23 (Mann et al. 1970) shows changes which have been recorded in spermatophores undergoing spermatophoric reaction in vitro with respect to volume and dry weight; concentration of sodium, potassium, and chloride; and biuret-reactive material of the spermatophoric plasma, at the following four stages: (1) at the onset of the spermatophoric reaction,

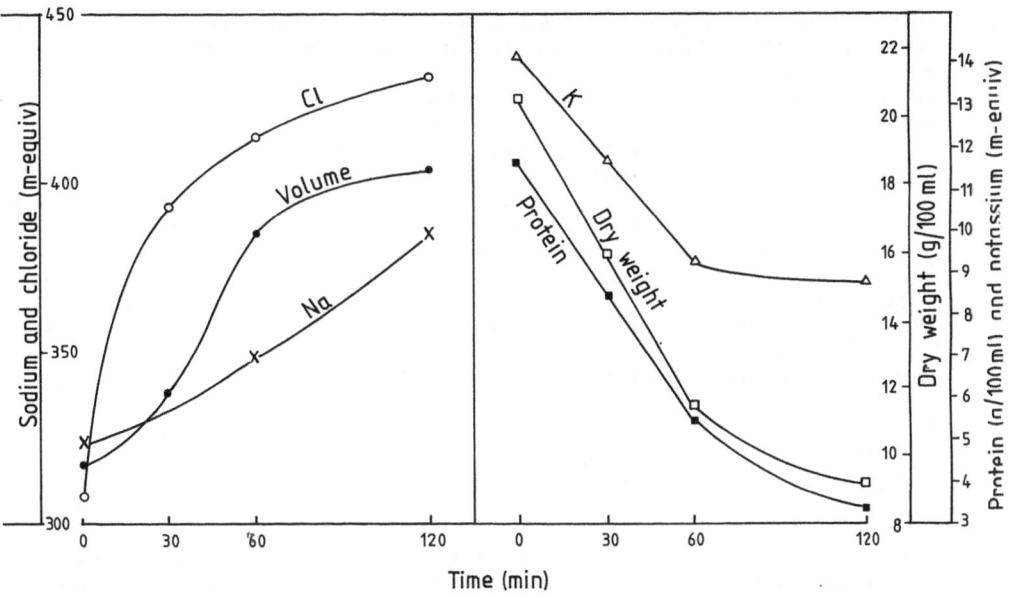

Fig. 23. Changes in the volume and content of dry weight, protein, sodium, potassium, and chloride in the spermatophoric plasma of octopus during the spermatophoric reaction. (Mann et al. 1970)

corresponding to placing the spermatophore in seawater; (2) 30 min later, when the sperm rope began to uncoil and started moving towards the distal end of the spermatophore; (3) 60 min later, after the spermatophoric bladder has just been formed, following the evagination of the ejaculatory apparatus; and (4) 120 min later, after the spermatophoric bladder had expanded further. As can be seen, concomitantly with the uptake of sodium chloride from seawater, the spermatophoric plasma became diluted, and hence its dry weight as well as the potassium and protein concentration were all drastically reduced. Yet, as mentioned earlier (this chapter), osmolality of the spermatophoric plasma undergoing dilution remains remarkably constant throughout the spermatophoric reaction. This is shown in Table 8, with reference to the same experiment as the one illustrated in Fig. 23.

The influx of sodium chloride across the outer tunic of spermatophores during the spermatophoric reaction in vitro has been also studied in other experiments (Hanson et al. 1973; Mann et al. 1981 b). The results of one such study (Mann et al. 1981 b) are shown in Tables 9 and 10. From Table 9 it can be seen that when a 2.7% NaCl solution (corresponding to the NaCl content of seawater in Puget Sound) was used as external medium, the level of NaCl in the spermatophoric-bladder plasma and the spermatophoric-tube plasma became nearly the same as that of the 2.7% NaCl solution, and when twice the concentration of NaCl was used to make up the external medium, then the level of NaCl inside the spermatophore also rose to twice as high. Table 10 shows that if phosphate or glucose, which are not normal constituents of seawater, are included in the 2.7% NaCl-external reaction medium, these two substances enter the internal milieu of the

Table 8. Octopus dofleini martini. Behaviour of osmolality and sodium chloride concentration in spermatophoric plasma during the spermatophoric reaction. (Mann et al. 1970)

Spermatophoric plasma	Osmolality determinations				NaCl determinations	
	Freezing point (°C)	Milli-osmols/ kg water	Corresponding NaCl concentration			
			(g/kg)	(mM)	(g/kg)	(mM)
Before the onset of the spermatophoric reaction	−1.254	674	22.21	379	10.20	174
20 min after the onset of spermatophoric reaction	−1.156	630	20.70	354	14.10	241
In the spermatophoric bladder	−1.257	676	22.57	384	20.20	346
In the tube behind the bladder	−1.246	670	22.17	378	19.60	335

Table 9. Octopus dofleini martini. Changes in pattern of electrolytes (in mEq l^{-1}) during spermatophoric reaction. Three spermatophores were removed from spermatophoric sac of the same octopus. One was used fresh, i.e., its spermatophoric plasma was immediately separated from the sperm rope. The other two were placed in trays containing 1 liter of 2.7% aqueous solution of NaCl and 5.4% NaCl, respectively, and allowed to undergo spermatophoric reaction in vitro, before spermatophoric-bladder plasma and spermatophoric-tube plasma were collected. (Mann et al. 1981b)

	Na	K	Cl	Ca	Mg
Fresh spermatophoric plasma	271	15	223	7	11
Spermatophoric-bladder plasma after spermatophoric reaction in:					
2.7% (462 mEq l^{-1}) NaCl	453	1	429	1	2
5.4% (924 mEq l^{-1}) NaCl	886	1	862	1	2
Spermatophoric-tube plasma after spermatophoric reaction in:					
2.7% (462 mEq l^{-1}) NaCl	452	1	436	1	1
5.4% (924 mEq l^{-1}) NaCl	920	1	880	1	2

spermatophore, and pass into the spermatophoric-bladder as well as into the spermatophoric-tube compartment. As pointed out already, the wall of the spermatophore does not provide a barrier to the entry of foreign chemicals, a fact of some significance in relation to the important question of effects of polluted seawater on reproduction of sea animals generally, and the function of spermatozoa in particular.

As also mentioned earlier (this chapter), the outer tunic of the spermatophore allows not only an influx of substances from the external medium, but enables some organic constituents of the spermatophoric plasma to pass into the surrounding medium. In the study referred to above, when a 2.7% NaCl solution was used as the external medium, 36 mg of anthrone-reactive carbohydrate and 26 mg of bound acid-hydrolysable amino sugar were found to have left the sper-

Table 10. Octopus dofleini martini. Entry of inorganic phosphate and glucose into a spermatophore undergoing spermatophoric reaction. Two spermatophores were removed from spermatophoric sac of the same octopus. In one, the concentrations of inorganic phosphate and glucose in the spermatophoric plasma were determined immediately. The other spermatophore was placed in a tray with 1 liter aqueous solution of 2.7% NaCl (462 mEq 1^{-1}) +0.1% inorganic phosphorus in the form of sodium phosphate buffer, pH 7.4 (32 mEq P 1^{-1})+1% glucose (55 mEq 1^{-1}); at the end of the spermatophoric reaction, analyses of inorganic phosphate and glucose were carried out in the external medium, the spermatophoric-bladder plasma and the spermatophoric-tube plasma. Results are in mEq $1^{-1.}$ (Mann et al. 1981b)

	Inorganic P	Glucose
Fresh spermatophoric plasma	0.1	1
After spermatophoric reaction in:		
external medium	36	53
spermatophoric-bladder plasma	23	37
spermatophoric-tube plasma	22	38

matophoric plasma during the spermatophoric reaction and to have entered the external medium. Corresponding amounts in the spermatophoric-bladder plasma of the same spermatophore were 440 mg bound carbohydrate and 420 mg bound amino sugar. In yet another experiment (Mann et al. 1981 b), when three spermatophores were placed in 150 ml seawater each, the total amount of organically bound nitrogen found in the external medium at the end of the spermatophoric reaction, was 6.1, 25.5, and 112 mg N. As regards the organically bound, acid-soluble phosphate of spermatophoric plasma, this, too, can pass to some extent across the outer tunic. About 13% of the acid-soluble P was shown to enter the seawater surrounding the spermatophore in the course of a spermatophoric reaction in vitro (Brooks et al. 1981). No doubt, a large part of this material must have been present in the form of small-molecular weight compounds, notably glyceryl-phosphorylcholine, but some of it may have been derived from the aforementioned galactophosphopeptide. This compound is known to pass a Cellophane membrane during dialysis, albeit at a relatively low rate.

Chapter 4

Annelida

Spermatophores are sporadically encountered throughout the entire phylum Annelida. They occur widely scattered in polychaetes, including the myzostomids and archiannelids (reviewed by Schroeder and Hermans 1975; Westheide 1984), and are also found in many Clitellata, that is, in both the oligochaetes and leeches (reviewed by Lasserre 1975).

4.1 Polychaeta, Including Myzostomaria, and Archiannelida

Asexually reproducing species apart, polychaetes are mostly gonochoric, though hermaphrodites are occasionally encountered. In their review, Schroeder and Hermans (1975) list a wide range of families, in which hermaphroditism has been definitely recorded. Male polychaetes release the spermatozoa mostly into the exterior, either by way of the coelomic gonoducts or following fragmentation (rupture) of the body wall.

The occurrence of spermatophores in polychaetes has been amply documented by studies on Capitellidae (Eisig 1887), Spionidae (Söderström 1920), Syllidae (Goodrich 1930), Myzostomidae (Jägersten 1934 a, b, 1939; Kato 1952), Arenicolidae (Okuda 1946), Hesionidae (Ax 1969; Westheide 1967, 1969, 1982; Westheide and Ax 1965), and Archiannelida (Ax 1968; Jägersten 1952; Jouin 1968, 1970). Polychaetan spermatophores are extruded either into the water (not necessarily in the vicinity of females) or they are deposited on the surface of the female's body (prior to penetration of the spermatozoa through the skin).

4.1.1 Early History of a Spionid Spermatophore

The first polychaetan spermatophore to be described was that of a *Spio*. An illustration of it appears in Kölliker's (1848) paper on "gregarine parasites" in the intestines of a *Spio*. In it (reproduced in Fig. 24), several details of the spermatophore structure are clearly discernible, including an outer and inner capsule, and the egg-shaped main part of the body fitted with a stalk and membrane. Unfortunately, the author of these observations failed to realize that he had discovered the spionid's spermatophores and regarded the "puzzling bodies" in *Spio* as possibly related to a parasitizing gregarine protozoon („rätselhafte Körperchen, die möglicherweise zu den Gregarinen in Beziehung stehen"). Twenty years had to elapse before this mistaken idea was refuted by Claparède and Mecznikow (1869),

72

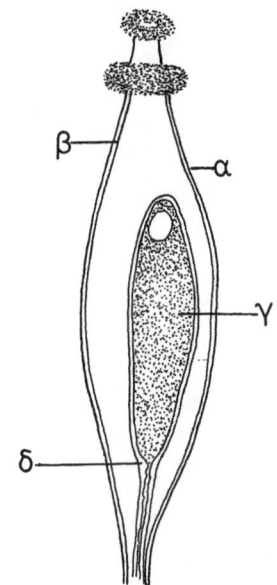

Fig. 24. The "puzzling body" described by Kölliker (1848) in a spionid polychaete; α outer capsule; β inner capsule; γ "egg-shaped body" fitted with a stalk and surrounded by δ a fine membrane. The contents of the egg-shaped body were later shown by Claparède and Mecznikow (1869) to consist of spermatozoa

who also deserve the credit for pointing out that the spermatophore of *"Spio Mecznikowianus"* is assembled from spermatozoa only after these have passed into the "middle and posterior portions of the male segmented organ" and there became engulfed by the special secretory materials.

4.1.2 Role of Nephridia in the Formation of Spionid Spermatophores

In due course, the modified nephridia of the male spionids were conclusively proven to be the "segmented organ" in which the spermatophores are formed. The occurrence of similar nephridial modifications was also demonstrated in the closely related family Disomidae. In addition, some spionid species were shown to possess special "seminal receptacles" or "genital sacs" (gamete-storing pockets on the sides of gonad-bearing segments). Species of *Spio, Pygospio,* and *Polydora,* amongst other spionids, have been studied in detail with respect to the mode of production and structure of spermatophores (Cerruti 1908; Greve 1974; Rice 1978, 1980, 1981; Rice and Simon 1980; Richards 1970; Söderström 1920). In the course of these investigations it became abundantly clear that within the family of Spionidae there are many distinct forms of spermatophores and several different ways in which they are deposited and transferred. Figure 25 shows a planktonic spionid spermatophore as actually observed by Grewe (1974) in a ctenophore culture vessel where *Spio filicornis* lives. When freshly emerging from the sand of that culture, this spermatophore looks like a transparent toy ballon, 1.6 mm in diameter, joined to a 1.5-mm-long pedicle. Close to the juncture is located

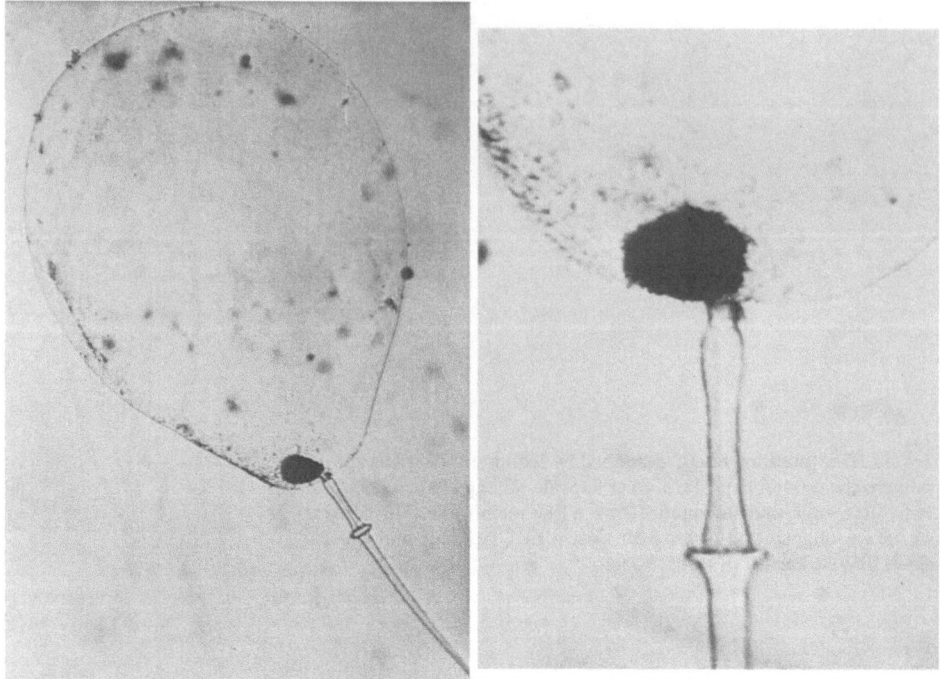

Fig. 25. The planktonic spermatophore of a spionid polychaete, shortly after emerging from the sand in a ctenophore culture vessel; total length about 3 mm. Details of the base of the spermatophore's bladder containing the mass of spermatozoa, are shown on the *right*. (Further details in Greve 1974. Courtesy of Dr. Wulf Greve)

the mass of spermatozoa which, when activated by light, swim out and aggregate at the wall of the balloon. In *Scolelepis squamata,* the spermatophores are leaf-shaped and they are deposited on the body surface of the female, but in *Polydora ligni* and *P. websteri* they are elongated, thin, and released by the male into the seawater, without contacting a female; in both *Polydora* species, the ultrastructure of the male nephridia and their role in the elaboration of spermatophores were meticulously explored (Rice 1980).

The mature nephridia are segmented organs composed of a nephrostome connected to a nephridial canal, ciliated throughout and composed of several cell types. The cells in the expanded region of the nephridia contain large urn-shaped depressions filled with long microvilli which are directly involved in the production of spermatophores. According to Rice (1980), the sequence of spermatophore formation and release in *Polydora ligni* is as follows.

The nephrostome acts as a mechanical sorting device to separate immature from mature spermatozoa, and directs the latter into the nephridial canal where, in the expanded area of this tract, they are packed together into what will become the head of the spermatophore. The glycocalyx-coated microvilli of the urn-shaped nephridial cells in this expanded area pinch off near their proximal ends and are moulded around the sperm mass by the cilia in the nephridial lumen. The newly formed spermatophore's head then moves along the dorsal nephridial canal and out of the nephridiopore while a tail is added to it. It takes about 3 min from the first appearance of the spermatophore's head in the dorsal nephridial canal to the passage of the tail through the nephridiopore. The final product is a spermatophore with a club-shaped head (0.2 mm long) and a tail (1 mm long). The mass of spermatozoa, centrally located within the head, is surrounded by a layer of randomly oriented tubules which make up a capsule and taper into a long tail. Since these tubules are identical in dimensions to the microvilli present in parts of the nephridium, they may be presumed to be of microvillar origin. The glycocalyx of the microvilli is thought to function as an adhesive to hold the spermatophore together.

At the time when they are released into the seawater, each two spermatophores of *P. ligni* appear coupled together into a pair. It is possible that apart from functioning as an adhesive, the microvillar glycocalyx provides, after the release, some kind of a species-specific recognition signal to the receptive female, thereby assisting in accurate transfer. A female of *P. ligni* will pick up with her palp the con-specific free-floating spermatophores notwithstanding the presence of other species.

4.1.3 Hypodermic Impregnation in Dinophilidae, Protodrilidae, and Hesionidae

An early indication that archiannelid spermatozoa can enter the body of a female through her "skin," came from Sidney Harmer (1889), who, having closely observed the sequence of copulatory events in *Dinophilus,* concluded that

> "the terminal, conical portion of the penis is protruded through the generative pore, and is passed into the skin of the female; spermatozoa are then seen to have passed from the vesiculae seminales, through the skin of the female, and to be accumulating themselves into a mass immediately beneath the perforation made by the penis.
> There seems to be no localization of the spot at which spermatozoa can be introduced into the female. The penis can obviously be inserted into the skin at any point, as shown by the fact that, in the cases actually observed, the point selected was sometimes in the region of the neck, in other cases far back in the body of the female, and in other cases near the middle of the body".

Half a century later, the curious events associated with copulation in *Dinophilus taeniatus* became the subject of a detailed study by Jägersten (1944), who described the lesion on the female's epidermis that serves as an opening for the entry of spermatozoa. Since the "penis" of the male lacks a sperm canal, it became obvious that it does not perform the role of an intromittent organ in the strict sense of the term, but merely breaks down the body wall, possibly with the assistance of special glandular secretions endowed with histolytic properties. Similar events were subsequently described in *Dinophilus gyrociliatus* (Traut 1966), *Dactylopodalia baltica* and *Trilobodrilus axi* (Ax 1968, 1969). In *Dactylopodalia* the formation of the spermatophore by the male (the individual in the male phase of development) is a two-step process. The first to be ejected is the empty, sac-like capsule and only some time later (up to 40 min) is the sperm mass ejaculated into this capsule. The spermatophore, about 30 μm in total length, is glued to the body surface of the female (that is, the partner in the female phase of development) at exactly the same spot at which it emerged from the male. In *Trilobodrilus,* the injection of the spermatozoa occurs within less than a minute after the male organ became attached to the female, probably as a direct result of rapid enzymatic dissolution of the female epidermis.

Many other spermatophore-producing polychaetes behave in an essentially similar manner, that is, they use the surface of the female's body as the site for spermatophore attachment. The archiannelid *Protodrilus* provides an instructive example (Jägersten 1952; Jouin 1970). In most species of this genus, paired sperm ducts, equipped with secretory glands, are present in a series of 3–5 anterior segments. As the spermatozoa pass through the ducts and are pressed out from them, they become coated by the glandular secretions and encased into spermatophores which have the form of balloons sitting on stalks. Spermatophore production has been closely followed in the viviparous *Protodrilus albicans* and in the non-viviparous *P. rubropharyngeus.* In the latter, Jägersten (1952) described the attachment of the spermatophore to a special receptor region on the surface of the female's body, consisting of middorsal, glandular pits which he named dorsal glands. However, unlike *Protodrilus,* some other Polychaeta tend to attach their spermatophores to the female in a rather haphazard manner, without discriminating between different regions of the recipient's body. In this category are some myzostomids and hesionids. In Hesionidae, the attachment of spermatophores and penetration of spermatozoa through the skin have been the subject of particularly thorough and detailed investigations by Westheide (1978, 1979, 1982, 1984).

In the hermaphroditic *Microphthalmus aberrans,* the spermatophores, which are single and equipped with special basal plates, are usually deposited on the posterior part of the body. The integument is pierced by the male organs, appropriately designated as copulatory stylets (rather than penis), and the unbundled spermatozoa are injected through the opening. It could be that the injection of spermatozoa is also facilitated by the male accessory secretions. These are known to contain special spherical granules which are discharged into the lumen of the ejaculatory ducts during sperm extrusion, and then dissolve.

In *Hesionides arenaria,* the cylindrically shaped and 160–175 μm long "double spermatophores" are deposited all over the body, and their attachment to the skin of the sexual partner is a matter of seconds or at the most, a few minutes. The formation of the sheaths of these spermatophores takes place immediately before transfer; the two sperm bundles in a double spermatophore are probably rapidly pushed through the ejaculatory duct and while passing the glandular projections of that duct they are surrounded by a sheath. Six different types of glands can be distinguished according to the ultrastructure of their secretory granules which, in addition to providing material for the sheath, probably contain lytic enzymes necessary for sperm penetration into the body of the female (Westheide 1982). It takes a day or so before the spermatozoa have forced their way into the coelom, leaving the empty wrappers behind. We will return to the question of how annelid spermatozoa pass from the epidermis to the site of fertilization later in this chapter, when dealing with the function of spermatophores in leeches.

76

4.2 Oligochaeta

Unlike Polychaeta, which are fundamentally gonochoric, Clitellata, that is Oligochaeta and Hirudinea, are regarded as fundamentally hermaphroditic. Clitellum is the name given to the special glandular development of epidermis, which extends over the region of genital pores in the clitellates. Asexual reproduction is seen in some oligochaetes, but is absent in leeches (reviewed by Lasserre 1975).

4.2.1 Main Features of Reproductive Function in Oligochaeta

In sexually reproducing oligochaetes there are usually present one or two pairs of testes and a single pair of ovaries. Spermatozoa pass from the testes into the so-called seminal vesicles where they undergo a maturation process. They then pass along the sperm ducts (vasa deferentia) before emerging through the gonoducts (coelomoducts). In some oligochaetes (naidids, tubificids, and phreodrilids) the sperm ducts expand into the so-called atrium, which may be connected to the male pore by means of a special ejaculatory canal. With the sperm ducts are also frequently associated various glands known under the general name of prostate. Spermatozoa are deposited in the spermatheca and there, in due course, they come into contact with the cocoon (secreted by the clitellum) at the time when this passes over the spermathecal pores.

4.2.2 Spermatophores of Tubificidae and Other Oligochaeta

Transfer of spermatozoa by means of spermatophores is not universal amongst oligochaetes, but it is highly characteristic of some, particularly the tubificids, in which the spermatozoa are aggregated in the atrium, probably under the influence of prostatic secretion, and enveloped into elongated spermatophores which after copulation are stored in the spermathecae. Among the early investigators of this subject, was Ray Lankester (1871), who in his paper *On the structure and origin of the spermatophores of two species of Tubifex* (p. 181) reported as follows:

> "The sperm-ropes of *Tubifex rivulorum* I have found in the copulatory pouches both in summer and winter, but especially abundant and well formed in the winter. They have a worm-like figure, with a curious conical head, an average from $\frac{1}{20}$ to $\frac{1}{15}$ of an inch in length, and from $\frac{1}{500}$ to $\frac{1}{200}$ of an inch in breadth, the narrowest part being that immediately succeeding the conical head, which has a breadth of about $\frac{3}{1000}$ of an inch.
>
> It appears that the material of which the sperm-ropes are formed, namely, spermatozoa and a cementing matrix, must be introduced in a viscid form from the male efferent duct, through the penis of one worm into the copulatory reservoir of another, and in the neck of that reservoir a 'setting' occurs; for the sperm-ropes, when fully formed, are very firm and compact bodies, of high light-breaking power".

Shortly thereafter followed similar observations on spermatophores in *Lumbricus* and other oligochaetes. The study by Katherine Foot (1898) on the cocoons and eggs of *Allolobophora foetida* was of particular importance on account of ob-

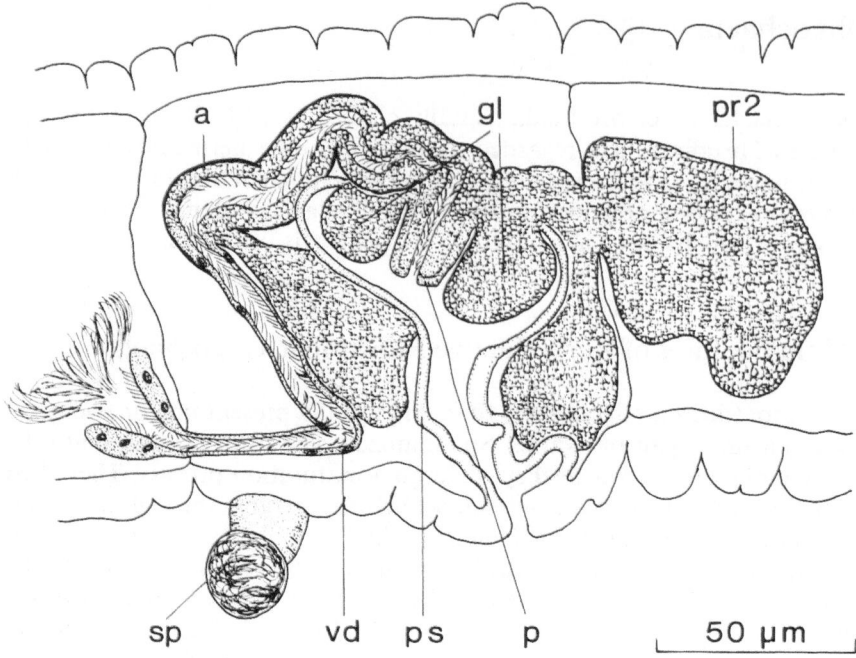

Fig. 26. Lateral view of the genital organs in the segments X and XI of the marine tubificid *Bacescuella mediterranea,* a new species described by Erséus (1980); *a* atrium; *gl* glandular ring; *p* penis; *ps* penial sac; *vd* vas deferens; *pr2* posterior prostate gland; *sp* the spermatophore (round capsule, stalked and filled with spermatozoa) attached to the body wall near the clitellar region. (Erséus 1980)

servations on the "slimy substance secreted by the two worms during union, which encircles the pair around those segments that are in contact" and the "measurements of the various parts of several hundred spermatozoa found in the seminal vesicles, spermathecae, spermatophores, seminal fluid of the slimy-tube and freshly deposited cocoons". More time had to elapse before convincing evidence could be brought forward that many earthworms, among Lumbricidae in particular, use exchange of spermatophores, along with spermathecal insemination, as a method of propagation; a recent survey of the genus *Dendrodrilus* provides an illuminating example of this type of investigation (Gates 1979).

Reproduction in Tubificidae, which, as mentioned earlier, already attracted Ray Lankester's attention, still continues to be extensively studied. Two marine tubificid genera of this kind are *Aktedrilus* and *Bacescuella,*in which the breeding patterns have been thoroughly explored by Erséus (1978, 1980). Both genera are characterized by having two pairs of prostate glands and a well-developed penis. While in *Aktedrilus* external spermatophores are absent and the spermatozoa are transferred by the penis directly into the spermatheca (storage organ), *Bacescuella* uses the penis to deposit external spermatophores on the body wall of the male (Fig. 26). In the recently described new marine species *Bacescuella labeosa,* a tubificid 3–4 mm long (with a clitellum 0.26–0.41 mm wide), the ovoid, unstalked spermatophores measure about 30 µm in the shortest dimension, and 43–51 µm in the longest dimension; their shape is probably determined by the lips of the male copulatory apparatus, and they are broadly attached to the epidermis of the mate, in or near the clitellar region (Baker and Erséus 1982).

Ryacodrilus arthingtonae is yet another tubificid which produces spermatophores; a special characteristic of this species is the occurrence of an unusually long vas deferens (Jamieson 1978). In general, external spermatophores are not very common among tubificids, but in addition to the genera already mentioned, they occur in *Bothrioneurum* and *Paranadrilus,* in which however, copulation and cross-fertilization do not represent the only method of reproduction; the first of the two genera reproduces also by fragmentation, and in the second, uniparental reproduction, possibly by parthenogenesis, predominates (Gavrilov 1955).

The secretory activity of the tubificid male reproductive tract has been recently explored in *Tubifex hattai* (Jaana 1982). In this freshwater tubificid, the prostate and atrium contribute a special kind of secretory granules which are held responsible for building the matrix of the peculiar tubular sperm-aggregates or "spermatozeugmata" found in spermathecae. The matrix which cements the spermatozoa in the wall of such spermatozeugmata reacts to histochemical tests for proteins and polysaccharides in exactly the same manner as the cytoplasmic granules of the atrial epithelium. After the release of spermatozoa, the matrix disintegrates and is probably absorbed by the spermathecal epithelium.

4.3 Hirudinea

Leeches, like oligochaetes, are hermaphrodites. They reproduce by internal fertilization, in many cases made possible by an exchange of spermatophores.

4.3.1 Main Features of Reproductive Function and Spermatophores in Hirudinidae

There are from four to ten pairs of testes in a leech, but only a single pair of ovaries. The spermatozoa pass from the testes along the vasa deferentia (which progressively enlarge, forming the sperm-filled ejaculatory bulbs) and then continue towards the atrium; during the passage, the spermatozoa often aggregate into dense bundles. In Gnathobdellida, the atrium connects with a spacial saccular bursa which by protruding externally forms an eversible "penis"; direct sperm transfer, as in oligochaetes, takes place during copulation. In Rhynchobdellida, which lack an eversible intromittent organ, sperm transfer is indirect, that is, by spermatophores deposited during "dermal copulation", similar in kind to that in some polychaetes. However, departures from these two basic patterns are not uncommon (Lasserre 1975; Mann 1962; Sawyer and Chamberlain 1972), and doubts are still being raised whether the intromittent organ of those leeches that have it, should be called a penis. Worth repeating in this respect is the view expressed by Whitman (1891, p. 363) nearly a century ago.

"The habit of discharging spermatophores directly into the vaginal office, presupposing direct union of the sexual pores, brings us to relations where such copulatory organs as we find in the Gnathobdellidae would become useful. The penis is here only an eversible end-piece of the *vasa deferentia* – a simple tubular elongation of what in its simplest form would be represented by a pore".

Friedrich Müller (1844), in his doctoral dissertation *De hirunidibus circa Berolinum hucusque observatis* (pp. 33 and 34) described the spermatophores of leeches as "organa singularia filiformia", a "phaenomenon, cujus neque analogon inter reliqua animalia reperire", and „Inter phaenomenon hoc et propagationem relationem quandam existere, nullus dubito". His observations were soon confirmed and extended by several distinguished investigators, among them Leydig (1849), who described the spermatophore of *Piscicola geometrica* as a double sac, surrounded by a membrane and filled with a mass of spermatozoa interlocked in the same bundle-like fashion as already present in ductus deferens. In due course, it became clear that species differ markedly in respect of the shape and size of spermatophores, most of them, however, equipped with a median chamber and two horn-like protrusions or diverticula. The chamber may be enlarged, as in *Trachelobdella punctata* or the diverticula may predominate as in *Pontobdella muricata*. Only one spermatophore is extruded from one leech at a time, and attached to the integument of the other, either in the clitellar area or in other regions of the posterior half of the body.

4.3.2 Dermal Copulation and Hypodermic Impregnation in Hirudinidae

Whitman (1891) was the first to demonstrate that the attachment of a spermatophore to the skin, or "dermal copulation", is followed in leeches by an "injection of spermatozoa through the body wall or hypodermic impregnation, as we may call it", and wrote as follows:

"Although it is well known that spermatophores are of very general occurrence among the invertebrates, and even among many vertebrates, the assertion that, as a perfectly regular and normal affair, in animals as highly organized as the leeches, *they represent an injecting apparatus, by means of which the spermatic elements of one individual are forced through the body-wall of another, at any point whatsoever,* may appear almost incredible, even when supported by direct observation many times repeated on different species. That such is certainly the case, however, is very easily demonstrated, and any one can verify it as often as he likes on almost any species of Clepsine that happens to be accessible."

One of the specimens of *Clepsina plana (Placobdella plana)* that Whitman experimented with was observed by him to extrude first a large spermatophore (8 mm long and 1 mm wide), which when placed on the skin of another leech stood "nearly perpendicular to the surface", but the spermatophores subsequently produced by the same specimen were much smaller: "Repetition of the act seemed to exhaust the individual's power of forming spermatophores". Upon placing under a magnifying glass the leech that has just received the spermatophore, he could "see the spermatozoa slowly flowing from the narrow mouth of the case through the skin."

Whitman's assertion that the spermatophores deposited during "dermal copulation" "represent an injecting apparatus" was soon verified by others, who also showed that in most instances the spermatophores are assembled in the atrium. The leeches on which these observations were made, included a number of Rhynchobdellida (*Glossiphonia, Helobdella, Piscicola, Pontobdella, Branchellion, Hemiclepsis*) and also certain Acanthobdellida and Gnathobdellida, *Erpobdella* (= *Herpobdella, Nephelis*) in particular (Brandes 1901; Brumpt 1900; Dall 1979; Damas 1968; Elliott 1973; Mann 1953; Nagano 1957). An updated record of the process of spermatophore transfer in leeches has been furnished by Westheide (1980, 1981) (Fig. 27).

Brumpt (1900), whose extensive treatise on *Reproduction des Hirudinées* included many observations on the mating behaviour of leeches, rightly described as a „prélude de l'accoupplement" the involved preparations by the intertwining partners, preceding the attachment of spermatophores to their bodies. He was also able to observe how the spermatophores that became attached to the body wall empty their contents, so that the released spermatozoa (by now "unbundled") can penetrate into the coelomic sinuses and eventually reach the ovarial sacs. In *Erpobdella* he found some spermatozoa around the ovaries within a few hours of spermatophore attachment. Brumpt favoured the view that a spermatophore that invades the cutaneous tissue does so owing to mechanical pressure by injected fluid. The concept of histolysis or proteolysis was evoked only some time later, and the view favoured by most investigators at the present time is that the penetration of the integument by the head of the spermatophore is the result of combined expulsion pressure and cytolytic action; protease and hyaluronidase have been found in the spermatophore of *Glossiphonia complanata* (Damas 1968).

However, even now it is far from clear how the spermatozoa released into the cutaneous tissue of the leech spread throughout the body cavities and make their way to the ovary. As a matter of fact, it is arguable whether the terms dermal copulation and hypodermic impregnation should be used at all, since direct contact between the genitals is not involved and the so-called penis which makes it possible for the male to place the spermatophore on the partner's body does not serve as an organ of intromission into the female reproductive tract. Various suggestions have been put forward at one time or other as to what the full significance of such a penis might be. Of all the ideas that have been put forward, the most bizarre came from Lang (1881) who, it will be recalled (Chap. 2), made the discovery of spermatophores in turbellarians. According to him, the penis of these particular platyhelminthes, serves the male as an effective "weapon for attack or defence" (Waffen zum Angreif oder zur Vertheidigung).

A related question, also in need of clarification, is how dermal copulation and hypodermic impregnation evolved into a method of propagation. Whitman's (1891, p. 363) view was that "this method of impregnation represents an important economical step in advance of the more primitive mode of setting the seminal elements free in water". However, the economy resulting from hypodermic impregnation cannot be all that important because many of the invading spermatozoa have no chance of reaching the ovary since they are disposed of by phagocytosis and possibly also other means. Brumpt (1900), contesting Whitman's view on "economy", preferred another hypothesis (p. 425):

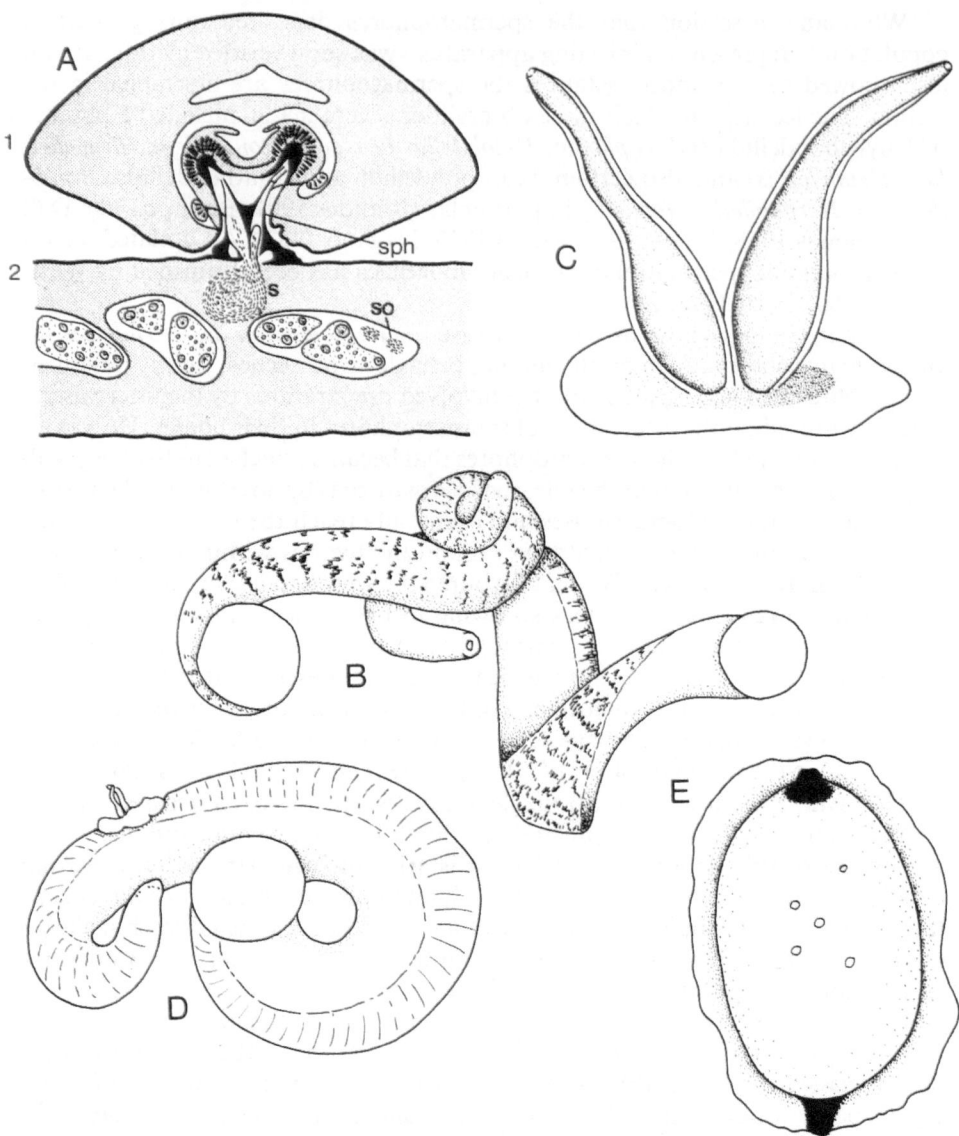

Fig. 27 A–E. Schematic presentation of the process of mating and transfer of spermatophore in a leech (*Erpobdella*). *A* Mating; *1* and *2* sections through the region of the male genital pore and the implanted spermatophore (*sph*); *s* the released spermatozoa invading the partner's body; *so* spermatozoa in the ovary. (After Brumpt 1900). *B* The intertwined mating partners (filmed). *C* The "double spermatophore." (After Brumpt 1900). *D* The leech with attached spermatophore (filmed). *E* A cocoon. (After Bennike 1943). (Reproduced from Westheide 1980)

«Il est probable que cette copulation, qui permet un accouplement plus facile, plus rapide que la véritable copulation, plus avantageux par conséquent pour l'espèce, a été fixée par la sélection et s'est définitivement établie dans les différents groupes où nous l'avons signalée. C'est la seule hypothèse qui permette d'expliquer les curieux phénomènes de la fécondation chez les Hirudinées et en particulier l'organisation du tissu vecteur parfois si hautement différencié et sur lequel nous avons plusieurs fois attiré l'attention.»

Brumpt's view on dermal copulation is much in line with other current ideas (Chap. 1) which favour selective adaptation of species to features of habitat, in preference to vaguely expressed notions on the economy or phylogenetic development, as an important factor most directly related to the whole question of the origin, purpose and function of spermatophores.

Chapter 5

Onychophora and Myriapoda

Of all the phyla of invertebrate animals, the Arthropoda certainly exhibit the greatest diversity of reproductive patterns and methods of insemination, spermatophores constituting only one of several devices that make the transfer of spermatozoa possible (Schaller 1979). This diversity is expressed not only in differences between the various classes which compose the phylum, but becomes even more evident when one considers the events associated with reproduction in separate orders, genera or species within each class. Nonetheless, there are certain features of sperm transit mechanisms and function of spermatophores than can be said to be bound up with individual classes and it seemed to us that it might be best, therefore, to deal with the material relating to arthropods in several chapters. The present one concerns Onychophora and Myriapoda. Insecta (Chap. 6), Crustacea (Chap. 7), and Arachnida (Chap. 8) will be treated separately.

5.1 Dermal Copulation and Spermatophores in Onychophora

Some onychophorans apparently insert spermatophores directly into the female genital opening, but most of them depend on dermal copulation similar in kind to that occurring in Annelida (Chap. 4) for transfer of spermatozoa to the eggs. Adam Sedgwick (1885), writing on the development of *Peripatus,* had this to say on the subject of sperm transit:

> "There seems to be no functional intromittent organ, but the male deposits little oval spermatophores quite casually on any part of the body of the female, and, for all that I know, of the male also; e.g. I have often seen them on the head. How these little packets of spermatozoa get into the vagina, and then up the uteri, which are always full of embryos, I cannot conceive.
> Fertilization is apparently effected in the ovary. I have never seen spermatozoa in any part of the female apparatus except in the ovaries, and in small numbers in the upper and of the oviducts at the time when the ova are entering the latter".

No wonder spermatozoa were invisible "in any part of the female apparatus except in the ovaries", since as we now know, the female tract is not the route along which the spermatozoa reach the gonads. How exactly they reach the ovaries remained something of a mystery until the following explanation was advanced by Sidnie Manton (1938) as a result of her observations on *Peripatopsis.* Having satisfied herself that the small spermatophores are indeed deposited by the males anywhere on the female's body, she proceeded to investigate the mechanism of sperm entry. She found that beneath a deposited spermatophore, leuco-

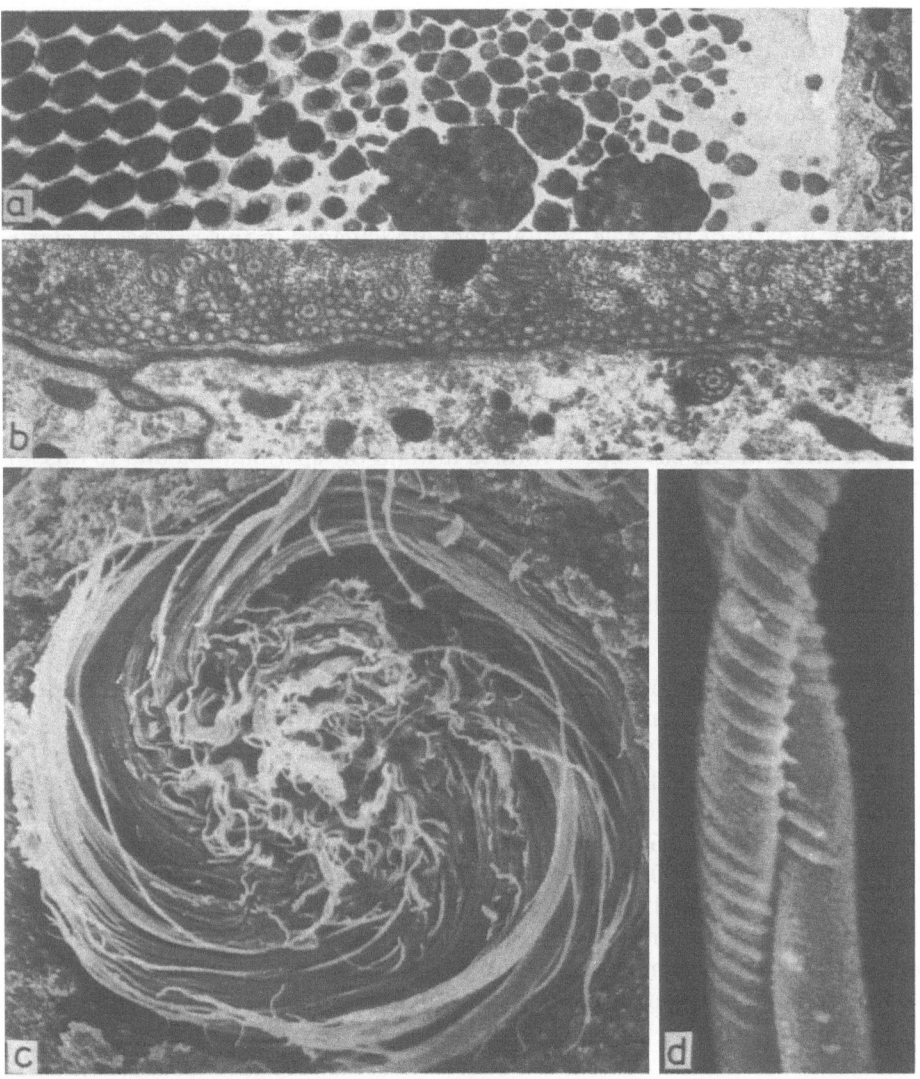

Fig. 28 a–d. Vas efferens with spermatophores of two Onychophora: *Opisthopatus cinctipes* (*a* and *b*) and *Peripatopsis moseleyi* (*c* and *d*). *a* Section through the periphery of spermatophore, showing sperm nuclei, the secretory coat which surrounds them, and vas-efferens epithelium. (× 5400). *b* Ciliated cells of the vas efferens. (× 18,000). *c* Spermatophore with centrally positioned zones of sperm nuclei and secretory coat. (× 850). *d* Heads of mature spermatozoa. (Storch and Ruhberg 1977)

cytes invade the subcutàneous region and break through the ectoderm thereby causing a rupture in both the body cuticle and the attached wall of the spermato-phore. Through the aperture thus formed, the spermatozoa are then able to swim into the haemocoele, and from there to the ovary, where they form a kind of a felt over the region bearing the egg follicles. Eventually, the spermatozoa force their way through the ovarian wall and reach the lumen. The empty cases of the spermatophores remain for some time attached to the cuticle, but in due course, the ectoderm regenerates and a new cuticle is formed. Typical onychophoran spermatophores, such as those of *Opisthopatus cinctipes* and *Peripatopsis moseleyi* (Fig. 28), are of a characteristically oval shape (Storch and Ruhberg 1977).

The process of spermatophore formation in Onychophora is now much better understood than at the time of the initial observations on dermal copulation. From the study of *Opisthopatus cinctipes* (Storch and Ruhberg 1977), it follows that several parts of the male genital tract are engaged in this process. It starts in the testis, where in the spermatocytes electron-dense secretory granules appear as products of the Golgi apparatus. These granules then merge into a droplet which is later extruded by the spermatids in the seminal vesicles. A second vesicular structure, probably an acrosomal vesicle, is produced from the Golgi apparatus near the plasma membrane of the spermatids, and this, too, is extruded into the seminal vesicle. It is from these secretory products that during the passage in the vasa efferentia an envelope is formed around the spermatozoa and to this, in the vas deferens, is added another multilayered outer covering.

5.2 Production of Spermatophores in Pauropoda

Pauropoda, like Onychophora, use spermatophores for sperm transfer, but the actual uptake of spermatozoa by the females has not been recorded. In *Stylopauropus pedunculatus* (Pauropodidae), the male spins a special web onto which he deposits the "sperm drop" in the absence of a female (Laviale 1964). Similar events have been lately described by Schuster and Hasenhütl (1983) in several Eurypauropodidae. In *Gravieripus latzeli,* the supporting web, measuring from 200×300 μm to 480×600 μm, consists of fine threads, and in its middle it has a special opening, about 130 μm wide, into which the sperm drop is placed. The round drop with an average width of 55 μm is surrounded by a thin membrane and contains several hundred filamentous spermatozoa, all completely immotile even after release into water or Ringer's solution.

5.3 Direct and Indirect Transfer of Spermatozoa in Diplopoda

Both direct and indirect transfer of spermatozoa have been observed among Diplopoda. *Polyxenus lagurus* (Pselaphognatha) provides a typical example of the indirect transference method (Schömann 1956). The male, having spun a double zig-zag thread, deposits on it two drops of semen and then proceeds to spin two thicker, parallel threads along which the female comes running to take up the semen. Other similar examples are to be found among Glomerida (Haacker 1968, 1969).

A characteristic property of the tailless, often crescent-shaped or hat-like millipod-spermatozoa is the ability to combine into pairs, each pair forming an in-

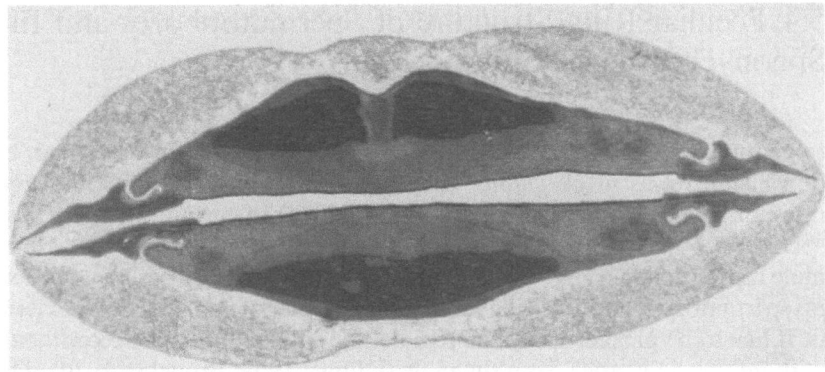

Fig. 29. Spermatophore with two spermatozoa of *Alloporus* (Diplopoda). (× 9,600). (Horstmann 1970. Courtesy of Dr. Heide Breucker)

dividual spermatophore. A binary structure of this sort, first observed under the light microscope, has been lately explored in more detail with the electron microscope in *Polydesmus* (Reger and Cooper 1968) and *Spirostreptus* (Horstmann 1970; Horstmann and Breucker 1969 a, b). As spermiogenesis advances, several organelles in the developing spermatids engage in a series of extensive migratory movements, the result being that the nuclei of two neighbouring spermatids assume a juxtaposition, and the crescent-shaped spermatozoa derived from them become attached to each other along their concave surfaces.

In some respect, the tailless mature spermatozoa of Diplopoda resemble similarly tailless sperm cells of certain other invertebrates, particularly those of Decapoda in the arthropodan class of Crustacea, but the hat-like form of milliped spermatozoa and the manner in which each pair merges into a spermatophore seems to be a special characteristic of the Diplopoda. With the decapods (Chap. 7), the diplopods share furthermore the extraordinary complex and forceful manner which the male uses to get hold and take a firm grip of the female so as to insert his spermatophores. In *Polydesmus,* according to one early authority (Latzel 1884), the "united animals do not part even after death and when force is applied, the copulatory legs of the males, though torn off, remain stuck in the vulvae".

An electron micrograph of the spermatophore of an *Alloporus* sp. (a near relative of *Spirostreptus*) is reproduced in Fig. 29 (Horstmann 1970). It shows the binary structure, completely enclosed in an envelope. The mechanism underlying the formation of this envelope, as indeed of the forces that make two spermatozoa join together along their concave surfaces, is still under investigation, but it appears highly probable that both the attachment and the subsequent formation of the distinct (PAS-positive) envelope around the binary structure is facilitated in some, as yet incompletely understood manner, by secretions produced in the spermatic duct. Of relevance in this respect may be that in *Polydesmus* species single spermatozoa are rarely observed, and then only in the upper portions of the spermatic duct where spermiogenesis is incomplete. Binary spermatozoa occur exclusively in the lower portions of the spermatic duct, the seminal vesicles and in the seminal receptacle of the female (Warren 1934; Reger and Cooper 1968).

5.4 Peculiar Ring Structure of Spermatophores and Indirect Sperm Transfer in Chilopoda

In centipedes, as in millipedes, spermatophores have been described, but in certain instances, as in *Geophilus*, they are of a texture and kind that remind one more of sperm drops (droplet spermatophores) in certain insects (Chap. 6) than spermatophores proper. The shape that a sperm conglomerate assumes in the male reproductive tract of *Geophilus* is that of a ring consisting of orderly rolled-up spermatozoa, and this ring is deposited in the form of a droplet externally, that is, it has to be picked up by the female and inserted into the receptacula seminis without any assistance from the male (Klingel 1959; Breucker 1970). The component parts of the male reproductive tract in which these peculiar ring structures are formed have been determined most thoroughly by Breucker (1970) in *Geophilus linearis*. Her study was preceded 2 years earlier by Horstmann's (1968) similarly exact electron micrograph analysis of *G. linearis* spermatozoa which are about 3 mm long.

Apart from Geophilomorpha, indirect transfer of spermatozoa has also been noted in other centipedes: Scolopendromorpha, Lithobiomorpha, and Scutigeromorpha (Schaller 1971). Typical examples are *Scolopendra cingulata* and *Scutigera coleoptrata* (Klingel 1957, 1960). In the former the male, having spun the web, deposits on it the bean-shaped spermatophore which is then taken up by the closely following female. In the latter, the male places the spermatophore on the ground and then pushes the waiting female towards it using his antennae and legs.

5.5 External Fertilization in Symphyla

Symphyla, a group of small, blind, soil-inhabiting arthropods, produce spermatophores of the sperm-drop type, which have to be taken actively by the females (Juberthie-Jupeau 1956, 1963; Schaller 1979).

Scutigerella immaculata and *S. silvestrii* can produce up to 24 spermatophores in a few days, even in the absence of a female. Each spermatophore is composed of a sperm drop and a stalk. Females collect the sperm drops and store them in special gnathal sperm pockets. Each deposited egg is smeared by the female with the sperm-containing drops and thus fertilized.

Chapter 6

Insecta

The diversity of reproductive patterns and sperm-encompassing devices, so highly characteristic of Arthropoda in general, is a particularly distinguishing feature of Insecta, even though taken as a whole class they show certain uniformity, being bisexual and reproducing mostly by internal fertilization. Propagation by parthenogenesis, however, occurs frequently, especially amongst Hymenoptera (Went 1982).

6.1 Insemination Routes and Sperm-Encompassing Devices

A great many insects produce and transfer semen in the form of ejaculates consisting of highly concentrated suspensions of free spermatozoa in seminal plasma. Semen of this kind is transfered to the females mostly by the intragenital route, but in some insects, such as *Cimex* (bed-bug), the transfer occurs extragenitally. In this instance, the male ejaculates the semen into a special, sac-like pouch, called Ribaga's organ (Ribaga 1897) or *spermalège* (Carayon 1959), which is located on the ventral surface of the female's body. Having broken through the wall of that sac, the spermatozoa scatter throughout the haemocoele of the female and reach the upper end of her oviduct within 2–10 h. Amongst insects, haemocoelic fecundation, similar in kind to that taking place in certain other Arthropoda (Chap. 5) and in Annelida (Chap. 4), has been observed in several Cimicidae, Anthocorinae, and Nabidae (reviewed by Engelmann 1970).

In contrast to those which ejaculate free spermatozoa, many insects employ a variety of pre-packaging devices. Spermatophores are only one of them. Other sperm-encompassing devices, to be described in more detail shortly, include sperm cysts, sperm-bundles (either naked or ensheathed into spermatodesms) and spermatozeugmata (large aggregates of spermatozoa). Spermatophores themselves can take various forms, the two most frequently encountered kinds being the largely naked sperm-drop or droplet spermatophore, and the properly encapsulated type to which we shall refer as the typical spermatophore. Spermatophores of these two kinds are distributed widely amongst many orders of insects, but with variable frequency.

The mode of spermatophore transfer is also highly variable, but species-characteristic. It can be either direct or indirect. The common form of direct transfer to the female is by the intragenital route. During the indirect transfer, the male deposits spermatophores externally, that is, either by attaching them to the surface of the female's body (by no means always proximally to her genital opening), or by placing them on a substratum (a leaf, for example), where they can be found and picked up by the female. But by whatever means the contents of a spermatophore may have reached the female's genital opening, her receptaculum seminis provides the ultimate site at which the spermatozoa are stored prior to fertilization.

6.2 Male Genital System

The material of which the spermatophores are built comes from the male genital system in the form of spermatozoa and accessory secretions.

In most insects, the male genital system is composed of paired testes, seminal ducts (vasa deferentia), seminal vesicles (often acting as storage organs for spermatozoa), a variety of accessory glandular structures, and an ejaculatory duct (ductus ejaculatorius), either single (simplex) or double (duplex) in part. The ejaculatory duct terminates in an opening, the gonopore, which forms the connection between the internal reproductive system and the intromittent organ. However, deviations from this pattern are not uncommon, as, for example, the presence of a single, rather than paired, testis, seminal vesicle, and other accessory structures.

External genitalia arise from two phallomeres, after each has divided to form a mesomere (medial lobe) and a paramere (lateral lobe). The mesomeres usually fuse to form an unpaired intromittent organ, the phallus, which may bear an aedeagus, typically a sclerotized tube around the inner part of the intromittent organ or endophallus. The parameres, on the other hand, depending on species, may develop as claspers or other parts of external genitalia.

The literature on the subject of male reproductive function in insects is extensive, but information concerning the role of male genitalia in contributing the material for building the spermatophores, and the ways in which the spermatophores are transferred, is scattered widely, often amongst publications that are difficult to come by. Early developments in this area have been described by Siebold (1848), Schneider (1883), Stitz (1901), Petersen (1907), and Godlewski (1910–1914), and as regards the subsequent advances, these have been expertly reviewed in respect of gonadal development and kinetics of spermatogenesis (Dumser 1980, Roosen-Runge 1977); participation of secretions produced by male accessory organs in the formation of spermatophores (Leopold 1976); and comparative aspects of male genitalia, sperm transfer mechanisms and function of spermatophores (Davey 1965; Engelmann 1970; Gerber 1970; Gupta 1946; Hinton 1964; Khalifa 1949 a, b, 1950 a, b; Landa 1960; Lipovsky et al. 1957; Schaller 1971, 1979; Scudder 1971; Wigglesworth 1972).

6.3 Relationships Between Sperm-Cysts, Sperm-Bundles, Spermatodesms, Spermatozeugmata, Droplet Spermatophores, and Typical Spermatophores

In insects generally, the testis is composed of a series of tubular follicles, each follicle consisting of cyst-like compartments filled with germ cells undergoing spermatogenesis. Since all the cells within a given germinal cyst divide synchronously and remain interconnected by cytoplasmic bridges, they constitute something of a clone and syncytium. As the spermatids present inside this germinal cyst undergo elongation, so does the cyst itself assume an elongated shape, and in the end it becomes a "sperm-cyst", filled entirely with spermatozoa, all perfectly aligned with each other. The literature abounds in descriptions of testicular follicles and germinal cysts in a variety of insects (Baccetti 1970; Dumser 1980; Roosen-Runge 1977; Wigglesworth 1972). As an instructive and carefully studied situation may be cited the one encountered in the caddisfly *Platycentropis* (order Trichoptera) (Phillips 1974). In some insects the cysts are released from the testes in an intact form, that is, with their envelopes closed, but in others they break open, often releasing whole clusters of spermatozoa in the form of sperm-bundles (Siebold 1836 a) varying in size, shape, number of spermatozoa, and nature of accessory contents. Sometimes, bundles of this sort persist naked, that is, devoid of any external sheath, but in other instances, partial ensheathing takes place, when, for example, the spermatozoa become fastened by the tips of their heads to a common hyaline cap, the bundle thus becoming a more tightly put together bunch of spermatozoa, often of brush-like appearance, since the sperm-tails stick out freely. Cholodkovsky (1913), being aware of the fact that in the locust, for example, a bunch of this kind bears little relation to a true spermatophore into which these spermatozoa are encapsulated only much later, proposed to call it a spermatodesm. Since then, sperm-bundles, either naked or ensheathed, have been observed in many insects amongst Orthoptera, Homoptera, Hymenoptera, Lepidoptera and other orders.

In Orthoptera, Cantacuzène (1968) reported the occurrence of two distinct types of sperm-bundle. One type, common in Tettigoniidae, is represented by a bundle composed of about ten mature spermatozoa, hooked together by acrosomes; it becomes a spermatodesm after acquisition of a special mucous coat in the seminal duct. In the other type, characteristic of Acrididae, the coat is laid down already at the stage of spermatids, that is, before completion of spermiogenesis, but later, this coat is reformed into a structureless, mucoprotein-containing cap, holding together about 200 bundled spermatozoa within a spermatodesm. In her subsequent study of spermiogenesis in *Locusta migratoria* the same author [Szöllösi (Cantacuzène) 1975] drew a clear distinction between ten developmental stages of which the differentiation process consists. Formation of the spermatid-bundle occurs during stage 8, but assembly of the spermatodesm, completion of the cytoplasmic sloughing off, and total condensation of the nucleus take place during stage 9. The transfer of spermatodesms into the seminal vesicle occurs during stage 10, when all microtubules other than the axonemal ones disappear.

Amongst Homoptera, the Coccoidea and Cercopidae have been studied with special care and in an exemplary critical fashion. The spermatozoa and sperm-bundles of coccids have been the subject of an extensive investigation by Robison and Ross (Robison 1966, 1968, 1970, 1972, 1977; Ross 1971; Ross and Robison 1969). Mature coccid spermatozoa, which are long and filamentous (200–300 μm × 0.25–0.3 μm), consist mainly of arrays of microtubules which surround elongated threads of nuclear material. They are packed into tightly ensheathed bundles of 16–256, in which they maintain a precise alignment, often in the form of multistranded helix. These bundles exhibit three-dimensional motility which persists within the female reproductive tract for periods varying from a few hours to several weeks. The motility of the bundle derives from the synchronous movements of the spermatozoa, but it may be complemented by structural specialization of the bundle sheath. In the male scale insects of the species *Parlatoria oleae,* each sperm cyst becomes a bundle with 32 spermatozoa. The bundles reach the vagina and spermatheca in an intact form before the spermatozoa are set free. Subsequently, the liberated spermatozoa pass along the oviducts to the ovarioles. In *Melanaspis smilacis,* the number of spermatozoa within each bundle is the same as in *Parlatoria,* but in *Melanaspis obscura* it is only 16, while in *Matsucoccus bisetosus* it exceeds 60 (Fig. 30). The sheaths can be thin or thick, and in some instances they are in complex shapes. In *Matsucoccus bisetosus* the sheath is modified into a thickened anterior cap (Fig. 30), and in *Stomacoccus platani* the anterior portion takes the form of a hardened corkscrew.

Spermatogenesis and the formation of ensheathed sperm-bundles or spermatodesms in a cercopid insect, *Philaenus spumarius,* were already described by Robertson and Gibbs (1937), but a more detailed investigation along similar lines, and extending over other Cercopidae, Cicadidae, Ledridae, and Ulopidae, followed some time later (Chevaillier 1962, 1963; Chevaillier and Maillet 1965; Folliot and Maillet 1970; Maillet 1959). These studies, which involved electron-microscopic and cytochemical observations, clarified the role of the special satellite cells within the cercopid testis in initiating the early stage of bundling and ensheathing the male germ cells; they also demonstrated that the content of the PAS-staining material increases as the ensheathed sperm-bundles pass along the male reproductive tract on their way to the seminal vesicle.

In a hymenopteran insect, the honeybee *Apis cerana indica,* the 72 spermatozoa constituting an average sized bundle are arranged in a hexagonal geometric array, and fastened to a hyaline cap rich in phospholipid and carbohydrate material which, however, disappears after the bundle has entered the seminal vesicle. The function of this material remains to be explored, but it has been suggested that the phospholipid within the cap might conceivably provide the honeybee spermatozoa with oxidizable substrate needed for respiration (Bawa and Marwaha 1975). If this supposition could be experimentally verified, this would indicate that the spermatozoa of the honeybee share with those of mammals the ability to respire at the expense of fatty acids derived from phospholipids (Hartree and Mann 1959, 1961).

Lepidoptera are yet another order of insects in which sperm-bundles occur with great frequency. Events associated with sperm packaging in the male reproductive tract of the cabbage looper, *Trichoplusia ni,* are particularly illuminating

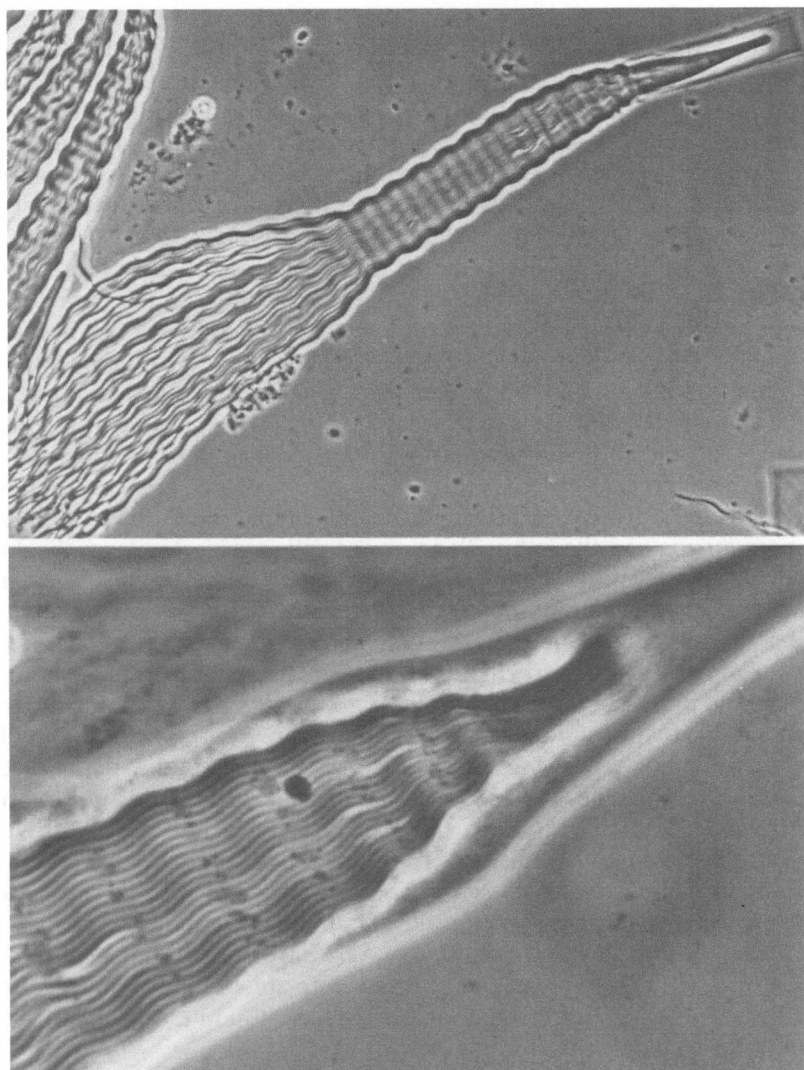

Fig. 30. Sperm bundle of the coccid insect *Matsucoccus bisetosus,* with 60 or more filamentous spermatozoa alligned in parallel, fully ensheathed, and forming a compact unit. *Top* bundle with living spermatozoa exhibiting synchronous, two-dimensional undulations which progress from anterior (*right*) to posterior (*left*) in a continuous train; the sheath of the bundle has been ruptured in preparation of the squash and the spermatozoa became deranged posteriorly; Ringer, phase contrast, (× 610). (Robison 1972); *bottom* region near the anterior end of a sperm bundle which shows the precise alignement of the spermatozoa, the sheath modified into a thickened anterior cap; this cap extends several microns beyond the tips of the spermatozoa; Acetocarmine squash. (× 2100). (Courtesy of Dr. W. Gerald Robison and the Rockefeller University Press)

in this respect. Two types of spermatozoa are produced in the testes of this noctuid moth: eupyrene (fertilizing) and apyrene (anucleate). Both become enclosed in bundles, together with some trophic cells, before passing to the seminal vesicles first, and then, by way of vasa deferentia, to the ductus ejaculatorius duplex. However, even before reaching the seminal vesicles, the sperm-bundles undergo some profound changes. In the bundles containing the eupyrene spermatozoa, following lysis of the trophic cells, a granular lysate is formed which engulfs the individual sperm cells, but as the lysate disappears and the interior of the bundles is thrown open, some new material appears in the form of characteristic strands that produce a network over the individual spermatozoa and a new cover for the bundles which thus revert once more from an open to a closed state. The rebuilt sperm-bundles then continue their passage to the ductus ejaculatorius duplex. In the apyrene bundles, lysis of the trophic cells is completed by the time they leave the testes, and although some strands of material are found to be associated with the apyrene cells in the seminal vesicles, they soon disappear, leaving the sperm cells detached from each other. At the time of mating, both the bundled eupyrene spermatozoa and the unbundled apyrene cells are ejected from the ductus ejaculatorius duplex into the ductus ejaculatorius simplex, and here they are encapsulated into a spermatophore by the male accessory secretions. The male deposits the spermatophore in the female's bursa copulatrix, from whence the spermatozoa pass into the spermatheca. In the contents of a spermatophore, dissected from the bursa copulatrix of a female about 20 min after the end of copulation, Riemann (1970) found the eupyrene spermatozoa to be still in bundles, but the supporting network of the strands was degenerating; in the spermatheca, both eupyrene and apyrene spermatozoa were present. Whether or not the transport of eupyrene spermatozoa to the spermatheca is somehow assisted by the apyrene cells is still a matter of dispute. Ever since their discovery by Meves (1903), the apyrene sperm-cells have been periodically credited with an ability to facilitate, in one way or another, the transfer of eupyrene spermatozoa of insects from the spermatophore to the spermatheca. As regards containment of eupyrene spermatozoa within a bundle, this is believed to shield the sperm cells from adverse effects which some of the male accessory secretion in the ejaculatory ducts might otherwise have exerted upon them.

Distinct from sperm-cysts and sperm-bundles are two other types of sperm agglomerates, namely, the spermatozeugmata and the spermatophores. A large sperm-aggregate designated as spermatozeugma may, to a casual observer, bear some resemblance to a spermatophore, but it lacks the proper capsule and the highly organized manner of sperm assembly which is so characteristic of typical spermatophores. The name *Spermatozeugma* (Greek zeugma, union), describing an aggregate of "yolked together numerous spermatozoa" (Zusammenjochung zahlreicher Spermatozoen), was introduced by Ballowitz (1890) who subsequently shortened it to *Spermozeugma* first, and then *Spermiozeugma* (Ballowitz 1895, 1916). In later years, when similar sperm-aggregates were also observed in animals other than insects, the use of the terms spermatozeugma or spermozeugma was extended to all such animals, invertebrates and vertebrates (fishes in particular) alike (Philippi 1909).

94

A detailed description of spermatophores and their function will follow later in this chapter, with reference to specific orders, genera and species of insects, but three points concerning spermatophores in general, must be raised now, in advance of detailed discussion. They relate to differences between (1) droplet spermatophores and typical spermatophores in apterygotan and pterygotan insects, respectively, (2) direct and indirect sperm transfer mechanisms, and (3) spermatophoric reaction and the passage of liberated spermatozoa in the female genital tract.

The name spermatophore is commonly applied not only to the typical, structurally highly differentiated and properly encapsulated sperm packages of pterygotan insects, but also to the naked, though mostly stalked, sperm drops or droplet spermatophores of the apterygotan insects. One may wonder if sperm drops of this kind deserve to be called spermatophores at all. However, since the function that they serve, namely sperm transfer, is not different from that of fully encased sperm packages, and the name droplet spermatophore (in preference to sperm drop) is now firmly established in the literature, there seems to be no overriding reason why this name should not be generally accepted. Moreover, the alternative name, sperm drop, might conceivably be confused with that of sperm droplet, which is commonly used to denote a special sperm organelle, namely the cytoplasmic or kinoplasmic droplet which is attached either to the neck or midpiece of individual maturing spermatozoa in a great many animals, including mammals (cf. Mann and Lutwak-Mann 1981).

Whereas pterygotan insects (with some exceptions) transfer spermatozoa during mating directly into the female genital system, sperm transfer in the apterygotan insects is indirect, that is, the spermatophores are deposited elsewhere, not necessarily even near the female's genital opening, and they have to be picked up by the females themselves. In insects that transfer spermatozoa directly into the females, the mechanism underlying the transfer varies, pending on how deeply the male intromittent organ can penetrate the female genital duct. While the male cowlouse *Linognathus vituli* is able to insert spermatophores as far as the spermatheca (Mukerji and Sarma 1951), other pterygotan insects deposit spermatophores in the vagina or bursa copulatrix, from where the liberated spermatozoa later pass to the spermatheca.

During the spermatophoric reaction the spermatozoa can escape from a spermatophore in several ways. In the house cricket, the spermatophore contains a pressure body and evacuation fluid; when the pressure body takes up the fluid, it swells and squeezes out the spermatozoa. In many Lepidoptera, the release of spermatozoa in the bursa copulatrix is probably helped in addition by the combined action of a chitinous structure, the lamina dentata or signum, and of a secretion produced by the bursa, which digests the spermatophore's wall. In *Rhodnius prolixus*, sperm passage to the spermatheca is made possible by rhythmic contractions of the female genital tract, elicited by the secretion of the opaque gland (to be described below), which had been previously incorporated into the spermatophore. After extirpation of the opaque gland, a spermatophore can still be transferred, but the spermatozoa do not leave the bursa (Davey 1965; Engelmann 1970).

6.4 Role of Male Accessory Secretions in the Production of Spermatophores

Over the years, increasing interest was taken in the chemical make-up of secretions that originate in the male accessory organs of insects, and the ways in which the secretory products participate in processes leading to the encapsulation of spermatozoa and formation of spermatophores (Leopold 1976). Proteins, glycoproteins, mucopolysaccharides, lipoproteins, neutral lipids, phospholipids, peptides, and amino acids are among the most ubiquitous constituents.

In *Blattella germanica* and certain other cockroaches, the moulding of spermatophores is made possible by a conjoint action of at least eight glycoprotein-rich accessory secretions (Ballan-Dufrançais 1968). In the domestic cricket, *Acheta domesticus,* the complex polytubular male accessory system secretes a number of distinct proteins, the individual tubules differing in their synthetic ability, as evident from the fact that in glands exposed to labelled amino acids in vivo, individual tubules from specific gland regions incorporate the label at different levels. Between 29 and 32 distinct protein-synthetic patterns have been detected amongst 300 such tubules, by using electrophoresis on sodium dodecyl sulphate-polyacrylamide micro gels (in capillary tubes of 1.1 mm internal diameter for casting the gels). It is of course highly probable that only some of these proteins serve as material for building the wall of the spermatophore, while others perform different functions, as for example, elaboration of spermatophoric plasma (Kaulenas 1976).

In the migratory locust, *Locusta migratoria,* the male accessory system consists of 16 pairs of distinct, delicate, blind-ended tubules, all opening into the ejaculatory duct (Mika 1959). Later in this chapter, we shall consider in more detail the function of such a polytubular complex in relation to the formation of the large spermatophore of the African subspecies, *Locusta migratoria migratorioides* Reiche and Fairmaire (Gregory 1965 a, b). In this insect, the 16 pairs of tubules form two masses, each 8–10 mm long and 4–5 mm wide, lying in the abdomen on either side of the hind gut. Each mass comprises four "white glands" filled with milky fluid; ten "hyaline glands" producing a colourless secretion; one "opalescent gland" containing minute granules; and one seminal vesicle in the form of a tightly coiled tubule which, when uncoiled, would measure 24–31 mm in length; the semen stored within this tubule gives it a whitish colour, but the sheath surrounding the coiled seminal vesicle is in the form of an oval, lipid-containing, orange- or yellow-coloured body. The ejaculatory duct itself is 4.4–4.6 mm long, and has a maximum diameter of 0.9–1 mm. It consists of an upper ejaculatory duct, into which are voided the contents of the accessory polytubular glands, and a lower ejaculatory duct which opens beneath the copulatory organ into the ejaculatory sac; this sac, in turn, opens through the gonopore into the spermatophore sac, an expansion of the base of the aedeagal canal of phallotreme (further details, this Chap. 6.10, Fig. 34).

Similar in composition to that of the migratory locust are the two masses of male accessory glands in the desert locust, *Schistocerca gregaria,* each mass containing 16 glands that secrete material necessary for the formation of spermatophores. Based on histological and histochemical findings, these 16 glands have been divided into several distinct types. One produces a proteinaceous, crystalline secretion; two, a minutely fibrous secretion; three other gland types, a globular secretion; one more, a lipoid secretion; but "gland 16", which is a typical, functional seminal vesicle, acts probably mainly as a storage organ for the spermatozoa (Odhiambo 1969 a, b).

The grasshopper *Gomphocerus rufus* is yet another acridid in which the tubular spermatophore is built from various proteins secreted by the 16 tubules of the paired accessory glands (Hartmann 1970). Based on their contents, these tubules have been subdivided into distinct types. Selective surgical ablations of the different gland types indicate that they perform different functions in relation to the spermatophore, as regards (1) conferment of shape, (2) pressing out the semen, and (3) formation of the spermatophore lumen. Decapitation of a male after the beginning of copulation, that is, when the formation of the spermatophore has already started, does not interrupt the process of spermatophore transfer (Hartmann 1970).

Distinct types of secretions participating in the formation of spermatophores have also been described in non-orthopteroid insects. Thus, in the blood-sucking bug *Rhodnius prolixus,* the jelly-like material which largely determines the size and pear-like shape of the spermatophore is formed from the fluid, protein-rich, neutral secretion discharged by the so-called transparent accessory gland, which turns into a jelly upon coming in contact with a second secretion, pH 5.5, at the level of the bulbus ejaculatorius and the intromittent organ. The bulk of the protein secreted by the transparent gland is actually synthesized in that gland directly, as was demonstrated in experiments on in vitro incorporation of [^{3}H]leucine (Barker and Davey 1982; Davey 1959, 1960 a, b, 1965).

The length of time needed by male insects to complete the assembly of spermatophores from the accessory secretions can be a decisive factor in determining the length of copulation. This follows clearly from observations on the lesser waxmoth *Achroia grisella* (Greenfield and Coffelt 1983): Females mated to males that have not been allowed to mate previously for 7 h 30 min, required only 11 min to copulate successfully and when dissected, they were found to contain spermatophores, but when the females were mated to males that have already copulated within the preceeding 1 h period, they required at least 6 h for copulation and when separated from the males forcibly before that time and dissected, they lacked any spermatophores. Obviously, the extended copulation period is needed because the male is incapable of producing a spermatophore within a shorter period of time.

Though much has been learned recently about the contribution made by individual male accessory secretions to the structure of a spermatophore, the interactions of these secretions during the process of moulding are still unclear. An answer to this fundamental question is unlikely to come until the different secretory proteins have been purified, and the enzymes that act upon them properly identified in the same manner as was done in much earlier studies of protein–enzyme interactions in mammalian accessory secretions responsible for coagulation or gelation of semen and the formation of copulatory plugs (cf. Mann and Lutwak-Mann 1981).

97

6.5 Some Unusual Chemical Constituents of Male Accessory Secretions, and Their Passage into the Semen and Spermatophores:

Indolalkyl Amine, Fructose, Glucose, and Trehalose,
Pyrophosphate, Uric Acid,
Cyclic Guanosine3′,5′-Monophosphate, and Prostaglandin Synthetase

In addition to the ubiquitous constituents such as some of the aforementioned proteins or lipids, the male accessory organs of certain insects are known to generate a number of much less common, in fact, in some instances quite unusual and highly specific secretory products.

Indolalkyl Amine. In addition to the transparent accessory gland, renowned for its role in the production of the transparent jelly for the spermatophore, one other male accessory organ of *Rhodnius,* the so-called opaque gland, is known to fulfil a special secretory function: the production of a substance capable of eliciting contractions of the oviduct, which has been identified as an indolalkyl amine with properties similar to 5-hydroxytryptamine (serotonin). A similar compound has also been shown to occur in the utriculi majores of the cockroach *Periplaneta americana* (Davey 1960a, 1965).

Fructose, Glucose, and Trehalose. In the honeybee, *Apis mellifera,* apart from the coagulable protein produced in the mucous gland, the semen contains an impressive number of amino acids and at least three sugars, fructose, glucose, and trehalose. The sugars are contributed largely by the penis bulb. Freshly ejaculated semen of the honey bee contains fructose at roughly the same concentration as human semen, that is, about 2.4 mg/1 ml. The content of fructose, glucose, and trehalose (μg/mg tissue of the penis bulb) in drones leaving the hive were given as 3.21, 3.02, and 0.51, but the corresponding values after return were only 0.02, 0.09, and 0.18, respectively (Taber 1977).

Pyrophosphate. In the hawk-moth, *Hyles (Celerio) euphorbiae,* the ejaculatory duct accumulates a large amount of pyrophosphate which, however, in its native state is not free, but occurs bound in the form of a loose complex from which it can be set free by extraction with trichloroacetic acid (Heller et al. 1960).

Uric Acid. In *Blattella germanica* and certain other male cockroaches, out of the eight male accessory secretions which participate in moulding a spermatophore, the one contributed by the utriculi majores is extraordinarily rich in uric acid, some of it in crystalline form as calcium urate. During copulation, the uric acid is poured over the spermatophore which later, when empty, is eaten by the female (Roth and Dateo 1964, 1965; Mullin and Keil 1980; Schal and Bell 1982).

Cyclic Guanosine 3′, 5′-monophosphate (Cyclic GMP). Earlier in this chapter, when describing the synthesis of proteins required for the formation of spermatophores, the male accessory gland of the domestic cricket, *Acheta domesticus,* was mentioned specifically as one in which a multitude of proteins is synthesized in different tubular regions. The same gland was also shown to contain exceptionally high concentrations of cyclic GMP, up to 500 pmol/mg glandular protein, and in individual spermatophores taken from the males, the content of cyclic GMP was about 100 pmol / spermatophore (Fallon and Wyatt 1975). Once more, an interesting analogy has thus been shown to exist between seminal characteristics of insects and mammals. Just as fructose, the chief seminal sugar of mammals (Mann 1946), is as characteristic of honeybee as of human semen, so is the high concentration of cyclic GMP a chemical feature that the domestic cricket and man share together; the levels recorded for human seminal plasma are 0.54–3.57 nM of cyclic GMP (Mann et al. 1981 a).

Prostaglandin Synthetase. The above analogy can be extended further. Just as the seminal vesicle of man, so are the testes, seminal vesicles and spermatophores of *Acheta domesticus* capable of synthesizing prostaglandin (PG). Moreover, prostaglandin synthetase has been detected in the bursa copulatrix, spermatheca, spermathecal canal, and oviducts of female crickets, but only after mating, and it would appear that the female receives this multienzyme from the spermatophore; a substance similar to PGE-2 has been detected in the reproductive tissues of both the male and the mated female, but not in virgin females (Destephano and Brady 1977).

Like those of *Acheta,* the spermatophores of the Australian field cricket *Teleogryllus commodus* carry the prostaglandin-synthesizing enzyme system, and in addition, they contain a substantial quantity of arachidonic acid (Loher et al. 1981; Stanley-Samuelson and Loher 1983). Prostaglandins as such, however, are virtually absent from these spermatophores. Arachidonic acid (5, 8, 11, 14-eicosatetraenoic acid) has long been recognized as the key substance in the biosynthesis of prostaglandins by mammalian vesicular tissue, and to function as a potent generator of PGE-2, and to a smaller extent, PGF-2α. Both these prostaglandins have now been shown to be formed upon incubation of the contents of *Teleogryllus* spermatophores with arachidonic acid (Tobe and Loher 1983).

Little is known as yet about the enzymes that make up the prostaglandin-synthesizing system in insect spermatophores, or its hormonal regulation. The three main unsaturated fatty acids in the spermatophore of *Teleogryllus* are arachidonic, linolenic, and eicosapentaenoic acid; their combined content is about 0.25 µg/spermatophore (Stanley-Samuelson and Loher 1983). However, since they are largely bound to phospholipids, arachidonic acid as such is not available for direct conversion to prostaglandins. It would have to be set free first by some phospholipase activity, possibly after mating, that is, in the female genital system. As regards regulation by hormones, it is possible that steroids are implicated in a manner similar to that described in the biosynthesis of prostaglandins in vertebrates. It is worth noting in this connection that ecdysteroids have actually been shown to be present in the spermatophores of the Mediterranean field cricket, *Gryllus bimaculatus* (Hoffmann and Behrens 1982); the mean concentrations, determined by radioimmunoassay, were (pg/mg fr wt): Ecdysone, free 129, conjugated 38; 20-hydroxyecdysone, free 40, conjugated 19. Further reference to prostaglandins will follow shortly, when discussing their role in oviposition.

6.6 Gustatory and Aphrodisiac Attributes of Male Secretory Products and Spermatophores

In addition to the usual auditory, olphactory (pheromonal) and tactile signals, mating behaviour of many insects involves other channels of communication, a most bizarre one being the habit of females to feed on male-derived secretions and spermatophores. Richards (1927), in his review of the early literature on sexual selection and other reproductive habits of orthopteran and neuropteran insects, listed a number of such instances, such as, for example, that "in the Locustid grasshoppers one lobe of the spermatophore, the so-called spermatophylax, is devoted to satisfying the female's appetite; by the time it has been eaten, the sperm has passed into the spermatheca" or that the female tree-cricket "as soon as she leaves the male for good, she eats up all that is left of the spermatophore". The same reviewer, however, was also fully aware of the notoriously ferocious appetite of females in general, and that "in a number of insects the male is eaten by the female during or after mating". Since then there have been many more observations on the nutritional value of spermatophores and "mate cannibalism" (Alexander and Otte 1967; Gwynne 1983), and most observers, both early and recent, concur with the idea that, apart from general nutritive value, the male-derived secretions and spermatophores can provide the female with substances that are directly beneficial to her reproductive function, particularly egg maturation or oviposition.

There can be hardly any doubt that in terms of caloric value and protein content, a spermatophore consumed by a female insect can indeed provide her with a substantial amount of food. Mormon cricket (*Anabrus simplex*) male has a very large spermatophore and loses up to 27% of his body weight as a result of spermatophore production, and in some other insects even higher values have been reported. No wonder that the ability of male insects to produce normal spermatophores is very much dependent on adequate nutrition (Chap. 1).

It has also been known for some time that proteins transferred to the female with the spermatophores are utilized in her tissues. In grasshoppers and butterflies, labelled spermatophore proteins are promptly and efficiently incorporated into the ovaries, and in some insects labelled amino acids from spermatophores have been shown to be used by both eggs and somatic tissues. Equally convincing are the numerous observations pointing specifically to the effect of spermatophore-derived substances on egg maturation and laying, such as, for example, that female crickets feeding on spermatophores and male glandular secretions substantially increase the number of their eggs and progeny, or that mating of female locusts without spermatophore transfer decreases their fecundity (Boggs and Gilbert 1979; Friedel and Gillott 1977; Gwynne 1981, 1984; Leahy 1973; Leopold 1976; Sakaluk and Cade 1980). Experiments involving artificial insertion or removal of spermatophores from the bursa copulatrix have been particularly helpful in clarifying the mechanism which controls ovarian function; an example is provided by the study of vitellogenesis in *Blabera fusca* (Dictyoptera) (Brousse-Gaury and Goudey-Perrière 1983).

There is, on the other hand, still some doubt as to whether male secretions and spermatophores act merely as enhancers or as true initiators of ovarian maturation, and whether the male-derived substances influence oviposition directly or indirectly, that is, by accelerating the rate of egg maturation (Itgi et al. 1982). From observations on the moth *Zeiraphera diniana,* it would apear that the accessory secretions transferred with the spermatophores during mating can, in fact, stimulate oviposition directly (Benz 1969). On the other hand, in cockroaches, the effect that the spermatophore exerts after deposition in the bursa is on oogenesis in the first place, most probably because of an activating influence on corpora allata (Engelmann 1970). Similarly, in *Acanthoscelides obtectus* (Coleoptera), a water extract prepared from spermatophores and injected into the female's abdomen stimulates oogenesis, but does not influence egg-laying. The active principle has been partially purified by column (Sephadex) chromatography, and on further purification two "paragonial substances", A and B, were separated; A stimulated oogenesis at 0.2×10^{-3} μg/μl concentration, but B, at a similar concentration, inhibited oogenesis and was generally toxic to the female (Huignard et al. 1977).

After young *Acanthoscelides* males had been injected with radioactive arginine or histidine, the C-isotope was incorporated into the male accessory glands and spermatophores from which it passed during mating to the female. Within 2 h of mating, radioactivity appeared in the haemolymph of the female, reaching a maximum at about 18 h; after some delay, the oocytes also became radioactive (Huignard 1983).

For the time being, the question concerning the chemical identity of the male-derived activators of oogenesis and oviposition remains wide open and in particular the relation to prostaglandins merits further attention. Earlier in this chapter, observations were described concerning the ability of cricket spermatophores to transfer the multienzyme system designated as prostaglandin synthetase from the male to the female genital tract. In the domestic cricket, *Acheta domesticus,* two prostaglandins, PGE-1 and PGE-2, and to a lesser extent also PGF-2α, have been shown to stimulate oviposition in virgin females, and from these observations the authors concluded that the post-copulatory biosynthesis of prostaglandin in the female genital system is directly responsible for triggering oviposition (Destephano and Brady 1977). Further clear indication that prostaglandins stimulate oviposition came from the studies of the Australian field cricket, *Teleogryllus commodus* (Loher et al. 1981). In the mated female of this species, oviposition follows shortly after the contents of a transferred spermatophore have reached the spermatheca, and it is probably triggered by PGE-2, freshly produced from arachidonic acid in the spermatheca itself, but under the influence of a vital component of the prostaglandin-synthesizing system, which has passed all the way from the testes, via the spermatophore, to the female. The transfer of a spermatophore from an orchidectomized, and therefore sperm-free male, does not stimulate oviposition, but a normal mating-induced release of eggs can be effectively stimulated by injecting nanogram quantities of PGE-2 into the oviduct.

Effects on egg maturation and egg-laying apart, certain male accessory secretions, including some that are incorporated into spermatophores, are credited with aphrodisiac attributes of direct relevance to female copulatory behaviour and fecundity. In his monograph on the *Physiology of Insect Reproduction,* Engelmann (1970) lists a number of such instances, among them the long-known alluring properties of the secretion produced by the metanotal glands (Hancock's glands) of the male cricket *Oecanthus fasciatus,* on which the female feeds prior to copulation. The functional significance of the occurrence of alluring glands in this and other insects is said to keep the female quiet during the transfer of the spermatophore and, perhaps, to restrain her from eating the spermatophore before the spermatozoa have been emptied (Gurney 1947). Of exceptional interest are in this respect some observations on the male decorated cricket *Gryllodes supplicans.* The spermatophore, which this cricket transfers to the female during copulation, consists a of a large gelatinous mass, the spermatophylax, attached to a much smaller portion, the sperm-containing ampulla. Immediately after mating the female detaches and eats the spermatophylax, but removes and consumes the ampulla only later, that is, after sperm transfer has been completed. It does look therefore, that the male crickets actually "feed females to ensure complete sperm transfer" (Sakaluk 1984).

Secretions with aphrodisiac properties are also offered by male cockroaches to their females even though these do not eat spermatophores. In this connection it is worth recalling, however, that prior to being discarded, a cockroach spermatophore undergoes at least partial digestion in the female genital tract. In some other insects, such as the beetle *Melolontha melolontha* or the moth *Galleria mellonella,* spermatophores are completely digested in the female tracts.

A particularly interesting example of close-range attraction of females by a male pheromone is the action of a blend of chemical substances formed by the scent-producing organs of the oriental fruit moth *Grapholitha molesta* (Baker et al. 1981). This moth is unusual among the Lepidoptera in that males attract females after they themselves have been attracted to the vicinity of a female by her sex pheromone. As in other species, the male scent-producing organs of *Grapholitha molesta* consist of groups of elongated hairlike scales (hair pencils) that are bundled into special pouches and then everted and splayed in the vicinity of a female during courtship. The "male lepidopterous courtship pheromone" emanating from the herbal-scented hairpencils of the oriental fruit moth, is a blend of (1) ethyl *trans*-cinnamate (similar to 2-phenylethanol, which has been found in the hairpencil of other lepidopteran species), (2) mellein (a fungal metabolite, also found in ants), (3) methyl jasmonate (a constituent of jasmine oil known in the perfume industry as the queen of aroma; chemically related to *cis*-jasmone in the hairpencils of the butterfly *Amauris ochlea*), and (4) methyl 2-epi-jasmonate. The formulae of these four compounds are shown in Fig. 31.

The selection by the female of one conspecific male in preference to another is yet another phenomenon connected with courtship behaviour of insects that might depend on the ability of an individual male to satisfy the female's appetite for adequate gustatory or aphrodisiac investments, most probably in the form of a glandular secretion (Bell 1980). Worth considering in this connection is the alluring effect exerted by the male black-horned tree cricket, *Oecanthus nigricornis* which, while courting the female for nearly an hour, performs courtship songs, vibrates, and raises his hind wings to provide metanotal gland secretion. However, this alluring effect diminishes as soon as the metanotal gland had been waxed over.

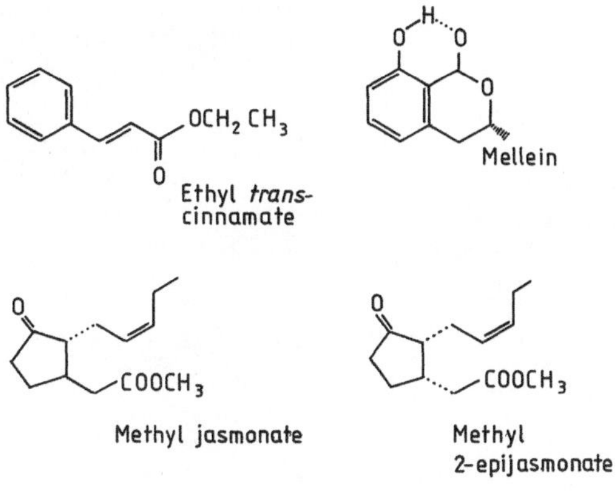

Fig. 31. Constituents of the male lepidopterous courtship pheromone emanating from the herbal-scented hairpencils of the oriental fruitmoth

Ethyl *trans*-cinnamate

Mellein

Methyl jasmonate

Methyl 2-epijasmonate

6.7 Droplet Spermatophores in Apterygotan Insects: Collembola, Diplura, and Thysanura

Droplet-spermatophores, or sperm-drops as they are often referred to by some investigators, are simple, bare, drop-shaped sperm-agregates, mostly stalked, which many male insects of the sub-class Apterygota deposit on a substratum where they can be collected by the females. In general appearance, they bear more resemblance to the analogous devices used for indirect sperm transfer by Arachnida (Chap. 8) than to the structurally complex spermatophores in Pterygota.

The literature dealing with indirect sperm-transfer mechanisms in Apterygota has been thoroughly reviewed by Schaller (1971, 1979). What follows is intended mainly to provide a series of examples pertaining to the function of droplet-spermatophores in three distinct orders.

Collembola. Spring-tails deposit droplet spermatophores of a variable size and number, mostly composed of a long stalk and an apical droplet filled with spermatozoa that may be in encysted form. In *Dicyrtomina ornata* (Fig. 32) the stem, made of fibrous material oriented along its major axis, is about 500 µm long and 8.5 µm wide at the apex, but much wider at the base where it adheres to the substratum; the apex, in the form of a swelling, provides the support for the droplet, usually elliptical in shape (35 × 25 µm), which encloses about 600 spermatozoa. No external membrane seems to exist; around the droplet there is only a thin (600–700 Å) layer of granular material similar in kind to that found between spermatozoa, only more electron-dense; each spermatozoon (in a cyst) has a diameter of 4.5 µm (Dallai 1975; Fig. 32).

The pick-up of spermatophores by the females can occur in a variety of ways, the body contact between the male and female ranging from non-existent to very close and highly involved (Betsch-Pinot 1974, 1977; Blancquaert 1981; Joosse et al. 1973; Klaver 1975; Poinsot-Balaguer 1976; Schaller 1971). Amongst collembollans in the sub-order Arthropleona, the males of the genera *Orchesella* and *Tomocerus* tend to deposit droplet-spermatophores in complete absence of the fe-

103

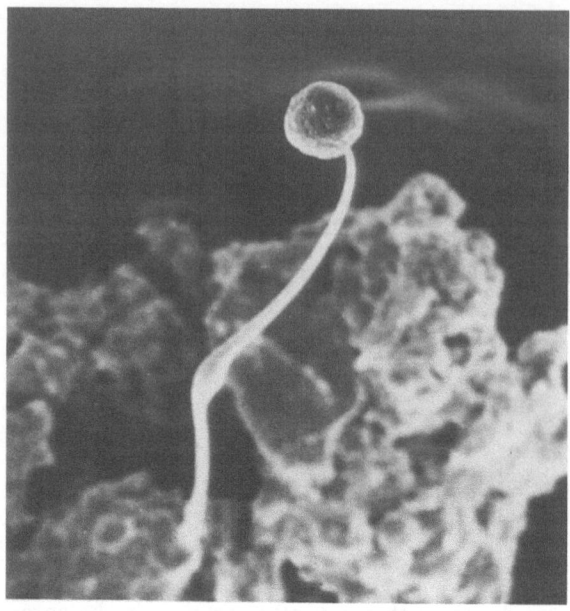

Fig. 32. Spermatophore of *Dicyrtomina ornata* (Collembola) seen with the scanning electron microscope; the sperm drop is at the top; the stem shows a swelling. (× 270). (Courtesy of Prof. Dtte Romano Dallai)

male, which has then to locate them and pick up the spermatozoa by brushing over with the vulva. But another arthropleon male, *Podura aquatica,* makes a thorough search for a female before placing around her a whole fence of sper-matophores and then pushes her towards that fence, thereby making it possible for the female to strip off properly the sperm-drops from the substratum. Similar behaviour is characteristic of certain spring-tails in the suborder Symphypleona, such as *Dicyrtomina minuta* and *Sminthurides aquaticus* (Schaller 1953, 1971; Schliwa and Schaller 1963), and three other *Sminthurides* species (Blancquaert 1981) in which mating behaviour extends over a period of a whole hour, the fe-males removing the sperm-drops in the end without bursting them. In at least one of these three species, *S. malmgreni,* the stripped-off sperm-drop can remain at-tached to the genital pore of the female for as long as 30 min, but throughout the period of attachment or following absorption into the genital pore, the female re-solutely refuses any further advances from the male, instead proceeding with a thorough clean-up of her body.

Diplura. These, in common with Collembola, produce droplet spermatophores of variable dimensions. In Campodeidae, for example, the stalk can be 30 to 300 µm long and 7–12 µm thick, and the round drops themselves 35–150 µm in diameter. Similarly variable can be the number of sperm-bundles enclosed within a drop: 1 in *Campodea rhopalota,* but 12 in *C. chardardi.* How, precisely, the spermatozoa are picked up by the females is not clear (Bareth 1966; Schaller 1971).

Thysanura. Bristle-tails, such as *Lepisma* (silver fish) or *Thermobia* (fire brat) use droplet spermatophores for indirect transfer, in very much the same manner as

104

Collembola and Diplura. The courting dance, in the course of which the spermatophores are deposited by the males on a substratum, can be long and involved (Sturm 1952, 1955, 1956). *Machilis germanica* provides an interesting example. Prior to depositing the sperm drops, the male spins a special thread which he fastens at one end to the ground while holding the other end with the "penis". He then attaches the sperm-drops to the thread, and already while spinning, pushes the female towards it with the help of his antennae, palps and legs, thereby enabling her to pick up the sperm-drops from the thread with her genital appendages.

6.8 Typical Spermatophores and Different Mechanisms of Their Assembly in Pterygotan Insects

The early observations on the occurrence of spermatophores in Pterygota (reviewed by Schneider 1883) were extended in the present century to a wide range of species, and sperm packages of diverse size and shape (rods, tubes, spheres, sacs) have been reported to occur in both Exopterygota (Odonata, Dictyoptera, Orthoptera, Dermaptera, Psocoptera, Hemiptera) and Endopterygota (Neuroptera, Coleoptera, Trichoptera, Lepidoptera, Diptera, Hymenoptera). With research steadily expanding it also became apparent that although in spermatophore-forming pterygotan insects generally (Odonata excepted) the spermatozoa are transferred during copulation directly into the female genital tract, the production of the spermatophores can proceed along several routes, depending largely on whether moulding occurs in the male or in the female genital system (reviewed by Gerber 1970; Scudder 1971).

In some insects, the process of spermatophore assembly is both initiated and completed within the male genital tract, most frequently in the anterior portion of the ejaculatory duct and the male copulatory organ and then, during copulation, the spermatophore is transferred to the female, usually by being placed between her external genital plates, so that one part of it can penetrate into the female genital duct, with the other part remaining outside the female genital opening. This method of formation and deposition operates in many orthopteroid insects (Tettigoniidae, Gryllidae, Phasmida, Mantoidea, and Blattidae), also in some Neuroptera and, probably, in the chironomid dipteran *Glyptotendipes paripes* (Chopard 1934; Du Bois and Geigy 1935; Gerhardt 1913, 1914; Gupta 1946; Khalifa 1949 a, b; Nielsen 1959). Acrididae are a special case (Gregory 1965 a, b). In locusts, where the spermatophore begins to be formed only after copulation has started, one part of it, sac-like in appearance, is retained by the male copulatory organ even after copulation has ended, but the other, tube-like portion, which acts as a conduit for the spermatozoa, extends into the spermathecal duct; when the pair separate, the two parts also separate. Another variant of the male-determined method of assembly operates in those pterygotan males that mould their spermatophore in the copulatory (spermatophore) sac while that sac is being everted into the bursa copulatrix of the female. After copulation the empty copulatory sac is withdrawn, but the spermatophore remains firmly anchored in the bursa. Amongst insects that employ this method are Hemiptera, such as *Rhodnius prolixus* (Davey 1959, 1965) and certain Coleoptera, namely, Dytiscidae and Scarabaeidae (Landa 1960).

A most distinguishing feature of the female-determined pattern of assembly is that the spermatophore is formed in the female reproductive system and its shape determined by that of the female ducts. This can happen when the male accessory gland secretions needed for spermatophore formation are ejected separately into the bursa copulatrix or vagina, either before or after transfer of semen, and encapsulation of spermatozoa takes place subsequently. This method is quite common amongst spermatophore-producing Trichoptera and Lepidoptera (Norris 1932, 1934; Srivastava and Srivastava 1957). It is also used by certain Diptera (Simuliidae), Coleoptera (Coccinellidae) and a few Hymenoptera (Davies 1965; Davey 1960 b; Fisher 1959). An interesting variant to the female-dependent pattern occurs in the psocopteran *Lepinotus patruelis,* where the spermatophore is probably moulded in the spermatheca, and not in either the bursa copulatrix or the vagina (Finlayson 1949). It seems likely that a similar mechanism applies also to the formation of the spermatophore-like structures in *Ibidoecus* and *Austrogoniodes metoecus* [Phthiraptera (lice): Philopteridae (Ischnocera)]. Cummings (1916), who described these structures in *Ibidoecus,* preferred to designate them as spermatodoses, rather than spermatophores, following Cholodkovsky's (1913) suggestion that sperm-lumps of this kind, when present only in female, but not in male insects, are in no way involved in sperm-transfer but serve to measure out "doses of semen" needed for fertilization of the eggs. Clay (1971), on the other hand, in her study of *Austrogoniodes* parasitic on penguins and ducks, described the bodies recovered from the dissected spermatheca as spermatophores, "broadly flask-shaped, the half with the pointed end having a thickened wall, the other half having a thinner wall and liable to collapse". The methods of spermatophore formation in lice generally seem to be in need of further explorations, particularly in view of Clay's (1968, 1971) other observation that in some Menoponidae (Amblycera) the spermatophores are formed in the male, "presumably inserted into the female genital chamber, their opening against the opening of the spermathecal duct and the contents discharged into the spermatheca, the empty spermatophore being ejected by the female".

Doubtful remains also for the moment the origin and function of the "sperm balls" occurring in the spermathecal lumen of certain Thripidae (Thysanoptera: Terebrantia). These, too, are surrounded by an enveloping structure, and have been categorized occasionally as spermatophores, "a problematic interpretation in some respects", as Bode (1975) rightly points out.

6.9 Mating Flights of Ephemeroptera and Odonata

Spermatophores are found in both Ephemeroptera (mayflies) and Odonata, the two supposedly most primitive orders of pterygotan insects in which mating takes place during flight. While in mayflies the spermatophore is transferred directly with the male and female genitalia pressing against each other, in Odonata the pattern of transfer is quite unique. A male dragonfly, either before or during the flight, places a spermatophore in a special cavity of his body, located on the 3rd and 4th abdominal segments, and equipped with the so-called prophallus. While the male is holding the female by her neck with the anal forceps, the female bends her abdomen forward, pressing against the spermatophore-containing pouch, thereby making it possible for the prophallus to introduce the sperm into her genitalia. Why the male, which has a penis, even though vestigial, does not use it as an intromittent organ is something of an enigma (Brinck 1960).

6.10 Spermatophores of Orthopteroid Insects: Blattidae, Mantoidea, Phasmida, Gryllidae, Tettigoniidae, and Acrididae

Orthopteroid insects provide excellent examples of specialization in formation, structure, and function of spermatophores. Even though in many of them the assembly of spermatophores is confined largely to the male reproductive tract, the time course of the moulding process varies between species. Whereas in some orthopteroid insects moulding is completed before mating, in others it continues during copulation.

Orthopteroid insects in which the structure, formation, and deposition of spermatophores have been extensively studied, in some instances as early as in the last century, include several blattids (cockroaches), mantids and phasmids (praying insects, stick- and leaf-insects), tettigoniids and gryllids (long-horned grasshoppers and crickets), and acridids (short-horned grasshoppers and locusts). Lately, special attention was given to conocephalids (Da Cruz-Landim and Ferreira 1977) and dolichopodids [Saltet (Boudou-Saltet) 1969; Boudou-Saltet 1972, 1980; Boudou-Saltet and Capolongo 1975]. Among the many interesting facts brought to light by studies of the Dolichopodidae inhabiting Italy and Greece is the finding that even very closely related species of *Dolichopoda* produce spermatophores which differ so much from each other as to provide useful species-specific markers. In the account which follows, the following orthopteroid insects will be dealt with specifically.

Blattidae: *Blattella germanica, Periplaneta americana.*

Mantoidea: *Mantis religiosa.*

Phasmida: *Baculum extradendatum, Extatosoma tiaratum, Phyllium bioculatum.*

Gryllidae: *Acheta (Gryllus) domesticus, Liogryllus (Gryllus) campestris, Nemobius sylvestris, Oecanthus pellucens.*

Tettigoniidae: *Requena verticalis.*

Acrididae: *Camnula pellucida, Chorthippus curtipennis, Chrysochraon dispar, Gomphocerus rufus, Locusta migratoria, Melanoplus sanguinipes, Schistocerca gregaria, Schistocerca vaga,* and *Taeniopoda eques.*

Blattidae. The production of spermatophores and secretion of glycoproteins, indolalkyl amine, and uric acid by the male accessory glands of cockroaches, has been mentioned earlier in this chapter. Details concerning two species are given below.

In *Blattella germanica* (Ballan-Dufrançais 1968; Khalifa 1950b), copulation lasts 2–3 h, during which time the pair will not separate, even if dipped into Carnoy's fluid or anaesthetized for dissecting purposes. On dissecting a female which has just mated, a fully formed spermatophore can be seen, an oval mass, about 2 mm in length and 1 mm in width, milky in general appearance and tough in consistency, its tip with the enclosed mass of spermatozoa inserted into the spermathecal groove. Normally, the spermatophore would remain in this position for about 12 h, during which time the discharged spermatozoa pass to the spermatheca. Eight groups of male accessory organs are held responsible for contributing the secretory material for building the spermatophore during the period of copulation, and at least four distinct layers are clearly discernible in the spermatophore's wall. The most internal layer, about 200 µm thick, has been contributed by the posterior accessory glands. It is surrounded by an elliptic mass, up to 300 µm long, of material derived from the median and ventral gland. Externally to it is the thinnest of all layers, no more than 150 µm thick, secreted by the dorsal gland. The outer cover, in the shape of a cupola, 400–700 µm thick and coated with uric acid, is derived from the utriculi majores.

In *Periplaneta americana* (Gupta 1946), the spermatophore dissected from a freshly mated female, has a pear-like shape, about 1.3 mm in diameter, its covering showing several projections and depressions. The wall of the spermatophore is about 400 µm thick and made up of three distinct acellular layers, the middle layer being the thickest (about 180 µm). So long as the spermatophore remains within the ejaculatory ducts, its wall has only two layers, the middle one derived from the secretions of the ejaculatory duct, and the inner one secreted by the utriculi majores. The outer layer, which is chitinous in character and only 60 µm thick, is deposited later at the time when the spermatophore becomes attached to the spermathecal papilla of the female cockroach, and it is probably formed by the secretion of the male's phallic gland. The central chamber, enclosed by the wall, is filled with a mass of spermatozoa floating in clear fluid which comes probably from the median accessory glands (utriculi breviores).

Mantoidea and Phasmida. The spermatophores of mantids and phasmids are mostly either sphere-shaped or bulb-like in appearance. In phasmids, the gelatinous bulb of the spermatophore which contains the sperm sac has been seen, in part at least, to remain after copulation outside the genital opening of the female. In the mantids on the other hand, the sub-genital plates of the female are modified in such a way that they conceal the spermatophore "so that it does not get eaten by other insects – a contingency of not infrequent occurrence among phasmids" (Cloudsley-Thompson 1976).

A spermatophore in the female genital tract of the mated *Mantis religiosa* is already mentioned by Przibram (1907) in his *Life History of Praying Insects [Die Lebensgeschichte der Gottesanbeterinnen (Fang-Heuschrecken)]*, and a more detailed description of it was given by Gerhardt (1914) who described the "whole structure as being of about the same size as hempseed", 3 mm in the longest diameter, and composed of a sperm-capsule enveloped by two membranes, and a duct conveying the spermatozoa to the female's receptaculum.

Amongst leaf- and stick insects, *Phyllium bioculatum* received early attention from Chopard (1934), who described its spermatophore as a small, pink-coloured sphere about 2 mm in diameter, and terminating in a tube-shaped conduit for evacuation *(conduit évacuateur)*. Similarly shaped spermatophores have been also found in other phasmids. In *Extatosoma tiaratum,* the giant Australian phasmid widely available as a laboratory insect, females have been seen after mating with a conspicuous cream-coloured sphere, 2–3 mm in diameter, attached to the side of the terminalia; similarly shaped detached spermatophores with long tubular extension were also occasionally found on the bottom of the laboratory cage (Clark 1974); an interesting statistically significant positive correlation exists at the time of oviposition between the size of the female Australian stick insects and the weight of the eggs laid by them (Carlberg 1984). *Baculum extradentatum* provides yet another example of a phasmid, in which spermatophores have been described (Carlberg 1981).

Gryllidae and Tettigoniidae. In gryllids and tettigoniids the spermatophore provides the usual vehicle for direct transfer of spermatozoa to the female; in some gryllids portions of the spermatophore can still be seen after copulation to protrude from the female's genital opening. Many female gryllids (*Gryllus, Oecanthus, Nemobius*) drop or remove the spermatophore after copulation and proceed to eat it. Female Tettigoniidae behave similarly. As mentioned earlier (Chap. 1), given the choice between two singing males of different weight, a female katydid mates with the heavier male which also produces a larger spermatophore, and subsequently she eats that spermatophore (Gwynne 1982, 1983, 1984). The size of the spermatophore varies greatly among the males of different species of katydids. In some, the contribution is as little as 2% of the body weight, but males of a few species can invest up to 40% of their body weight in one huge spermatophore. Spermatophylax is a common component of the katydid spermatophore. Near the end of copulation, the male contracts his abdominal muscles and produces first the sperm package which is inserted into the female's genital opening. He then slowly squeezes out of his abdomen the white proteinaceous mass of the spermatophylax which remains attached to the sperm package. After the pair uncouple, the female proceeds to consume the spermatophore, a feat that may last several hours. The first to be consumed is the spermatophylax which contains no spermatozoa. The female then removes and eats the empty sperm package, that is, the ampulla. Experiments with *Requena verticalis* (a katydid common in Western Australia) have shown that consumption of the spermatophylax prevents premature removal of the sperm ampulla. Females with the spermatophylax experimentally removed eat the ampulla prematurely, but when the females are given an additional spermatophylax, the time until the ampulla is removed is significantly extended. Evidently, feeding on the spermatophylax prevents the female *Requena verticalis* from removing and eating the ampulla before its sperm contents have been evacuated (Gwynne et al. 1984).

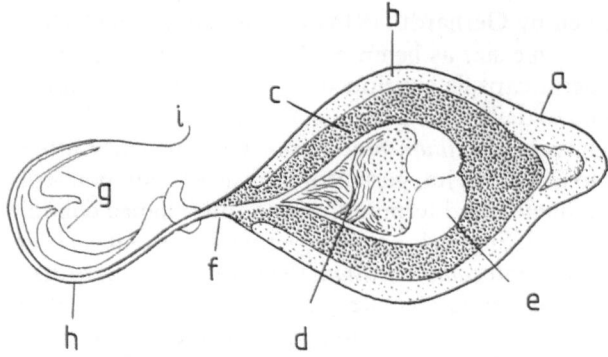

Fig. 33. Diagrammatic presentation of the spermatophore of *Liogryllus (Gryllus) campestris* (Orthoptera); (*a–c*) three layers of the capsule (outer, middle, and inner, resp.); (*d*) spermatozoa; (*e*) "pressure body" (material which presses out the spermatoza); (*f*) neck of the capsule; (*g*) the hook-like "fixing apparatus;" (*h*) spermatophore tube; (*i*) closure of the tube. (Redrawn from Regen 1924)

Lespès (1855a, b) was the first to point out that a cricket spermatophore consists of three distinct parts: a "vésicule", "lamelle", and "filet". His observations were later confirmed and extended by other authors, among them Boldyrev (1915), whose study also included the tettigoniids, Gerhardt (1913, 1914), Regen (1910, 1924), Hohorst (1936), and Khalifa (1949a).

Regen (1924), working with *Liogryllus (Gryllus) campestris,* described the spermatophore of a mature cricket as measuring about 5 mm in total length and composed of three parts (1) a bulb-shaped, sperm-containing capsule (Lespès's *vésicule*), 2.3 mm long and 1.4 mm wide, (2) a thin, 2.5-mm-long filamentous spermatophore tube (Lespès's *filet*), and (3) adjacent to the tube, a hook-like "fixing apparatus" (its convex surface in close apposition to the concave surface of the tube), serving as a device for anchoring the spermatophore in the female tract (Fig. 33). He also noted that after a male had been castrated, it will produce and transfer during copulation spermatophores of the same type, but free of spermatozoa.

Three distinct parts to the structure of spermatophore are also clearly discernible in the domestic cricket, *Acheta (Gryllus) domesticus;* they have been named ampulla, handle and spermatophore tube (Khalifa 1949a). The ampulla, a globular capsule (approximately 1.6×1 mm) with a projection at its posterior end, has a wall composed of (1) a thin, transparent outer membrane, (2) a certain amount of so-called evacuating fluid underneath, (3) an internal layer, most conspicuous of all, of very firm material which, owing to its hardness and thickness (0.17 mm) maintains its form even in an empty spermatophore, and internally to it, (4) the thin, transparent inner membrane which envelops both the sperm mass and the so-called pressure body consisting of highly condensed proteinaceous matter. The handle (corresponding to Regen's *fixing apparatus*), an extension of the outer membrane at the anterior end of the ampulla, has two projections which fit exactly on the base of the female's ovipositor; with the two lateral processes fully extended, the handle measures 1.5 mm in length and 1.25 mm in width. The spermatophore tube, a capillary tube originating from the inner layer, extends through the middle of the handle and about 3.65 mm beyond it; it tapers towards the end and terminates in a fine-pointed closure.

110

When observed under conditions in vitro, under a binocular microscope, the evacuation of spermatozoa from the cricket's spermatophore is a relatively slow process. Under these conditions, it takes nearly 20 min after the spermatophore tube had been cut between the handle and the ampulla for the whole mass of spermatozoa to flow out. The process of emptying the spermatophore is believed to be due to osmotic pressure exerted by the evacuating fluid and the pressure body; when the closure at the tip of the tube is removed, the pressure body swells, the evacuating fluid presses inwards, and the sperm mass is expelled from the ampulla through the spermatophore tube. As already mentioned (this chapter), egg release in the mated *Acheta domesticus,* which follows copulation, is believed to be triggered by a prostaglandin formed with active participation of a prostaglandin-synthesizing system transmitted to the female with the contents of the spermatophore (Destephano and Brady 1977; Loher et al. 1981).

No less complex is the structure of the spermatophore and the events leading to its evacuation in *Oecanthus pellucens* (Hohorst 1936). In this cricket, too, the spermatophore consists essentially of three parts, an ampulla, an anchoring device with two lateral lamellae, and the terminal filamentous tube for evacuating spermatozoa. In this cricket, as in *Gryllus,* the wall of the spermatophore is built by the male accessory secretions. The cover for the spermatozoa is provided mainly by the secretions that have entered the distended atrium, but it is only after the still soft mass had been moved by peristalsis from the atrium to the cavity of the penis, that it assumes the form of a round sac narrowing towards the neck. Next, starting at the neck, are added the lamellae and the tube (formed by secretory material flowing into the groove of the penis). Hardening of the whole spermatophore follows immediately. During copulation the neck of the ampulla is inserted into the vagina and there it becomes firmly anchored, partly by the lamellae, but partly also with the help of some additional male-derived secretory material. The process of evacuating spermatozoa starts immediately and is completed in about 15 min, during which time the female feeds on the metathoracic secretion of the male. Subsequently, she removes the empty spermatophore and devours it. *Nemobius sylvestris* is another gryllid in which copulation and spermatophores have been investigated meticulously (Gabbutt 1954; Gerhardt 1921; Richards 1952, 1953). The male of this species bears the distinction of extruding during copulation, not just one, but two spermatophores, each with an ampulla, handle and tube. The first spermatophore, however, is much smaller than the second one. The average diameter of the ampulla in the first spermatophore has been given as 0.303 mm, and that in the second one as 0.946 mm; on occasions when a male has failed to effect the transfer of a spermatophore, he has been seen to scrape it off and eat it (Gabbutt 1954).

Acrididae. In acridids, the male intromittent organ penetrates during copulation as far as the spermathecal duct, while pressing the spermatophore into the female. Siebold (1845), as part of his early study of *Locusta viridissima,* described the spermatophore which he recovered from a female mated some time before, as being oblong in appearance, but later Schneider (1883) reported that if the spermatophore is observed early during copulation, at the moment when it is just emerging from the male orifice, it still lacks a tube and consists merely of a round, soft

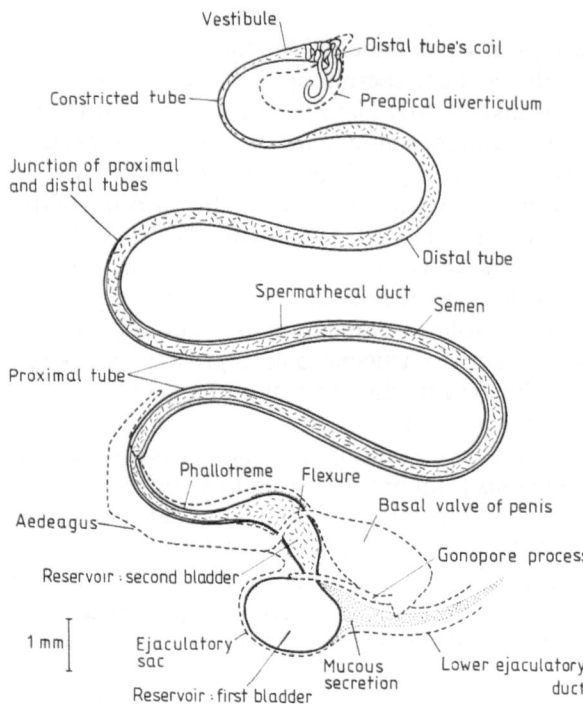

Vestibule

Distal tube's coil

Constricted tube

Preapical diverticulum

Junction of proximal
and distal tubes

Distal tube

Spermathecal duct   Semen

Proximal tube

Phallotreme   Flexure

Basal valve of penis

Aedeagus

Gonopore process

Reservoir : second bladder

1 mm
Ejaculatory
sac
Mucous
secretion   Lower ejaculatory
duct
Reservoir : first bladder

Fig. 34. Fully formed
spermatophore of the African
migratory locust, *Locusta
migratoria migratorioides*
(Orthoptera) in situ, with male
copulatory organ holding the
reservoir (the two "bladders"), and
the female spermatheca (uncoiled)
holding the tubular portion of the
spermatophore. (Redrawn from
Gregory 1965a)

vesicle, comparable in size and shape to a pea. Other investigators, Russian in
particular (Boldyrev 1927, 1929; Iwanowa 1925; Sokolow 1926), became subsequently engaged in extensive studies of the migratory locust, the spermatophore
of which has been reported by Boldyrev (1929) to consist of a tube with a swollen
sac-like structure at one end, only the tubular part being transferred to the female,
the sac-like part remaining within the male. It took many more years, however,
before the remaining mysteries concerning the formation, transfer and fate of the
locust spermatophores were cleared up by Gregory (1965a, b) in a detailed investigation of the African migratory locust, *Locusta migratoria migratorioides,* based
on dissections of over 400 pairs, at different times during the 5–18-h-long period
of copulation.

As mentioned earlier (this chapter), the spermatozoa of the migratory locust
are first bundled and then bunched into spermatodesms, before transfer to the
seminal vesicles, where they remain in storage until the time of copulation. The
formation of the spermatophore under active participation of the male accessory
secretions, which reach the ejaculatory duct from the polytubular accessory
glands, does not begin until early during copulation, that is, after the introduction
of the aedeagus of the male between the ventral ovipositor valves of the female.
However, once it has been initiated, the production of the spermatophore becomes a continuous process.

Figure 34 shows a completely assembled spermatophore of the African migratory locust, at the time when it extends fully into the spermathecal duct, and is
35–45 mm long. At that time, it consists of (1) a dilated "reservoir" (which during

copulation remains within the male copulatory organ) in the form of two "bladders": the first filled with secretion, and the second with semen, that is, the spermatodesms in a fluid suspension, and (2) a long, blind-ended tube which passes into the female. The first to be formed is the reservoir. Its formation begins usually less than 2 min after the start of copulation, but externally it cannot be recognized until about 10 min later, when the first reservoir-bladder has expanded sufficiently to become visible as a swelling at the base of the aedeagus. Then, 20–25 min later, the tube of the spermatophore begins to be extruded under the pressure of the musculature of the male copulatory organ. The extrusion is an eversion process, the tube turning inside out at its tip; thus the secretion of the distal tube precedes that of the proximal one. When the tube has been fully extended, its tip is ruptured and it becomes filled with semen. Copulation continues until almost all the spermatozoa have passed into the female.

At the end of copulation the male and female separate and at this point the spermatophore is broken, the tubular part remaining in the female but the sac-like part kept by the male until it is ejected altogether. As regards the spermatophore tube, its distal portion is dissolved in the spermatheca within 24 h of the start of copulation by proteolytic enzymes derived from the spermatheca and possibly also from semen (spermatophoric plasma); the proximal tube is not dissolved, but ejected by the female within a few days.

The presence of the female is certainly indispensable for the formation of spermatophore, for this cannot take place until copulation has begun. As regards the spermatophoric reaction as such, that is the extrusion of the tube, unless the female is present and copulation continues, the tube of the spermatophore is never extruded more than a few millimetres from the male aedeagus, possibly because of exposure to air and rapid drying. However, by inducing a male to extrude the tube of the spermatophore along a fine glass tube filled with saline and fitted over the tip of the aedeagus at the moment when extrusion has started, it could be shown that as long as the glass tube reproduced within narrow limits the internal dimensions of the spermatheca, extrusion of the spermatophore tube was able to continue, almost to completion (Gregory 1965 b).

No less complex and intriguing than in migratory locusts are the events associated with the function of the spermatophore in the grasshopper *Gomphocerus rufus*, the male accessory glands of which have already been described (this chapter). Just as in the migratory locust, the spermatophore formation in the grasshopper begins as soon as the copulating organs of the male and female become hooked together, and it then proceeds in a similar fashion, but much more rapidly, so that the whole process is completed within 3–4 min. In its final form, after penetrating deeply into the ductus receptaculi seminis, the spermatophore stretches over a distance of 18–25 mm, but the stretching during transfer is not done by turning the inner side out, but by expansion. Finally, it bursts at its tip, releasing into the receptaculum the semen with spermatozoa still bunched in spermatodesms (Fig. 35) (Hartmann 1970). The output of the male accessory secretions required for the formation of spermatophores must be very efficient, since as many as eight spermatophores have been observed to be turned out by a single male grasshopper within 6 h. After copulation, the entire empty spermatophore is digested and resorbed (Loher and Huber 1966).

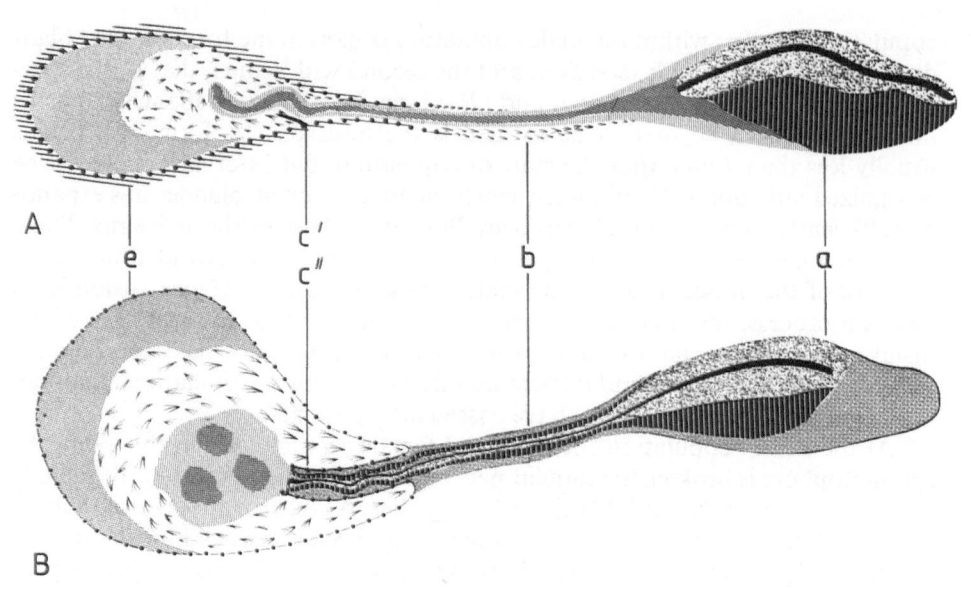

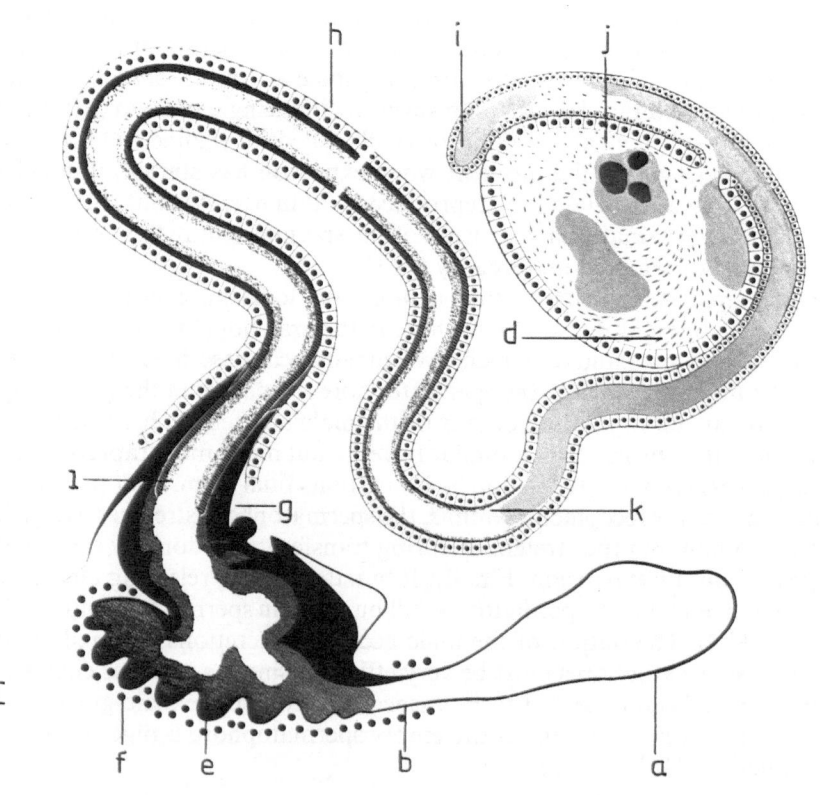

In addition to those mentioned already, the structure and function of spermatophores has been explored in other acridids such as *Camnula pellucida* (Ewen and Pickford 1975), *Chorthippus curtipennis* (Hartmann and Loher 1974), *Chrysochraon dispar* (Renner and Kremer 1980), *Melanoplus sanguipines* (Friedel and Gillott 1977; Pickford and Gillott 1971), *Schistocerca gregaria* (Odhiambo 1969 a, b; Pickford and Padgham 1973), and *Taeniopoda eques* (Whitman and Loher 1984). Acridoid spermatophores in general, are typically composed of two parts, a bulb-like sac which remains in the male, and a long tubular portion that obtains its final shape within the spermathecal duct of the female. While in some grasshoppers only a single spermatophore is produced during each mating, in others several spermatophores are passed to the female while the copulating partners remain locked together, but after emptying each individual spermatophore is removed from the genital tracts of the male and female before the next one is produced and transferred. In the desert locust *Schistocerca gregaria,* sperm transfer is accomplished by means of 1.5 mm-long spermatophores, of which as many as 14 can be manufactured by the male during the 4 h long copulation period. In *Schistocerca vaga,* the release of spermatozoa from the receptaculum seminis can be experimentally induced by electrical stimulation of the receptaculum nerve or the ductus aperture nerve (Okelo 1979). Multiple spermatophore transfer is equally characteristic of *Taeniopoda eques,* a grasshopper in which the male accessory glands consist of 17 separate tubules belongings to 8 distinct categories. In this grasshopper, the gelatinous, 7 mm-long and S-shaped spermatophore is a cast of the combined genital tracts of the joined male and female. Each spermatophore is emptied of its spermatodesm contents and subsequently extracted from the male and female genital tracts through the motions of the aedeagal valves while the copulating partners remain firmly locked together (Whitman and Loher 1984).

Fig. 35 A–C. Development of the spermatophore in the grasshopper *Gomphocerus rufus* (Orthoptera). *A* and *B* Two successive initial stages associated with the formation of the first bladder and proximal tube in the male reproductive tract. *C* The final stage, about 4 min later, when the spermatophore has been completely lodged in the receptaculum of the female. (*a*) Upper ejaculatory duct; (*b*) lower ejaculatory duct; (*c'*) and (*c''*), at the start of formation and after evagination, respectively; (*d*) spermatophóric semen with spermatozoa still in spermatodesms; (*e*) ejaculatory sac; (*f*) musculature instrumental in pressing the spermatophore into the female tract; (*g*) aedeagus; (*h*) ductus receptaculi seminis; (*i*) appendix of the bulbous end-portion; (*j*) the main part of the bulbous end-portion of the receptaculumn seminis [with spermatophoric semen (*d*) inside]; (*k*) and (*l*) the two ends of the solid tubular part of the spermatophore [*shaded area* above (*k*) contains the fluid secretory material of the non-solidifying part of the spermatophore]. (Further details: Hartmann 1970. Courtesy of Dr. Rüdiger Hartmann)

6.11 Spermatophores of Hemiptera, Neuroptera, Coleoptera, Mecoptera, Trichoptera, Lepidoptera, Diptera, and Hymenoptera

Hemiptera. Earlier in this chapter, Hemiptera were mentioned on several occasions, most particularly as regards the significance of sperm-bundles in the scale insects (suborder Homoptera) and the special secretory products of the male reproductive tract in the "assassin bug" *Rhodnius* (suborder Heteroptera). In both suborders, spermatophores have been described. In the large European lantern-fly, *Dictyophora europaea,* pear-shaped spermatophores have been recovered from the bursa copulatrix, and similar structures were shown to occur in other Fulgoroidae; there is also some evidence that in these Homoptera, the actual release of spermatozoa from the spermatophore and subsequent passage to the receptaculum seminis is facilitated by secretions produced in the female genital system (Strübing 1955).

In *Rhodnius prolixus,* about which so much has been learned owing to Davey's efforts (Davey 1959, 1965; Barker and Davey 1982), the bulk of the spermatophore is made up by the "transparent jelly" contributed by the aforementioned male transparent gland. The formation of the spermatophore takes place during copulation inside the copulatory (spermatophore) sac, that is, within the male intromittent organ. As the penis is inserted into the bursa copulatrix and the secretion of the transparent glands is filling the spermatophore sac, the sac is everted into the bursa. It takes about 20 min from the moment the male copulatory organ has penetrated the bursa until the eversion process is completed and the spermatophore fully extruded, its anterior narrow neck-portion extending farthest into the female. After a further 10 min, the male withdraws the intromittent organ together with the empty sac, leaving the spermatophore in the bursa. About 8 h later, the used-up spermatophore is ejected by the female and falls to the ground.

A spermatophore fully extruded into the bursa is pear-like in shape and consists of (1) the gelatinous matrix contributed by the transparent jelly, (2) a slit in the narrow neck-like portion of the jelly, which contains the semen, and (3), on the same surface, but on the broader part of the spermatophore, a small region containing the opaque accessory secretion (Fig. 36). Once the spermatozoa have been released from the slit, they are transported to the spermatheca along the oviducts; a stimulus for the contractions of the oviducts is provided by the opaque accessory secretion incorporated in the spermatophore. As already mentioned, the active principle present in that secretion is a serotonin-like substance. The bulk of the spermatophore functions merely as a supporting plug which holds the semen in place while the spermatozoa are being moved by the contracting oviducts to the spermatheca, there to be stored until the time of fertilization. In the spermatheca of *Rhodnius* the spermatozoa can survive for at least a month.

Neuroptera. With a few exceptions (Coniopterygidae, Planipennia) spermatophores are widely distributed among Neuroptera. *Osmylus chrysops* is one of them. Copulation in this beautiful insect with a wing expanse of nearly 5 cm that

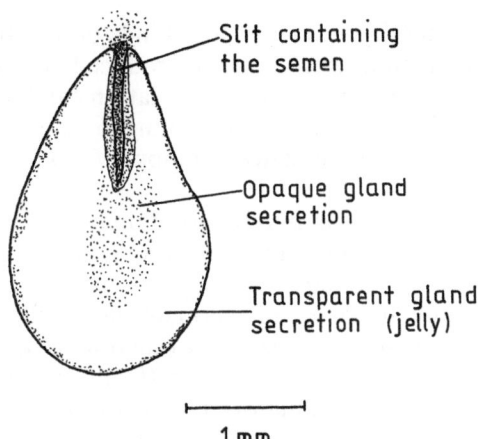

Slit containing
the semen

Opaque gland
secretion

Transparent gland
secretion (jelly)

1 mm

Fig.36. Spermatophore of *Rhodnius prolixus* (Hemiptera). (Redrawn from Davey 1959)

"cannot be mistaken for any other of the British Neuropters" has been described by Withycombe (1922) as follows:

"The male everts his scent glands, and almost immediately females within a foot or two become agitated and wave their antennae vigorously. Next, they walk or fly towards the male and commence caressing the scent glands with their antennae and palpi.
The whole performance of pairing is very leisurely, and occupies several minutes. From time to time, peristallic contractions of the male's abdomen may be noticed, and the female levers with the rod-like valves as though to extract something from the male's abdomen. This indeed is the case, for in from ten minutes to an hour a large white spermatophore is withdrawn and remains projecting forward from the tip of the abdomen of the female, the rod-like valves lying just below it".

The spermatophore itself has been described by Withycombe as about 4 mm long, white in colour, but yellowish in the centre, and as consisting of four rounded lobes and a short stem. Within a few minutes of parting from the male, the female bends her head, brings the mouthpart close to the genital opening and begins to devour the nearest lobe, the large spermatophylax, while the spermatozoa are already proceeding to the spermatheca, thus escaping a similar fate (David 1936). The spermatophylax, à rich source of proteinaceous food, may remain, in part at least, outside the female genital opening for periods of time ranging from 10 min to 2 days.

No less complex is the act of copulation and spermatophore extrusion in *Sialis lutaria*, as observed by Du Bois and Geigy (1935), who described the spermatophore of this neuropteran insect as being composed of two distinct parts, one, a large, gelatinous, heart-shaped body with two cavities containing masses of spermatozoa, and the other, a thin neck, inserted into the genital opening of the female. Soon after insertion, the male has been seen to leave the female with the main body of the spermatophore still protruding from her genitalia. Further details about the structure of the *Sialis* spermatophore have been given by Khalifa (1949b), who described the body as being 0.6 mm long and 1 mm wide in its widest dimension, and the neck as 1.5 mm long and 0.4 mm wide at the base. He also noted that after the spermatophore had been firmly lodged inside the bursa copulatrix, it breaks into pieces and consequently the spermatozoa are set free and migrate into the spermathecae (one on each side of the bursa).

117

In *Chrysopa perla,* four kinds of secretion are needed for the formation of the spermatophore. These are produced in a male accessory gland consisting of seven paired pouches. Before copulation can take place a cover barring the entrance to the female genital opening must open first. The spermatophore is then deposited in the bursa copulatrix (Philippe 1972).

Coleoptera. On purely evolutionary grounds, it would be difficult and unwise to speculate why some Coleoptera should deliver ejaculates containing free spermatozoa, while others, equally well-developed and often members of the same family, make use of spermatophores. Similarly perplexing is the diversity in the morphology and secretory function of the female spermathecal complex in which the spermatozoa are stored (Hallberg 1984). In beetles that produce spermatophores, the copulatory (spermatophore) sac, that is, the bladder-like structure forming part of the male intromittent organ which evaginates during copulation, provides the preferred site at which the spermatophore is assembled. Several aspects of this process, and the ways in which the spermatophores are deposited and function in Coleoptera will be discussed below, using the following species as examples.

Acanthoscelides obtectus
Diabrotica virgifera
Dytiscus marginalis
Hydrophilus olivaceus
Lytta nuttalli
Melolontha melolontha
Popillia japonica
Pimelia angulata confalonierii
Tenebrio molitor

Cockchafer, *Melolontha melolontha* (Scarabaeidae), has been a favourite object of studies on reproductive behaviour and mode of insemination in Coleoptera ever since Straus-Durckheim (1828) published his extensive treatise on the anatomy of this insect, containing a detailed description of male genitalia and the exceedingly long filamentous glands to which he gave the name of spermatic vessels (vaisseaux spermatiques). The subsequent developments in this area of research were reviewed by Landa (1960) in his important publication on the mode of formation, function, and fate of the cockchafer's spermatophore. His findings were briefly as follows.

The copulatory sac, which the male inserts into the bursa copulatrix during copulation, begins to fill with semen containing the sperm-bundles and the accessory secretions, at about ½–1 h after copulation has started, at the time when it is already well spread inside the bursa. It takes another 2 h or so before the process of filling and mixing has run to completion and until, following the withdrawal of phallus, the spermatophore is left behind in the bursa. Following deposition in the bursa, characteristic changes take place in the spermatophore when the components of the accessory secretions begin to differentiate into islands of granular and hyaline material, and the clusters of spermatozoa, now highly motile, start their passage to the receptaculum seminis. For a period of about 12 h, the membrane of the spermatophore remains largely unchanged, but then it swells and begins to disintegrate. The period of time necessary for complete digestion, probably under the influence of a secretion produced by the bursa itself, continues for 5–8 days. The shortest interval between two copulations associated with transfer of a normal spermatophore is 5 days (Landa 1960, 1961).

Some Coleoptera, such as the water beetle *Dytiscus marginalis* (Blunck 1912) and the dung beetle *Phanaeus* (Halffter and Lopez 1977) deposit spermatophores in the "vagina" (Blunck's *Spermatophorentasche*). When this happens, then the spermatozoa must presumably be squeezed out from the spermatophore, perhaps by pressure from the terminal sternites of the abdomen, before they can penetrate into the female genital tract. Blunck (1912, 1913) who studied extensively the reproductive organs and spermatophores in Dytiscidae, has also included in his investigation the Hydrophilidae, showing that they, too, deposit large spermatophores during copulation. More recently, two distinct pairs of male accessory glands have been shown by Gundevia and Ramamurty (1977) to participate in the elaboration of the spermatophore of *Hydrophilus olivaceus*. One, the outer pair of glands, secretes muco- or glyco-protein which avidly binds ribonucleic acid, and provides material for the formation of the massive eosinophilic ring and pigment granules that surround the helically coiled sperm column in the spermatophore. The other, inner pair produces in addition to muco- or glycoprotein some free glycogen which, after incorporation into the spermatophore, is believed to provide a nutrient substrate for the spermatozoa in the same manner as suggested earlier by Anderson (1950) for the Japanese beetle, *Popillia japonica*. If this supposition could be substantiated further, glycogenolysis may well prove to be as important to the spermatozoa of this insect as it is to those of the octopus which are already known to metabolize efficiently glycogen to D(−)lactic acid (Mann et al. 1974).

The fact that, as in orthopteran and neuropteran insects, the spermatophores in Coleoptera serve not merely the purpose of sperm transfer, but provide the females with essential nutrients with gustatory or aphrodisiac properties (already mentioned in this chapter), has been repeatedly stressed by many investigators, including Landa (1960). A particularly instructive example has been brought to light by studies of *Acanthoscelides obtectus* (Bruchidae) (Huignard 1969, 1974, 1983; Huignard and Lamy 1972; Huignard et al. 1977), which showed that certain proteins disappear from the spermatophore after it had been deposited in the bursa copulatrix, while some others, those presumably that compose the wall, remain unchanged for at least 24 h. Esterases and aminopeptidases which pass from the male accessory secretions into the spermatophore are credited by Huignard and Lamy (1972) with a role in the disintegration of the spermatophore structure and the ability to form degradation products which eventually reach the haemolymph and the oocytes. The authors regard these products as being of paramount importance in stimulating oogenesis, a conclusion supported by evidence that when the "paragonial substance A" extracted from the spermatophores is injected into the abdomen of a virgin female, it elicits oogenesis (but not egg-laying).

A prolonged period of copulation is characteristic of many Coleoptera: at least 3 h in the cockchafer (Landa 1960) and *Diabrotica virgifera* (Chrysomelidae) (Lew and Ball 1980); and 8–10 h in the blister beetle *Lytta nuttalli* (Meloidae)

119

(Gerber et al. 1971). The spermatophore of *Lytta* consists of a mass of jelly-like material and a tube; the spermatozoa are located in the gelatinous part at the junction with the tube. The jelly is secreted by the vasa deferentia, while the wall of the tube, consisting of three layers, is built from secretions coming from the male accessory glands. After completed copulation, in about 2 h the tube is rejected by the female. During the next 24 h most of the spermatozoa pass into the spermatheca. Sufficient numbers are transferred to the female during copulations preceding the first oviposition to fertilize all the eggs produced by her. Males can copulate as frequently as every 24 h, but the most frequent interval between copulations in virile males is 2–3 days. Copulation is an important factor in stimulating oocyte maturation and probably also egg-laying; the large quantities of spermatophoral materials from repeated copulations are obviously important to the female (Gerber and Church 1976). A spermatophoral tube similar to that of *L. nuttalli* is produced by other Lyttini, and probably also by Meloinae and Nemognathinae, in which the morphology of the reproductive systems is strikingly similar (Beauregard 1890; Gupta 1965, 1967). Whereas in those species of Meloinae that have been observed copulation generally lasts for several hours, in Nemognathinae it can be as short as a few seconds.

The biochemistry of secretory proteins engaged in the construction of spermatophores in beetles has so far received only scanty attention, but a welcome contribution to the knowledge of this subject has been the result of amino acids analyses performed on the reproductive tissues and spermatophores of *Tenebrio molitor* (Frenk and Happ 1976). A typical tenebrionid spermatophore, such as the one described by Fiori (1954) in the bursa copulatrix of *Pimelia angulata confalonierii,* is composed of two parts, a vesicle filled with spermatozoa, and a long tube linking that vesicle to the penis, both parts surrounded by a three-layer wall. A great deal of information about the formation and function of spermatophores in the mealworm beetle, *Tenebrio molitor,* has been provided by Happ and his associates (Gadzama and Happ 1974; Happ and Happ 1975; Frenk and Happ 1976; Grimes and Happ 1980; Dailey and Happ 1983).

The male genital system of the mealworm beetle consists of paired testes, vasa deferentia, and seminal vesicles which converge upon the ejaculatory tract, and in addition, two pairs of accessory glands located at the junction of the seminal vesicles and the ejaculatory tract. The larger gland type is bean-shaped, and the other tube-like in form. The bean-shaped glands produce secretions that form the outer layers of the spermatophore. Acid hydrolysates of the spermatophore and of the bean-shaped gland are strikingly similar in their content of amino acids. Proline, glutamic acid, aspartic acid, and alanine predominate in both. Moreover, the two antisera against extracts of the spermatophore and the bean-shaped glands, share common cross-reactive male-specific components as shown by both complement fixation tests and immunodiffusion procedures. The terminal differentiation of the tubular accessory glands in the postecdysial male mealworm beetle is characterized by the synthesis of four major groups of proteins; the results of crossed immunoelectrophoretic analyses indicate that at least two of the tubular accessory gland antigens contribute to spermatophores (Black et al. 1982).

Mecoptera. In scorpion flies generally, the spermatozoa are transferred during copulation in the form of liquid semen. Boreidae are a notable exception. The spermatophore of *Boreus westwoodi,* described by Mickoleit (1974), is about 1.4 mm long, spindle-shaped, and divided by a median partition into two chambers, each extending into a tube. It is produced in the genital duct of the male during copulation, but except for the tip which is inserted into the female, the bulk of it is retained by the male inside the aedeagus. Copulation takes 20–30 h (even longer therefore, than in Coleoptera) and during that time, the spermatozoa are pressed out, most probably under the influence of the swollen granular secretion contained in the spermatophore. At the time when most of the spermatozoa have already been pressed out into the receptaculum seminis, no more than one-quarter of the spermatophore is present in the female, the rest still firmly lodged inside the aedeagus.

Trichoptera. The external genitalia of all Trichoptera are essentially alike, yet whereas some families possess well-developed male accessory organs and produce spermatophores, in others the accessory organs are small, and the spermatozoa are delivered in free form. Of the twelve species of caddis flies examined by Khalifa (1949b), six, belonging to the families Sericostomatidae, Molannidae, and Limnophilidae, had spermatophores. In this type of caddis fly, the vasa deferentia are characterized by high secretory activity, and they carry a pair of accessory glands filled with either milky, granular secretion (in *Sericostoma* and *Silo*) or a transparent fluid (in *Molanna*). There is a positive correlation between the amount of secretion found in the paired accessory glands and the size of the bursa copulatrix which receives the spermatophore. At the point where the vasa deferentia merge into the single ejaculatory duct, there is a third accessory gland which secretes a transparent jelly. During copulation, which lasts several hours, the secretion of the unpaired gland is poured into the bursa first, and that of the paired glands next. As the two kinds of secretion come into contact with each other, a mass of coagulated protein is formed, for which the term *spermatophylax* has been proposed, in analogy to a similar body in certain other insects (this chapter). The next step in the formation of the spermatophore is the ejection of the contents of the vasa deferentia.

In its final form, the trichopteran spermatophore is composed of two parts, the coagulated mass of protein and a sac containing the spermatozoa. In *Sericostoma,* the protein mass is white, weighs about 0.2 mg; and the spherical sperm sac, when empty, weighs 0.02 mg, and measures 0.9 × 0.83 mm. In *Limnophilus,* the protein mass is pink, and the sperm sac 1.16 mm long and 0.33 mm wide. Twelve hours after copulation the protein mass becomes softer, and after another 12 h it is nearly dissolved. In *Anabolia,* when a female copulates for a second time, the old sperm sac is squeezed aside; up to three sperm sacs may be found in the bursa.

Lepidoptera. Omnipresence of spermatophores in moths and butterflies had been recognized by the end of the last century (Petersen 1907), and similarly evident was at that time the striking diversity in the anatomy of the lepidopteran male and female genital systems. Since that time, a great deal more has been learned not only about the morphology of the reproductive organs and spermatophores, but other aspects of reproduction, such as the ejaculatory process, that is, the sequence with which the accessory secretions and spermatozoa are voided by the male during copulation, and the ways in which the spermatophore is first assembled and later evacuated and disposed of in the bursa copulatrix of the female.

The species listed below are intended to provide representative examples of the various facets of spermatophore function which will be discussed later:

Acrolepia assectella
Adoxophyes orana
Anagasta (Ephestia) cautella, A. elutella, A. kühniella
Artogeia (Pieris) rapae crucivora
Bactra verutana
Bombyx mori
Choristoneura fumiferana
Colias eurytheme, C. philodice
Diatraea saccharalis
Diparopsis castanea
Dryas julia
Euxoa auxilliaris
Galleria melonella
Grapholitha molesta
Heliconius charitonius, H. erato
Heliothis virescens, H. zea
Hyalophora cecropia
Laspeyresia pomonella
Leucania separata
Leucinodes orbonalis
Manduca sexta
Ostrinia nubilalis
Pectinophora scutigera
Plodia interpunctella
Plutella xylostella
Pseudaletia unipuncta
Sitotroga cerealella
Spodoptera littoralis
Trichoplusia ni
Zygaena trifolii, Z. filipendulae, Z. fausta

In general, the lepidopterous spermatophore consists of one or several sperm sacs embedded in an ovally shaped gelatinous body ("corpus") narrowing into a neck ("collum", "stalk"), at the tip of which a solid transparent horn ("frenum", "hook") can usually be recognized, corresponding with the diverticula from the male ejaculatory duct of the species in question. Near the horns there is the aperture through which the spermatophore is placed during copulation in the bursa in such a way that the aperture at the tip faces the opening of the seminal duct. Precise positioning of the spermatophore in the corpus bursae is a prerequisite for further successful passage of spermatozoa from the spermatophore to the spermathecal duct and spermatheca. In some lepidopterous families, the seminal duct is equipped with a special bulb-like bulla seminalis. Other groups, like sugarcane borers, lack a bulla. The sperm mass emerging from the spermatophore of the

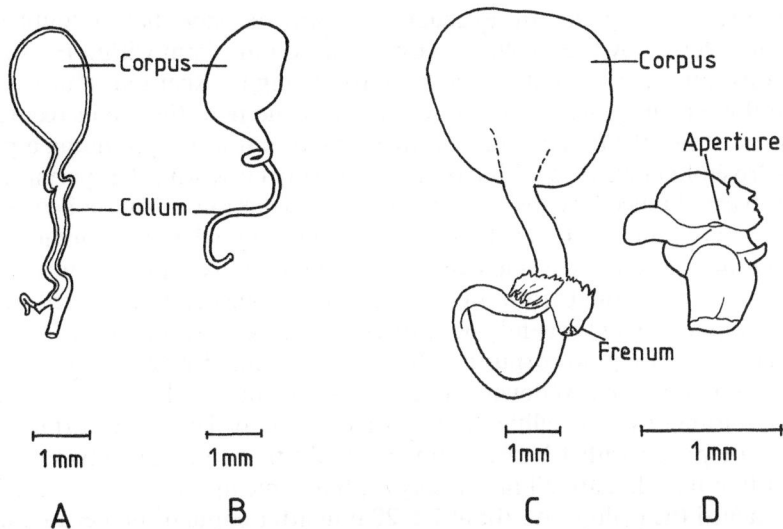

Fig.37 A–D. A and *B* Spermatophore of the European corn borer, *Ostrinia nubilalis* (Lepidoptera). *A* Inside the female reproductive tract; the corpus lodged in the bursa copulatrix, and the collum in the ductus bursae (the bursal gland is not shown). *B* The spermatophore by itself. *C* and *D* Spermatophore of the army cutworm, *Euxoa auxilliaris* (Lepidoptera). *C* Complete spermatophore. *D* Enlargement of the frenum, showing also the aperture. (Redrawn from Drecktrah and Brindley 1967 and Drecktrah 1978)

sugarcane borer *Diatraea saccharalis* includes eupyrene sperm bundles along with non-functional motile apyrene sperm cells, but after arriving at the spermatheca, the apyrene and eupyrene cells are segregated (Miskimen et al. 1983).

The size and shape of spermatophores in Lepidoptera are largely determined by the internal form of the bursa. Hence the characteristic differences between species (Williams 1939, 1941; Callahan and Chapin 1960). Two typical spermatophores, one of the European corn borer, *Ostrinia nubilalis* (Pyraustidae), and the other of the army cutworm, *Euxoa auxilliaris* (Noctuidae) are shown in Fig. 37 (Drecktrah and Brindley 1967; Drecktrah 1978). In both insects, the female reproductive system is of the ditrysian type. It has two separate openings: the ostium bursae, which receives the aedeagus and the contents of the spermatophore during mating, and the oviporus, through which the eggs pass during oviposition.

The spermatophore of the corn borer has an oval corpus which fills the lumen of the bursa completely, and a tubular collum of exactly the same coiled appearance as the ductus bursae in which it is lodged. Multiple matings are common in the corn borer and when this happens, the additional spermatophores are crowded into the same bursa, but are smaller than the first spermatophore, since there is less space for them (Drecktrah and Brindley 1967). Similarly oval-shaped is the corpus of the army cutworm spermatophore, which extends into a somewhat flattened, coiled collum terminating in small, flared wing-like projections, the frenum (Drecktrah 1978).

Special spines on the cuticular lining of the bursa, called signa or cornuti, are commonly credited with being directly responsible for tearing open the spermato-

123

phore and triggering the evacuation of spermatozoa, but undoubtedly, the pressure exerted by the swollen accessory secretions from within the spermatophore constitutes a contributory factor in triggering the spermatophoric reaction. This follows from many observations, both old and new, the events recorded in another noctuid, the tobacco budworm *Heliothis virescens,* providing a good example (Proshold et al. 1975). Like that of the army cutworm, the spermatophore of the tobacco budworm consists of a bulbous corpus and a long coiled collum with an aperture and frenum at the tip. During mating, when the contents of the future spermatophore are ejaculated into the bursa, the corpus inflates and the semen, that is, the spermatozoa and spermatophoric plasma, is deposited within the spermatophore. Consequently, in a spermatophore dissected out of the bursa just after copulation, the corpus can be seen to be filled with a milky white fluid, and the collum with a yellowish material. As the material from the collum flows out of the aperature, it jellifies, forming a cone-shaped cap. The formation of this cap is completed with 1 h after copulation. At that time the eupyrene spermatozoa, still in bundles, are all in the corpus, but some apyrene sperm cells have already reached the collum. At about 2 h 20 min after copulation, because of an increase in pressure within the spermatophore, the spermatozoa, eupyrene following closely after the apyrene ones, are suddenly ejected from the spermatophore. A similarly sudden release takes place in vitro, that is, in a spermatophore which, having been dissected out of the bursa earlier, at 0, 1 or 2 h, was placed in a saline solution and left there until the end of the 2 h 20 min period. The reason for the sudden increase in pressure is unknown, but two events that occur just prior to ejection of spermatozoa seem to be relevant: the milky material within the collum becomes more transparent; and the spermatophore undergoes additional coiling. Whether the pressure is the result of coiling alone, or of an enzymatic reaction, perhaps mucolytic or proteolytic in kind, remains to be determined.

Just as the size and shape of spermatophores, so is their number in the bursa subject to species, and indeed in some instances, individual, variations. In many species only one spermatophore is found in the bursa; in *Spodoptera littoralis* (Noctuidae), an average of 1.7 has been recorded (Takeuchi and Miyashita 1975); and in the wax moth *Galleria mellonella,* up to 7 can be present (Khalifa 1950a). Variable is also the rate at which the males can produce spermatophores, largely depending on the speed with which they are able to replenish their genital system with accessory secretions and spermatozoa. Males of *Pectinophora scutigera* (Gelechiidae) pass one spermatophore in 24 h (Vickers 1982). Males of *Acrolepia assectella* (Plutellidae) can refill their genital system with secretions within 12 h after the last copulation (Thibout 1971), and are able to produce spermatophores at a rate of 1 every day for 9 days, all perfectly fertile; but the size of the spermatophores decreases and the sexual activity of the female remains inhibited for several days by the presence of a spermatophore in the bursa (Rahn 1971).

Mechanical or stretch stimuli exerted on the walls of the bursa by the spermatophore, concomitantly with afferent nervous responses to the inflated state of the bursa, are the most likely reason for the post-copulatory behavioural change of the female, no longer willing to accept a male. This follows most particularly from experiments on cabbage white butterflies, in which either saline solution (Obara et al. 1975) or silicone oil (Sugawara 1979) was used to inflate the copulatory pouch. The afferent nervous impulses are normally conducted from the bursa of *Artogeia (Pieris) rapae crucivora* by a pair of bursal nerves. While in virgin females the frequency of spontaneous afferent impulses conducted by these nerves is 1–3/s, in the bursa filled with a spermatophore during copulation, the impulse frequency increases tenfold; a similar enhanced afferent activity can be achieved by inflating the bursa with silicone oil. In two other pierids, *Colias eurytheme* and *C. philodice,* the probability that the mated females would approach males was shown to be directly related to the degree of depletion of the contents of spermatophores received in previous matings (Rutowski et al. 1981).

The period of time during which a spermatophore remains in the bursa is also highly variable, largely dependent upon the rate of disintegration which in some species is probably assisted by secretions originating in the bursa itself. In the diamondback moth, *Plutella xylostella,* once a spermatophore has been fully formed, it can last for 96 h, which makes it very convenient to identify each successful mating (Yang and Chow 1978). On the whole, a count of spermatophores provides a reliable method for determining the number of matings in individual females (Clarke et al. 1983). It is a method widely employed for defining not only differences between species, but individual intra-species fluctuations attributable to season and other environmental factors. As an example may serve the results of a study conducted on two noctuid moths, *Heliothis virescens* and *Heliothis zea,* collected from traps equipped with blacklight lamps during a 6 months' period (Hendricks et al. 1970). Both species mated most frequently once only, but some individuals as many as six times; a definite decrease in the percentage of mated *Heliothis zea* occurred (in Texas) between June and August. Dye-marking of the spermatophores provides an additional means of identification (Cantelo 1973); male moths of the tobacco hornworm *Manduca sexta,* reared by using a larval rearing medium containing a red dye, produced spermatophores which when removed from mated females were red-coloured.

Of the other moths that have been investigated, the Mediterranean flour moth, *Anagasta (Ephestia) kühniella* (Phycitidae), has received particular attention in the past. Its reproductive organs, both male and female, conform on the whole to the pattern featured by a great many other Lepidoptera, as described by Cholodkovsky (1880, 1884), Stitz (1901) and others, but they possess also certain special characteristics discovered by Norris (1932, 1933, 1934) in the course of her fundamental study on fertility of insects, and later expanded by Musgrave (1937) by histological observations. The male reproductive tract of this moth consists of (1) an unpaired testis, (2) two vasa deferentia leading into the seminal vesicles, (3) paired accessory glands joining basally to form the unpaired gland system composed of four glands, the lowest of them dilated to about twice the width of the other three and forming a reservoir distended by granular, opalescent secretion, and (4) a single ductus ejaculatorius starting at the base of the reservoir gland, subsequently widening slightly to form the bulbus ejaculatorius, and finally continuing into a thin-walled chamber, the lower part of which is wrapped round the inner side of the aedeagus, and forms two distinct, curved hollow horns. During copulation, when the entire penis is pushed out, the terminal part of the ductus ejaculatorius, till now coiled up inside the aedeagus, is uncoiled and thrust through the end of the aedeagus; the evaginated tube is known as the vesica. Copulation lasts 2 h or more, and by the end of this time the contents of the male reproductive tract and accessory glands have all been transferred into the bursa copulatrix of the female to form the spermatophore.

Two distinct basic types of secretory cells compose the epithelium of a major part of the paired accessory glands and the adjoining segment of the ductus ejaculatorius in the Mediterranean flour moth (Riemann and Thorson 1979). The granule-secreting cells of one type deliver their secretion as PAS-positive granules that had previously been packaged in the Golgi system. The foliate cells of the other type form characteristically shaped elongate apical projections that are composed mostly of membranes. These membranes fuse in pairs, forming multilayered whorls, and when during copulation they are split off to become part of the ejaculate, they contribute a major part of the material from which the wall of the spermatophore is built. Both the granule-secreting and foliate cells occasionally contain mycoplasm-like organisms similar to those encountered in the tissues of the moth. The wall surrounding the nearly spherical body of the spermatophore has been described as a cuticle with a thin amorphous layer on the outside, and a much thicker layer of membranes underneath, no doubt derived from the apical projections of the foliate cells which also contain dense granules, probably representing particles of glycogen.

Hemel (hexamethylmelamine), a chemosterilant highly effective in moths, when injected into the male flour moth, induces striking mating abnormalities, expressed as permanent copulation, sterile mating and spontaneous production of a spermatophore sac (Tan 1974, 1977). The cause of permanent copulation is the inability of the male to dislodge the horns of the spermatophore from the horns of the ejaculatory duct, possibly owing to a stiffening of the horns before the spermatophore is duly transferred into the bursa copulatrix. Consequently, the copulating partners become locked together, the male being often dragged around by the female. A similar aberration in which males cannot be parted from the females has also been observed in other Lepidoptera, such as the male red bollworm *Diparopsis castanea,* treated with Hemel (Campion 1971), and the male cabbage looper *Trichoplusia ni,* fed 1% Tepa (trimethylenephosphoramide); Tepa-fed males were also less responsive to the female sex pheromone than untreated males (Henneberry et al. 1966). Tepa, incidentally, has been well known for some time to be highly effective as an antispermatogenic agent in rats (cf. Mann and Lutwak-Mann 1981).

Less obvious is the cause of the Hemel-induced sterile mating in which the male flour moth manages to deposit the spermatophore, but the female fails to respond by laying eggs. Similar sterile matings have been previously reported in the same species after the males were reared at excessively high temperatures and consequently became sterile, probably as a result of either arrest or abnormalities of spermatogenesis (Norris 1933). Another possible cause of the sterile matings could be an abnormality in the spermatophore itself. Sterility associated with spermatophore malformation has been described in the male noctuid *Spodoptera littoralis,* orally treated with D-glucoascorbic acid (Navon and Levinson 1976) or D-araboascorbic acid (D-isoascorbic acid; D-erythorbic acid) (Navon and Marcus 1982; Navon et al. 1983). The feeding of either of the two analogues of L-ascorbic acid causes the production of deformed spermatophores which fail to inflate their bodies during mating. D-Glucoascorbic acid feeding to adult males leads in addition to permanent copulation. In the males fed D-araboascorbic acid, in spite of the grossly malformed spermatophores, the mating activity seems to be unaffected, but the fecundity of the mated females is markedly reduced. However, 2 days after cessation of feeding of this nutrient analogue egg fertility was restored. When L-ascorbic acid is fed to the moths that had been previously treated with D-araboascorbic acid, the effect of spermatophore malformation is not completely reversed. It remains to be shown which of the male accessory secretions needed for inflating the spermatophore was missing in the treated males, a question of special interest in relation to the males of *S. littoralis* which elaborate several distinct secretions including one, derived from the ductus ejaculatorius simplex, which contains a red pigment; some of the pigmented secretion is known to be normally incorporated into the spermatophore (Haines 1981). As yet, the function of this secretion, which occurs also in other noctuid species and can vary in colour from yellow or orange to deep purplish-red or even brownish-black, remains unknown. Some time ago, it was suggested (Callahan and Cascio 1963) that the function of the pigmented secretion in the corn earworm, *Heliothis zea,* is to harden the spermatophore as it is being transferred during copulation. In this connection, our own observation may be mentioned that in the giant octopus of the North Pacific, the rod-like ejaculatory apparatus of the spermatophore, which is normally colourless or yellowish-coloured, assumes an orange, red or brown appearance in some older, thinner, and more hardened spermatophores stored inside the spermatophoric (Needham's) sac (Chap. 3).

Spontaneous spermatophores formation, that is, the induction of spermatophore-sac production in the unmated, Hemel-treated males of the Mediterranean flour moth, is again a phenomenon not solely restricted to one species, but has been noted in other Lepidoptera, such as the aforementioned red bollworm. The cause of it, in *Anagasta (Ephestia) kühniella* at any rate, is, according to Tan (1977) a disturbance in the neuro-endocrine mechanism that normally controls spermatophore production. In the flour moth, he has been able to induce spontaneous spermatophore formation by treatment with cholinesterase inhibitors, the effect being most pronounced after topical application of organophosphorous insecticides, and also by injections of certain nerve stimulants, notably caffeine or amphetamine.

Before closing the above outlined account, largely based on observations concerning the reproductive behaviour of the Mediterranean flour moth, it should be added that apart from this species other Phycitidae have been receiving due attention. *Plodia interpunctella* is one of them; its spermatophore, together with those of *Anagasta (Ephestia) cautella* and *glutella,* has been already investigated by Norris (1932, 1934) as part of her fundamental study of insect fertility. An interesting recent contribution to the knowledge of spermatophores in the almond moth, *Anagasta cautella,* concerns the effects of irradiation (Brower 1979). The irradiated females showed a curious tendency to mate more often and to contain more spermatophores than control females. The dose required to sterilize 50% of both sexes was approximately 20 krad, and the doses for 99% sterility were 50 and 60 krad in females and males, respectively.

Males of several lepidopterous insects have been extensively used in exploring the sterilization effects of irradiation, largely as part of attempts to counteract pests. Three typical examples involving observations on the function of spermatophores concern the European corn borer, *Ostrinia nubilalis* (Nabors and Pless 1981), the codling moth, *Laspeyresia pomonella* (Olethreutidae), a worldwide pest of many pomes, drupes and nuts (Robinson 1974), and the cabbage looper, *Trichoplusia ni* (Holt and North 1970).

The role of spermatophores in the reproduction of *Laspeyresia pomonella* and serveral other tortricid moths, such as *Adoxophyes orana, Bactra verutana, Choristoneura fumiferana,* and *Grapholitha molesta,* has been for some time, and continues to be, a topic of concentrated research (Ferro and Akre 1975; Frick and Wilson 1982; George and Howard 1968; Howell et al. 1978; Noguchi 1981; Noordink 1970; Outram 1970). Normally, within 15 min of the start of copulation, the male codling moth begins to force the spermatophore into the bursa copulatrix and after about 40 min, when the body of the spermatophore is almost completely filled with accessory secretion, he ejaculates into the spermatophore the milky semen, at first containing mainly apyrene sperm cells, and later also the eupyrene spermatozoa (still in bundles); sperm transfer is completed within 1 h after the start of mating. Upon entry into the spermatophore, the apyrene sperm cells become highly motile, while the eupyrene sperm-bundles remain inactive; it takes another 3 h or so before the sperm mass, made up of apyrene and eupyrene spermatozoa, begins to enter the seminal duct (Ferro and Akre 1975). Irradiated (15 and 35 krad) male codling moths do not differ from the non-irradiated ones in their ability to deposit the spermatophore in the bursa, but their inseminating potential, as determined by the presence of spermatozoa in the spermatheca, is significantly reduced and, moreover, the ovipositional response of the mated females is also smaller (Robinson 1974).

Likewise, gamma irradiation of the male cabbage loopers (30 krad) results in a failure to transfer spermatozoa to the spermatheca, the main cause being an inability of the irradiated males to incorporate the spermatozoa properly into the spermatophore (Holt and North 1970). Normally, both apyrene and eupyrene are included in the spermatophore, but in the irradiated cabbage loopers, the number of the eupyrene ones (still in bundles and immotile) incorporated into the spermatophore, is drastically reduced. Moreover, when eupyrene sperm-bundles are deposited directly in the bursa copulatrix, they fail to migrate to the spermatheca. It does look, therefore, as if the motile apyrene spermatozoa are needed to provide a transport vehicle for the immobile eupyrene sperm-bundles, a conclusion in line with that reached by Iriki (1941), who claimed that in the silkworm, *Bombyx mori,* normal sperm passage from the bursa copulatrix to the spermatheca is made possible by the mobility of the apyrene sperm.

Bombyx mori, along with a few other insects such as *Anagasta (Ephestia) kühniella* (Norris 1932) and three species of *Zygaena (Z. trifolii, Z. filipendulae, and Z. fausta)* investigated by Hewer (1934), has been in the past a source of invaluable information about the mechanism underlying the passage of spermatozoa in the female reproductive tract, and has helped a great deal in tackling the vexed question:

Is sperm motility essential for the migration of insect spermatozoa from the site of spermatophore deposition to the spermatheca, or are the spermatozoa propelled forward passively, that is, by concomitant contractions of the female genital tract?

It is a question to which we shall return once more in this chapter when discussing sperm passage in certain Diptera, but perhaps it might be appropriate to mention already at this point that similar doubts regarding active versus passive sperm passage existed at one time in respect of transit of mammalian spermatozoa in the uterus, until it was conclusively proven that their own progressive motility is of paramount importance to spermatozoa for their entry into the ovum, that is, the final act of fertilization, but in the uterus, and to a large extent also in the oviducts, sperm passage is substantially assisted by ciliary and contractile movements of the female reproductive tract, the contractions being strictly controlled by a neuroendocrine mechanism (reviewed by Mann 1964; Mann and Lutwak-Mann 1981).

Notwithstanding the presumed role of apyrene sperm cells as carriers for the eupyrene spermatozoa, experiments on *Bombyx mori* indicate that in this insect at any rate the formation of the spermatophore and the transport of spermatozoa in the female are under neural control. It has been shown by MacFarlane and Tsao (1974) that the removal of the head of the male silkmoth immediately after the start of copulation does not significantly alter either the number or the fertility of the eggs laid by mated females. However, when the nerve cord of the male was cut between the 2nd and 3rd abdominal ganglia, this resulted in significantly reduced egg production and egg fertility. Thus, it would appear that the passage of the male accessory secretions is under at least partial control of the ganglia located anterior to the 3rd abdominal ganglion. As in many other insects already mentioned, it is highly probable that the male accessory secretions transferred during copulation with the spermatophore provide the female with an essential nutrient needed particularly for oogenesis and/or oviposition.

- *Leucinodes orbonalis* (Pyraustidae) is another moth in which the transference of spermatozoa from the spermatophore to the receptaculum seminis has been shown to depend on rhythmic, muscular contractions of the ductus seminalis (Srivastava and Srivastava 1957). As in *Bombyx mori* (Omura 1938), *Anagasta kühniella* (Norris 1932), and *Galleria mellonella* (Khalifa 1950a), in which the sequence of ejaculatory events was previously determined, the individual secretions of *Leucinodes orbonalis* are voided by the male into the bursa according to a definite sequential pattern, emission of the secretions from the ejaculatory duct preceding that from the accessory glands. If the spermatophore is removed from the bursa about an hour after copulation, its neck is already found broken, and in a female dissected at that time, the bursa, ductus bursae, ductus seminalis, ductus receptaculi, and the receptaculum itself can all be seen undergoing rhythmic contractions. There can be little doubt that the main factor responsible for the pumping of spermatozoa into the receptaculum is the muscular contraction affecting the entire route from bursa to the receptaculum, though of course, motility of the spermatozoa may also be of some assistance.

Acquisition of ability to move by themselves is acquired by the *Leucinodes* spermatozoa relatively early. If shortly after copulation had been accomplished the spermatophore is removed from the bursa and dissected in saline solution on a microscopic slide, motile spermatozoa can already be seen. In *Sitotroga cerealella* (Gelechiidae) activation of spermatozoa also takes place within the spermatophore directly after mating, but the rupture of the spermatophore's sheath and subsequent sperm passage along the seminal duct towards the spermatheca is caused mainly by muscular contractions of the bursa, possibly facilitated by a lubricating action of secretions produced in the female reproductive tract (Stockel 1973).

As regards the nutritive value of the spermatophore's contents to the mated females, and in particular, the stimulation of oogenesis and/or oviposition, this appears to be as important in Lepidoptera as it is in the earlier described Neuroptera or Coleoptera. Apart from the already mentioned reduction in the ovipositional response of female codling moths that copulated with irradiated males (Robinson 1974) and of the female silkmoths mated to males with cut nerve cords (MacFarlane and Tsao 1974), there are other observations plainly indicating that the role of the spermatophore extends far beyond its function as transport vehicle for spermatozoa, mainly in two directions, as stimulant to egg production and barrier to remating, the latter related to the earlier mentioned effects of sphragis (spermatophragma) and copulatory plugs.

In Lepidoptera, mating has long been known to stimulate oogenesis and increase the rate of oviposition, most investigators favouring the view that the stimulation is attributable to material released from the spermatophore. It is by no means clear, however, even now, whether the stimulant originates in the male accessory secretions or in the spermatozoa. Nor can we be certain whether the spermatophore's contents are acting alone or in conjunction with an agent produced in the bursa or some other part of the female reproductive tract. Of special significance in this respect are experiments carried out on female Cecropia silkmoths, *Hyalophora cecropia* (Riddiford and Ashenhurst 1973). These have shown that (1) after implantation of the bursa copulatrix (minus the spermatophore) from a mated female, the virgin female behaved as if it had copulated, that is, it oviposited eggs in the typical mated pattern, and (2) similar implantation of bursae from females which had mated with castrate males did not alter the virgin behavioural pattern. Furthermore, injections of haemolymph from mated females into virgin females caused an increase in oviposition rate, but blood from virgin females had no effect on oviposition. Altogether, it would appear that the switchover from virgin to mated behaviour in the female moth, that is, a change in the ovipositional response caused by mating, must be ascribed to a combined effect of a blood-borne factor which is secreted by the bursa copulatrix, but only after contact with either spermatozoa or some other material of testicular origin that has been transferred with the spermatophore to the female. It may not be out of place to recall at this point the previously mentioned observations on mated crickets, showing that oviposition is triggered by a prostaglandin produced in the spermatheca from arachidonic acid, under the influence of a prostaglandin-synthesizing agent, most probably of enzymatic nature, which has passed all the way from the testis, via the spermatophore, to the mated female.

Apart from those already mentioned, absorption of male-derived nutrients and their utilization for egg production has been explored in other Lepidoptera.

130

Some examples have already been described earlier in this chapter, when dealing in general, with the gustatory and aphrodisiac attributes of male secretory products and spermatophores. Among the lately explored instances of a similar kind relating specifically to Lepidoptera are additional observations on some Lycaeninae (Ehrlich and Ehrlich 1978), two heliconiine butterflies, *Dryas julia* and *Heliconius charitonius* (Boggs 1981; Brown 1981), and the moth *Leucania separata* (Chao 1981). In the last-cited investigation, radioisotope-tracer measurements and autoradiography of ^{32}P were used to demonstrate that after spermatophore transfer, the constituents of semen, including the secretion of the yellow gland, are absorbed into the ovarioles.

Before closing the discussion on the various aspects of spermatophore function in moths and butterflies, two additional points seem to be worth raising. One concerns the possible influence of the male lepidopterous courtship pheromone such as the one, already mentioned, which is produced in the hairpencils of the oriental fruit moth, *Grapholitha molesta,* on the formation of spermatophores (Baker et al. 1981). Apart from some vague suggestions relating to Orthoptera, such as that the secretions of the "alluring glands" help to keep the female quiet during the transfer of a spermatophore, or that they dissuade her from eating the spermatophore, little is known as yet about the possible influence of male sex pheromones on the production and fate of spermatophores, not only in Lepidoptera, but in insects generally.

The second point concerns the possible role of so-called anti-aphrodisiac pheromones in controlling mating behaviour. The existence of this sort of relationship is borne out by several observations. The male armyworm moth *Pseudaletia unipuncta* (Noctuidae) emits a distinct almond-like odour due to benzaldehyde released by a pair of abdominal scent brushes, which acts as a male-to-male pheromone preventing other males from approaching a mated female (Hirai et al. 1978). A direct involvement of anti-aphrodisiacs released from male accessory secretions may provide an alternative mechanism for preventing re-mating. As already mentioned (Chap. 1), male insects have evolved a variety of plug-like devices, such as copulatory plug, sphragis or spermatophragma, which are formed from hardened male accessory secretions during copulation and seal off effectively the genital region of the females. Such structures, commonly occurring in Lepidoptera, are generally held to function by preventing the backflow of semen and at the same time providing a mechanical barrier against a second insemination (Parker 1970). There are indications, however, that apart from equipping the female with something in the nature of a chastity belt, the male may enforce monogamous behaviour in the female by introducing with his accessory secretions an odorous substance that repels other males. *Heliconius erato* depends apparently on a deterrent of this kind, in the form of an anti-aphrodisiac pheromone with the odour of witch hazel, exuded during copulation by glands located in the male clasper (Gilbert 1976).

Diptera. In spite of early evidence (Pomerantzev 1932) for the presence of spermatophores in *Culicoides nubeculosus* (Ceratopogonidae), their occurrence in Diptera continued to be denied until Nielsen (1959) showed that males of *Glyptotendipes (Phytotendipes) paripes* (Chironomidae) frequently carry between

131

their claspers a peculiarly pink-coloured spermatophore consisting of a gelatinous mass "about the shape of the seed of a birch tree", its base housing "two cavities in which the thread-like spermatozoa can be seen". He also found that no more than 5 s are needed by the male to place the spermatophore in the female, and that it takes something like 20 min for the spermatozoa to reach the spermatheca. Subsequently, the observations on spermatophores were extended to a third family, Simuliidae. Davies (1965) showed that the spermatophore of *Simulium (Wilhelmia) salopiense* has a spherical, but flattened body, 150–170 μm at its greatest diameter, packed with spermatozoa with a "considerable degree of parallel bunching". The spermatophore is formed when first the viscous male accessory secretion and next the spermatozoa are ejected into the cavity opened up within the female terminalia in the course of a 2–3-min copulation. Immediately after ejaculation, the contents become encapsulated within a wall composed of two layers, an inner thin, hyaline layer, and a thicker outer layer of more spongy appearance, often carrying some irregular external projections. Wenk (1965), apart from demonstrating that in several other Simuliidae, including *Boophthora erythrocephala,* sperm transfer occurs by means of spermatophores, provided a great deal of information about the peculiar pre-copulatory recognition signals exchanged during flight between the two sexes, and the mechanism of copulation itself. Linley and Simmons (1983) found that in the black fly, *Simulium decorum,* the total number of spermatozoa ejaculated by the male into a spermatophore can vary from 572 to 6892.

Behavioural interactions, spermatophore formation and sperm passage to the spermatheca have been extensively explored by Linley and his associates in *Culicoides melleus* (Linley 1975, 1981 a, b; Linley and Adams 1972; Linley and Hinds 1975), and the question of active versus passive sperm passage in Diptera generally, has been thoroughly discussed by Linley and Simmons (1981, 1983). Much of the case for non-involvement of sperm motility arises from the fact that complete transfer from spermatophore to spermatheca in the *Culicoides,* and from the copulatory bursa to the three spermathecae in the *Aedes,* is extremely rapid. For both the *Culicoides* and the *Aedes,* the speeds (109 μm/s and 414 μm/s, respectively) seem much too high to be attributable to motility of spermatozoa alone. Within 10 min of emergence, male *Culicoides melleus* is capable of copulating and inseminating a female. Male potency increases over the next few hours and declines thereafter. Copulation between virgin insects occupies about 10 min, and very soon after genital union is established the male begins to secrete a spermatophore, flask-like in shape, with a long neck penetrating into the spermathecal duct of the female. The mean numbers of ejaculated spermatozoa, recorded with virgin, once-mated, and twice-mated females, were 857 ± 24, 732 ± 23, and 673 ± 25, respectively. The spermatophore's capsule is formed during copulation and subsequently held in the male genitalia, while the neck is formed within the spermathecal duct. As copulation ends and the insects separate, the spermatophore may remain either in the male's genitalia or attached to the female by the neck still lodged within the spermathecal duct. Both young and old males, and old females, remove a retained spermatophore usually within 40 s of separation, but very young females sometimes require a much longer time.

In addition to Ceratopogonidae, Chironomidae, and Simuliidae, sporadic occurrence of spermatophores has also been recorded in the family Bibionidae (Leppla et al. 1975; Pollock 1972). In *Plecia nearctica,* semen is transferred by extrusion of a filament from the distal part of the large bilobed spermatophore to the bursa; branches of the filament extend into the spermathecal ducts, but the spermatophore proper remains in the male and is expelled within seconds after disengagement. As regards suborders other than Nematocera, it seems that the tsetse flies still remain the sole properly authenicated instance of spermatophore-carrying insects (Pollock 1970; Tobe and Langley 1978).

Tsetse flies, sole members of the genus *Glossina* and well-known vectors of trypanosome diseases, are peculiar in that they reproduce by adenotrophic viviparity. Their male reproductive system, consisting of paired testes and paired accessory glands which empty into the ejaculatory duct, conforms to the pattern in other cyclorrhaphous Diptera, but the spermatozoa are transferred by means of spermatophores. These are constructed in the female late during copulation, from several male accessory secretions; the spermatozoa begin their migration to the spermathecae along the spermathecal ducts before the end of copulation. Ultimately, substances that are released from the spermatheca of the mated female are held responsible for stimulating egg production and inhibiting receptivity to subsequent matings. A study of the spermatophore in the tsetse fly *Glossina morsitans morsitans* (Kokwaro and Odhiambo 1981; Odhiambo et al. 1983) has shown that in this insect the spermatophore is a highly organized sac-like body, its wall comprising two distinct layers, whose components are similar to those of the secretions of male accessory glands; dense aggregates of material, matrix filaments, and clusters of granules predominate in both layers. The spermathecae are a pair of orange, spherical structures containing a secretion in which the spermatozoa may last for several months. The secretory material appears to consists mainly of mucopolysaccharide and protein (Kokwaro et al. 1981).

Though other Diptera do not seem to be capable of producing proper spermatophores, in many of them the mating plugs formed during copulation appear to function in a similar fashion, at least as regards stimulation of egg production and reduction of female receptivity to subsequent matings. In fact, perusal of the literature on the subject of plug-like structures in insects such as *Aedes* and *Anopheles* species leaves one with the impression that these plugs are imperfectly formed spermatophores. To some extent, this may perhaps apply to male accessory secretions that appear to stimulate oviposition or decrease female receptivity to remating in other dipteran insects, such as the cabbage maggot *Hylemya brassicae* (Swailes 1971), fruitfly *Drosophila melanogaster* (Bairati 1966; Perotti 1971; Leopold 1976), or housefly *Musca domestica* in which no less than 20 distinct protein fractions have been identified in homogenates of the ejaculatory ducts alone (Terranova et al. 1972). In *Drosophila mojavensis,* the formation of the copulatory plug is followed by transfer of the male-derived nutrients to the oocytes and somatic tissues of the female (Markow and Ankney 1984).

Hymenoptera. As in Diptera, properly encapsulated sperm packages in the form of spermatophores are rarely encountered amongst Hymenoptera. Flanders (1939), the first to record the presence of a spermatophore in *Calliephialtes extensor,* also provided later a detailed description of the role played by spermatophores in *Macrocentrus ancylivorus* (Flanders 1945), pointing out that malposition of the spermatophore in this parasitic insect may adversely affect the chances of the female becoming impregnated and furthermore, that an excessive number of spermatophores may cause the pre-oviposition period of the females to last longer than the time period during which the host, that is, the potato tuber worm, *Gnorimoschema operculella*, remains susceptible to the attack. Normally, the empty spermatophore is ejected by the female before oviposition takes place. Lack of oviposition resulting from a prolonged pre-oviposition period may be caused by the extra time required for breaking up the supernumerary spermatophores before they can be cast out of the female's body.

Pimpla instigator is another hymenopteron which, when dissected shortly after copulation, can be seen to cary a spermatophore in the vagina (Khalifa 1949 b). The fully formed spermatophore is a transparent egg-shaped body, 0.66 mm long and 0.61 mm wide. There is a small milky portion at the anterior end containing a double sperm sac with a common opening which has a conspicuous rim fitting into the opening of the spermathecal duct. The spermatozoa start to migrate into the spermatheca soon after copulation and within 2–3 h the sperm sacs become empty, but it takes 16–20 h before the empty spermatophore is ejected.

In contrast with those enumerated above, the honeybee, *Apis mellifera*, is a hymenopteron insect in which the male accessory secretions produce a copulatory plug instead of a spermatophore. The coagulable protein required for this purpose comes from the secretion of the mucous glands, which forms part of the normal ejaculate. This is a fact of considerable importance as regards the practice of artificial insemination (Taber 1977). The special syringe which one uses routinely for collecting spermatozoa for insemination has to be handled carefully in such a way as to prevent the inclusion of the mucous gland secretion as otherwise the coagulated protein would quickly clog up the syringe.

Chapter 7

Crustacea

Just as in other classes of Arthropoda, transfer of spermatozoa by means of sper-matophores is a common feature of reproduction in Crustacea; it is characteristic of many Ostracoda, Copepoda, and Malacostraca. With insects the crustaceans also share some general characteristics of spermatozoa, such as long survival in the female body. A well-known example is the American lobster, *Homarus ameri-canus,* in which the spermatozoa that have been transferred as a spermatophore to the thelycum of the female (a receptacle located on the ventral surface of her body), can remain there in storage for several years before being released for fer-tilization (Herrick 1911; Kooda-Cisco and Talbot 1982).

Apart from spermatophores, crustaceans, again in common with insects, use other sperm-packing devices, one, particularly characteristic, being the so-called extracellular tubule. Tubules of this kind, which originate as secretory products of the Golgi complex in nurse cells surrounding the spermatids, are found around sperm-bundles in the vas deferens of various Copepoda, Cumacea, Isopoda, and Amphipoda. A good many of these tubules disintegrate in the lower parts of the vas deferens, but it is possible that the remaining ones "may function in a capacity analogous to typical spermatophores, in bundling spermatozoa for transport dur-ing fertilization" (Reger and Fain-Maurel 1973).

7.1 Ostracoda

A spectacular feature of ostracod spermatozoa and spermatophores, which did not escape the attention of G. W. Müller (1894) in the course of his extensive sur-vey of Ostracoda in the Gulf of Naples, is their extraordinary length, sometimes exceeding that of the whole body. In *Propontocypris monstrosa,* he found the sper-matozoa to be 5–7 mm long, thus exceeding by 8–10 times the length of the ani-mal; this is, one should add, still less than of the spermatozoa in an insect, *Notonecta glauca* (backswimmer, Heteroptera), reputed to be 12 mm long (Pantel and De Sinéty 1906). A great deal of information on ostracod spermatozoa and spermatogenesis has accumulated since the time of Zenker's (1854) general mono-graph on Ostracoda, and Stuhlmann's (1886) treatise on the male genital organs and spermatogenesis in *Notodromas monacha* (Lowndes 1935; Gupta 1964, 1968; Zissler 1966, 1969 a, b; Reger 1970).

The spermatophore which Müller (1894) recovered from the receptaculum seminis of an ostracod has been described by him as consisting of three distinct parts, a ball, neck, and tail; the ball filled with a balloon-shaped sperm-containing bladder, and connected to a tube-like channel running through the neck and continuing all the way to the end of the tail. Puzzled by the dimensions of the bladder, which he found to be far in excess of the diameter of the copulatory channels, he concluded that although the production of the sheath engulfing the sperm may have started in the male reproductive tract, the formation of the bladder could not have been accomplished without the active participation of some material contributed by the receptaculum seminis. Similar reasoning stressing the presumptive role of female genital secretions in the completion of spermatophore development has also been subsequently advanced by other investigators of crustacean spermatophores. Admittedly, as pointed out earlier, there are situations where secretory material coming from the female genital tract can contribute to the formation of a spermatophore. However, as regards the development of so-called bladders in spermatophores generally, in our view the argument that the large dimensions of a spermatophore's bladder as observed within the female tract are due to female secretions alone carries little conviction. A bladder of this kind may have been formed during the terminal phase of spermatophoric reaction, as a result of evagination of the spermatophore, in the same way as in octopus (Chap. 3).

The extraordinary length of ostracod spermatophores may be more a reflection of the length of the sperm tails encased inside than that of the encasing sheath. However, not much is known as yet about the relative contributions of the testis, vas deferens and other parts of the male genital system towards the formation of the sheath as such. In this respect, some material produced by the so-called nurse cells of the spermatic duct seems to be of special importance.

7.2 Copepoda: Harpacticoida, Calanoida, Cyclopoida, and Caligoida

The formation of copepod spermatophores, made largely possible by secretions derived from the vas deferens, has been known for a long time. *Cyclops castor* was one of the first copepods thoroughly explored in this respect, not only as regards transfer of spermatophores, but male mating behaviour as a whole (Siebold 1839; Wolf 1905). Extrusion and transfer of spermatophores in Calanoida have also been subject of an early study by August Gruber (1879), later extended by other investigators to include various members of other copepodan orders. In most of these, the spermatophores have been described as tube- or flasklike in shape. The present terminology commonly used for designating the different parts of either the copepodan male reproductive system or the spermatophores is based on the classification used by Heberer (1926, 1932, 1937, 1955 b), Lange (1948), and Fahrenbach (1962). In many instances, the studies of the spermatophores also included observations on copepodan spermatozoa which are distinguished by a number of unusual features, one of them being the occurrence within the spermatophore, of two quite distinct types of spermatozoa, variously described, mostly as the fertilizing and swelling, or the B-(Befruchtung) and Q-(Quell) kind (Heberer 1955a; Gupta 1964; Roosen-Runge 1977).

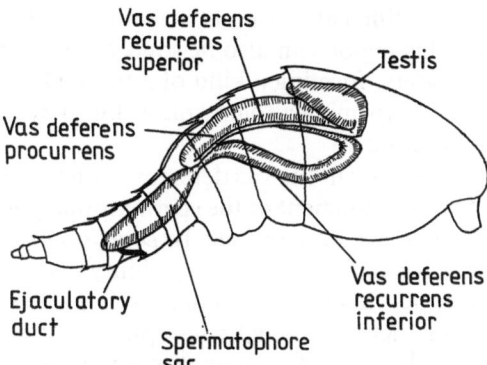

Fig. 38. Male reproductive tract of harpacticoid copepod. (Redrawn from Fahrenbach 1962)

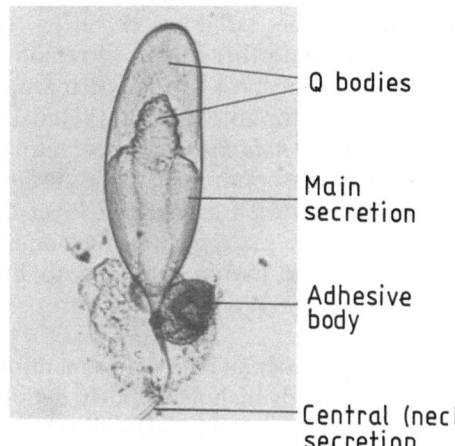

Fig. 39. Spermatophore of the harpacticoid copepod *Diarthrodes cystoecus* Fahrenbach, during the initial stage of discharge from the spermatophoric sac; the central secretion has formed the elongated neck. (× 250). (Fahrenbach 1962)

Harpacticoida. As a particularly instructive example of studies on the reproductive function of harpacticoid copepods, can serve the detailed investigation made by Fahrenbach (1962), of *Diarthrodes cystoecus,* a marine copepod (males 650–780 µm long) which inhabits red algae. Figure 38 shows the male reproductive tract of this animal; the main parts are the testis, vas deferens, spermatophore sac and ejaculatory duct. Figure 39 shows an unfixed spermatophore (in the initial stage of discharge) removed from the spermatophore sac.

The flask-shaped spermatophores of this copepod are 130–152 µm long and 37–41 µm wide. The distal part of each spermatophore (located at the posterior end of the spermatophore sac) is packed solidly with a granular mass of socalled Q-bodies (with swelling properties), filled with PAS-staining material that originated in the secretion of the vas deferens. When during copulation a spermatophore is inserted into the female gonopore, it contracts, and following the discharge of the contents, its dimensions shrink to 38 × 25 µm. Under in vitro conditions, that is, when it had been dissected in seawater, the spermatophore undergoes rapid swelling, largely, it seems because of water imbibition by the Q-bodies which in the course of this process become able to push both the main secretion (M-secretion) and the spermatozoa in the direction of the neck of the spermato-

137

phore. When at this point a slight pressure is applied to the neck, the fluid which fills this region can also be squeezed out. In contact with seawater it hardens immediately, forming a kind of a tube. During all this time, the Q-bodies continue to swell, forming a foamy mass, but they themselves rarely emerge from the spermatophore.

Apart from the vas deferens, which produces the Q-bodies, an important role in the development of the spermatophore is played by material secreted in the wall of the spermatophore sac. In *Diarthrodes cystoecus,* at least two regions of the sac are endowed with secretory ability. The product of one of them is the granular material which later becomes compacted into the spherically "adhesive body" (at the neck of the spermatophore); after copulation this secreted mass can be found between the spermatophore and the ventral body wall of the female close to the copulation pore, and because of its sticky consistency, it is supposed to function as the adhesive secretion (Heberer's *Klebesekret*). The material produced in the other secretory region of the spermatophore sac lacks precise localization in respect to the spermatophore; it is thought to function as some kind of spermatophore lubricant (Heberer's *Gleitsekret*). In another harpacticoid copepod, *Tisbe holothuriae,* the secretory material from which the stratified wall of the spermatophore is made up consists of a loose fibrillar network, sandwiched between a granular layer and a set of thin dense plates; all three segments of the male genital duct and the seminal vesicle are involved in the secretion of the material needed to build the spermatophore (Gharagozlou-van Ginneken 1978).

Calanoida. Gruber's (1879) aforementioned publication on the reproductive organs of copepods includes a fairly detailed description of the steps leading from what he calls *Spermatophorenanlage* to the extrusion of the spermatophore by male calanoids. For successful copulation, proper positioning of the spermatophore on the female urosome is essential and to achieve this the male must first of all assume the correct position. His attempts to do so are associated with some quite impressive gymnastic efforts. Upon encountering a female, he squeezes her frequently by the urosome and then, if successful, executes an upward somersault, grasps the genital segment of the female's urosome with his fifth limb, in which he is already holding the freshly extruded spermatophore, and places it on the body of the female. A secretion escaping from the neck of the capsule serves to fix that capsule first to the limb and then to the genital pore of the female. No less impressive are the copulatory events associated with spermatophore transfer in *Diaptomus gracilis,* as recounted by Wolf (1905; reviewed by Schöne 1961). The male of this species first clasps the female's abdomen with the right fifth foot, and then uses his left foot, which encloses the neck of the spermatophore, to push it towards the female genital orifice; according to Wolf (1905) the final push leading to the expulsion of spermatozoa comes from some special quick-swelling material (Gruber's *Austreibstoff*) contained within the spermatophore.

Copulatory behaviour in different calanoid copepods varies in details, but the basic sequence of events in terms of male acrobatics seems to be similar. Some of the recent behavioural studies also contain indications that the quick, complex swimming pattern which forms part of the mate-seeking behaviour of the male is elicited by a special sex pheromone emanating from the female (Gauld 1957, 1966; Katona 1973; Blades 1977; Dunham 1978; Hayward 1981). Among the recent studies well illustrating the highly ritualized mating behaviour among calanoid copepods, one of *Labidocera aestiva* (Blades and Youngbluth 1979) deserves special attention on account of structural observations concerning the area of spermatophore attachment to the female's body. Scanning electron microscopy revealed the presence of about 20 special pore structures (pit-pores) on the right ventrolateral surface of the female's abdominal segment, which corresponds to the area of spermatophore attachment. Prior to the transfer of a spermatophore, the male vigorously strokes this area with the modified endopod of his left fifth leg to prepare the pit-pore region for the attachment of the spermatophore; transmission electron microscopy of the cells underlying these pit-pores strongly suggests that they perform a secretory function; it is possible that the material secreted by these cells helps to dissolve the cement which binds the coupling device of the spermatophore to the female's abdomen, thereby enabling her to remove a discharged spermatophore from the urosome. Structural features of coupling devices have been described in several calanoid copepods belonging to the genus *Centropages*; the form of the plates which compose these devices is important for the accurate location of the spermatophore on the female urosome (Lee 1972).

A great deal of additional information has accumulated lately with regard to the structure of spermatophores in a number of Calanoida, the exact location of the sites to which these spermatophores become attached, and also the respective contributions by the various male genital secretions to the formation of spermatophores. The aforementioned *Labidocera* has been receiving special attention in this respect (Blades and Youngbluth 1981; Fleminger 1979; Turner et al. 1979). The male reproductive system of *L. aestiva,* studied by Blades and Youngbluth (1981), includes a sinuous duct composed of a ductus deferens and seminal vesicle. The glandular, columnal epithelium of the ductus deferens produces a flocculent substance and two granular secretions which together constitute the future spermatophoric plasma. In its terminal portion, the ductus deferens synthesizes another secretion that is chiefly responsible for the formation of the spermatophore's wall, but the final assembly of the wall, which eventually will encase both the spermatozoa and spermatophoric plasma, is accomplished within the seminal vesicle. Contiguous to the seminal vesicle is an elongate, highly glandular spermatophore sac. The chitin-like coupling apparatus, by means of which the spermatophore will be able to attach itself later to the female's abdomen, is formed in the anterior region of the spermatophore sac, while the posterior region functions as a storage organ for the flask-shaped spermatophore and in addition produces a secretion that aids the extrusion of the entire spermatophore complex.

Calanus finmarchicus (Raymont et al. 1974) and *C. hyperboreus* (Brown 1970) are two other calanoid copepods which have received special attention. The latter produces a club-shaped spermatophore, about 200 μm long, consisting of an outer cylinder made up by densely packed spermatozoa, and a central core composed of PAS-staining granular material. Among the recently investigated species are also *Acartia tonsa* (Hamner 1978), *Pseudodiaptomus acutus, P. cokeri*, and *P. coronatus* (Jacoby and Youngbluth 1983), *Pleuromamma piseki* (Hayward 1981), and *Euchaeta norvegica* (Ferrari 1978; Hopkins 1978, 1982; Hopkins and Machin 1977; Schweder 1979). *Euchaeta* has been studied with special care as regards breeding intensity and mode of spermatophore attachment. The total output of spermatophores by the male during his lifetime is probably around three, and the ejection of spermatozoa is triggered by seawater infiltrating the spermatophore contents; unlike in the aforementioned *Diaptomus* (and many other calanoids), the spermatozoa of *Euchaeta* are all of the same structural type.

Cyclopoida and Caligoida. These are two further copepodan orders in which the participation of spermatophores in reproduction was established some time ago.

Among the recent investigations on Cyclopoida, one of special interest is the comparative study of 12 species belonging to the family Oithonidae in the eastern coastal waters of Central and South America; in *Oithona oculata* and *O. plumifera* spermatophores were shown to be located ventrally on urosome segments, and it has been suggested that tactile recognition of the male "pore signature" is accomplished by the modified endopod of the female's fourth leg (Ferrari and Bowman 1980). There has also been renewed interest in the spermatophores of Caligoida inhabiting the Eastern Pacific Ocean; in the eight species of *Alebion* investigated, the spermatophores attached to the female were shown to provide a very useful character in recognizing individual species (Cressey 1972).

7.3 Malacostraca: Isopoda, Cumacea, Mysidacea, Euphausiacea, and Decapoda

Among members of Malacostraca, the orders of Isopoda, Cumacea, Mysidacea, Euphausiacea, and Decapoda provided some valuable information about the ways in which spermatophores are formed and transferred.

Isopoda. With some exceptions (*Cymothoa*), the isopods are characterized by the presence of two separate sexes, and in both the parasitic and non-parasitic isopods sperm transfer occurs as a rule by means of spermatophores. A detailed account of the structural and functional features of the spermatophores and spermatozoa of several isopods has been compiled by Fain-Maurel (1970). Her article also outlines the ways in which the spermatids are transformed into spermatozoa, and the latter enclosed into spermatophores before passing into the seminal vesicle.

Related in function to spermatophores proper are probably so-called extracellular tubules which are found surrounding bundles of spermatozoa in the vas deference of several species of isopods, and also certain amphipods, schizopods, copepods, and cumaceans (Reger and Fain-Maurel 1973). They are believed to originate as secretory products of the Golgi apparatus in nurse cells surrounding the developing and elongating spermatids, the mucopolysaccharide material of the secretions probably acting as a glue that keeps the bundled spermatozoa together. Somewhat similar filamentous structures are found in certain insects, such as

Drosophila, for example (Bairati 1966; Perotti 1971). More information about the ways in which isopod spermatozoa are bundled and ensheathed is given by Itaya (1979) in her paper on woodlice. The resulting spermatophore-like structure consists in *Armadillidium vulgare* of two distinct component parts: a cone-shaped assembly of longitudinally aligned tubules, 400–450 Å in diameter, extending from the area rostral to sperm acrosomes to the region of sperm nuclei, and a matrix surrounding the spermatozoa, all along the entire length of the bundle. Morphological evidence suggests that the two component parts are produced from material originating in the endoplasmic reticulum and Golgi apparatus of the testicular follicle cells.

Cumacea and Mysidacea. The cumaceans and mysidaceans are two other peracarid orders in which prior to extrusion the spermatozoa are prepackaged into spermatophores (Duncan 1983; Wittmann 1982). Females of the marine mysidacean *Leptomysis lingvura* produce a pheromone which attracts the males, and during copulation, lasting 1 s, a female receives from the male a spermatophore which is about 2 mm long, tube-like in shape, and which contains thousands of spermatozoa.

Euphausiacea. As regards the order of Euphausiacea, special importance attaches to ecological observations on the relationship between the production and attachment of spermatophores on one hand, and seasonal changes in the phytoplankton on the other; in at least two *Thysanoessa* species (*T. inermis* and *T. raschii*) of Balsfjorden in Norway, the annual phytoplankton cycle appears to be the dominant factor in the regulation of spawning activity (Falk-Petersen and Hopkins 1981). In three rare giant *Thysanopoda* species (*T. cornuta, T. egregia,* and *T. spinicaudata*) which spawn in Japanese deep sea, each species carries differently shaped spermatophores, their size varying characteristically according to the size of the females (Nemoto et al. 1977).

Decapoda. Kölliker (1841), the first to take note of the semen tubes *(Samenschläuche)* in the vas deferens of *Pagurus bernhardus,* described them as being "sufficiently large to the recognized with the naked eye", and equipped with a double membrane ("outer" and "inner"). Over the years that followed, spermatophores were found to be present in many decapods, among Reptantia and Natantia alike, most investigators commenting also on the distinct bisexual characteristics and in some instances, a most bizarre or at times highly aggressive, copulatory behaviour of these animals.

In fresh-water crayfish, Andrews (1910) noted:

"The phenomena of conjugation are, in brief outline, the following: The aggressive male seizes the comparatively inert female and mounting upon her ventral surface becomes firmly attached both by the large chelae and by special clasping-hooks on the bases of the legs. The male has three sets of organs concerned in the transfer of sperm, the papillae and the two pairs of specialized limbs. These limbs we shall call the stylets, as they are firm, calcified, tapering organs that are thrust with force into the narrow slit of the calcified receptacle in the shell of the female. By means of these stylets the sperm is transferred some half-inch or so through the water from one animal to the other without being exposed to contact with the water, which would, it is believed, destroy the sperm.

With the male and the female firmly locked together and the stylets held at an angle of about 45 degrees, the transfer of sperm can take place after the muscular efforts of the male have forced the tip of the first stylet into the firm orifice of the annulus of the female".

Since then, mating behaviour has been closely observed in other Decapoda, a recent addition being the spiny lobster *Panulirus argus* (Lipcius et al. 1983). The extensive literature on decapodan spermatophores has been reviewed by Dudenhausen and Talbot (1983) who list 47 anomurans, macrurans, and brachyurans, in which spermatophores have been investigated: 44 by light microscopy, and 3 (*Libinia emarginata, Homarus americanus,* and *Pacifastacus leniusculus*) by electron microscopy. In general, the assembly of decapod spermatophores takes place during sperm transit from the testis to the distal part of the vas deferens; the newly formed spermatophores are then retained in either that part of the vas or in a special ampulla until the time of mating. There is still little precise information available about the physicochemical nature of the secretions produced in the decapod vas deferens, but the ease with which they undergo coagulation (Mouchet 1931) must be taken as an indication that in order to form the wall of the spermatophore, the secretions interact with each other in a manner resembling coagulation of semen and formation of copulatory plugs in a great many other animals, mammals included.

In the spider crab *Libinia emarginata* (Hinsch and Walker 1974) the spermatozoa entering the vas deferens from the testis are still free, but as they pass along the vas they become engulfed in secretions produced by three distinct types of cells, all rich in rough endoplasmic reticulum and Golgi apparatus, but poor in mitochondria. Another secretion, produced by cells with a similar type of cytoplasm, is added to the surface of the spermatophore in the middle portion of the vas, before the spermatophore passes into the posterior part of the vas, where it remains until the time of mating. In *Penaeus kerathurus* and other prawns (Malek and Bawab 1974 a, b), four regions of the vas deferens are clearly discernible: proximal, medial, distal, and terminal. The proximal part provides a viscous secretion which, on mixing with the free spermatozoa, converts them into a compact mass. This process continues in the next part of the vas, a blind pouch, in which the sperm mass is retained for some time, before being moved to the interior of the ascending limb of the vas. This portion of the vas, which plays a key role in the production of spermatophore, consists of two separate ducts. The wider one, named spermatophoric duct, forms the main body, while the narrow one is named wing duct because of its role in the formation of the accessory structure, the so-called wing of the spermatophore. Five layers of secretory material produced by the ascending limb are deposited over the surface of the spermatophore; some of them contain protein, carbohydrate as well as lipid. At the time of transfer, the body of the spermatophore becomes lodged within the female's thelycum, while the wing remains fully exposed to seawater at the ventral surface of the female. Broad wings and the manner in which spermatophores are lodged in the thelyca have also been described in several American white shrimps of the penaeid subgenus *Litopenaeus* (Farfante 1975). In some penaeids, it is possible to transfer the spermatophores manually for the purpose of artificial insemination. A special continuous gill irrigator has been developed, in combination with a restraining device, to keep the female alive and reduce stress during the spermatophore transfer (Tave and Brown 1981).

Extrusion and transfer of decapod spermatophores can occur in several ways. In some brachyuran crabs, the spermatophores are extruded through the gonopores (at the base of the fifth walking legs) directly into the female genital ducts, prior to internal fertilization. However, in other brachyurans, and in penaeids and nephropsideans with external fertilization, the spermatophore is often deposited in a ventral thoracic groove or pouch (thelyca of penaeids, annulus ventralis of some nephropsideans). The male hermit crab *Pagurus prideaux* sticks the spermatophore to the internal wall of the snail's shell in which a crab has settled down. Carideans plaster the spermatophores on some location of the female body where they can be in the path of eggs. The male caridean shrimp *Heptacarpus pictus,* which produces a spermatophore composed of two U-shaped cords, attaches it to the underside of the female's first abdominal segment (Bauer 1976; Chow et al. 1982). Electroejaculation, followed by artificial insemination, can be used to collect caridean shrimp spermatophores (Sandifer and Lynn 1980).

Three types of decapodan spermatophores have been described (reviewed by Dudenhausen and Talbot 1983). The simplest is the small, spherical or ellipsoid spermatophore of the brachyuran crabs, composed of a sperm mass surrounded by a thick acellular wall (Hinsch and Walker 1974). The second type is the pedunculate spermatophore, common to anomurans, in which the sheath is broken into small separate spermatophores, elevated on stalks which are attached to a common gelatinous base (Subramoniam 1977). The number of spermatophores forming a group is highly variable; the hermit crab, *Pagurus novizealandiae,* for instance, releases spermatophores in groups of 2–14, attached to a common pedestal (Greenwood 1972). The third type is the tubular spermatophore equipped with several distinct investing layers surrounding the sperm mass arranged in a cord-like manner. Tubular spermatophores are produced by most macrurans (crayfish, lobsters); they are characteristic of the astacids (*Pacifastacus leniusculus*), the homarids (*Homarus americanus*), the nephropsids (*Enoplometopus occidentalis*), and the panulirids (*Panulirus interruptus, P. penicillatus*). Although anomuran spermatophores are generally pedunculate, a macruran, non-pedunculate type occurs in a few members of the family Hippidae (Matthews 1956; Subramoniam 1977, 1984). In the sand crab *Albunea symnista* for example, the spermatophore consists of a highly convoluted tube surrounded by a firm membrane containing a neutral mucopolysaccharide; the fully formed spermatophoric mass is composed of the spermatophoric tube, a supporting basal gelatinous cord and a protecting gelatinous matrix (Subramoniam 1984).

An excellent account of tubular spermatophores in homarids has been presented by Kooda-Cisco and Talbot (1982, 1983) in their papers on the lobster *Homarus americanus.* The technique employed by these authors for collecting spermatophores from live lobster depends on applying an electric stimulus (12 V) to the base of the fifth walking leg. This stimulates the contraction of the vas deference and induces spermatophore extrusion without harm to the male. A spermatophore freshly obtained in this manner is shown in Fig. 40, and a transverse section through it is diagrammatically depicted in Fig. 41. The spermatophore measures 1–2 cm in length, with a circumference of 0.5–1 cm; the mass of spermatozoa (which can be seen through the translucent wall) is located within the first half to be extruded. The tubular spermatophore has the shape of a partially flattened cylinder with a prominent central bulge on the ventral surface and two

143

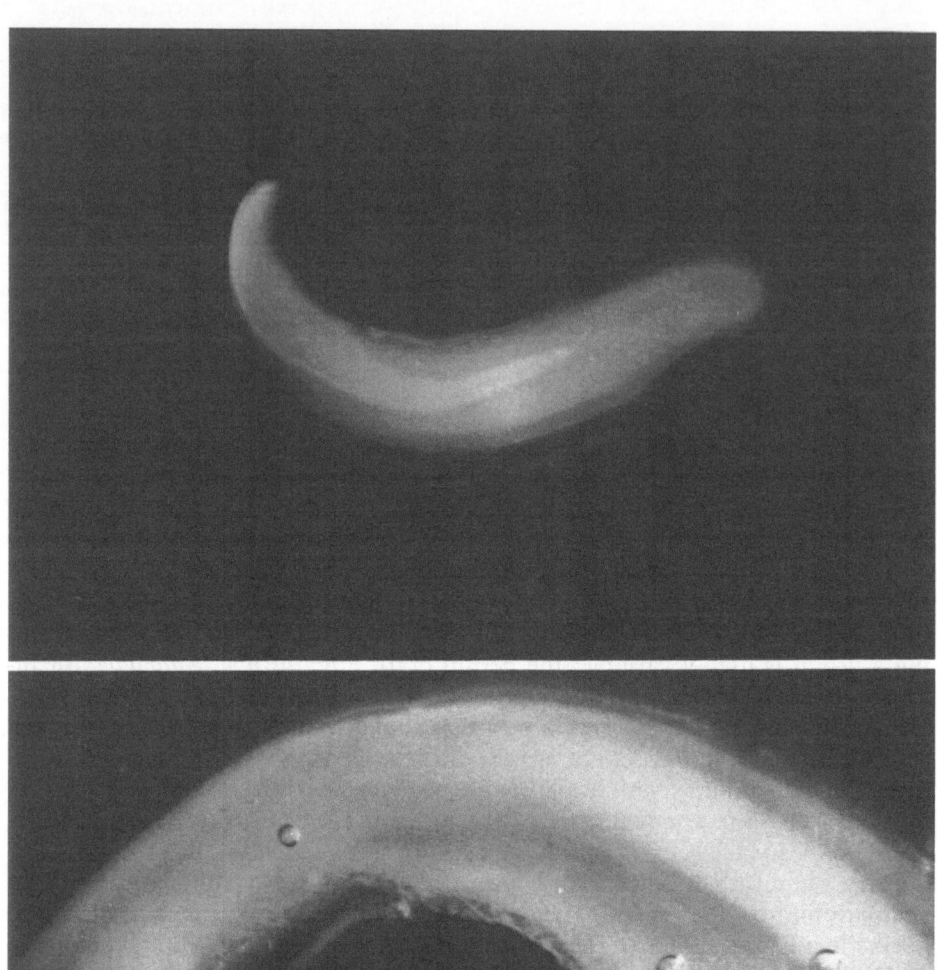

Fig. 40. Freshly extruded spermatophore of the lobster, *Homarus americanus*. *Top* the entire spermatophore; the first portion to be extruded is on the left; *bottom* part of a spermatophore freshly extruded in seawater, showing the release of gas bubbles from the surface. (Courtesy of Drs. Marcia J. Kooda-Cisco and Prudence Talbot)

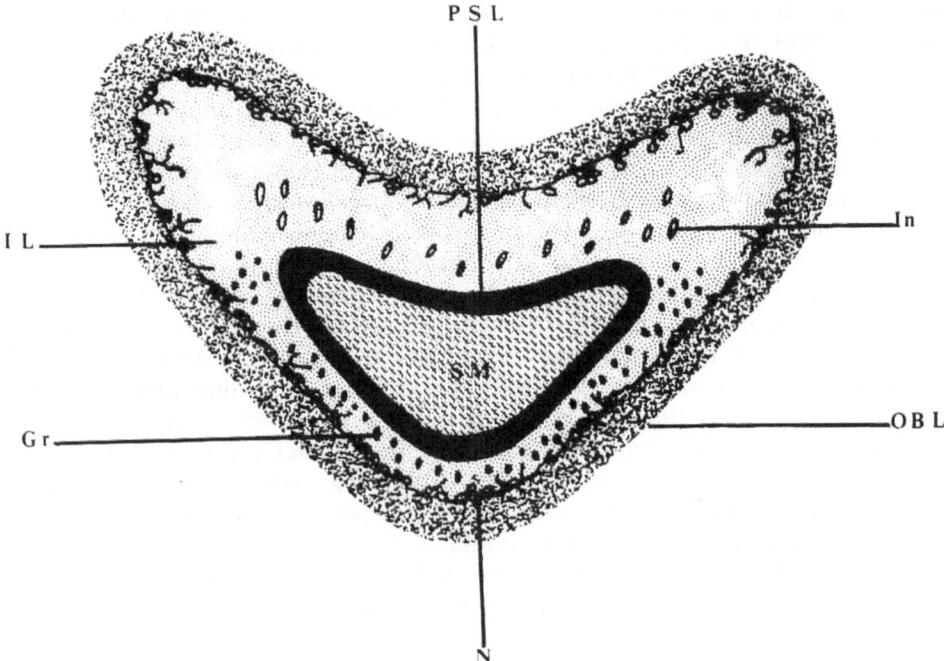

Fig. 41. Diagrammatic presentation of a transverse section through the lobster's spermatophore, showing the sperm mass (*SM*), the primary spermatophore layer (*PSL*), the intermediate layer (*IL*), and the outer bounding layer (*OBL*). Indicated are also the granules (*Gr*), inclusions (*In*), and the network of acellular processes (*N*). (Kooda-Cisco and Talbot 1982)

lobes tapering laterally. In the transverse cross section four regions can be identified. These are, from the inside to the outside, (1) the mass of randomly oriented spermatozoa, supported by a matrix of secretory material, (2) the primary spermatophore layer, PAS-positive, 1 mm thick, amorphous, but containing at the periphery peculiar ring-like structures (approximately 0.14 μm in diameter) and an array of crystals, (3) the intermediate layer, thicker dorsally than ventrally, with dense, PAS-positive granules dominating the ventral half, and (4) the outer bounding layer, 13.3 μm thick, comprised of small filaments and some flocculent material which presumably imparts stickiness to the freshly extruded spermatophore. All three layers of the homarid spermatophore are probably secreted by the vas deferens.

When placed in seawater, a freshly extruded lobster spermatophore does not swell (as would have been the case in the octopus, for example), but on the contrary, it becomes less sticky and more rigid, and at the same time, numerous small gas bubbles appear on its surface (Fig. 40) in a manner resembling the phenomenon of bubbling activity caused by CO_2 evolution in tick spermatophores (Feldman-Muhsam et al. 1973; see Chap. 8).

Hardening of freshly extruded spermatophores has also been observed in other decapods. Extensive histochemical analyses of the layers that make up the spermatophore wall of *Scylla serrata* suggest that some sort of polymerization of

145

chitin may be responsible for the hardening phenomenon (Uma and Subramoniam 1979). In *Scylla,* both the thick, outer "chitinous layer" and the thin, inner "non-chitinous layer" are rich in acid mucopolysaccharide, but while that of the outer layer appears to contain sulphated groups, that composing the inner layer has carboxylic groups. Both layers have proteins which contain tryptophan. A special characteristic of the outer chitinous layer is that it is readily permeable to low-molecular dyes. In *Penaeus kerathurus (trisulcatus),* 1 hardening of the spermatophore has been reported to involve phenolic tanning, an enzymatic cross-linking reaction associated with hardening of chitin complexes (Malek and Bawab 1971). In the crayfish *Pacifastacus leniusculus,* the hardening of the spermatophore involves several structural changes in the wall (Dudenhausen and Talbot 1983). In the soft and sticky unextruded spermatophore (4–9 mm long), the wall is composed of three concentric layers: a thin primary PAS-positive layer around the mass of spermatozoa; a thick middle layer composed of electron-dense, spherical granules; and a thick outer layer formed from a dense globular secretion (PAS-positive). Following extrusion and hardening, the wall shows the following changes: (1) division of the middle granular layer into a compact inner region and a highly reticulated outer region, (2) the loss of the outer globular layer, and (3) the formation of a thickened ridge, fibrillar in structure, along one side of the spermatophore wall. These changes have been interpreted to indicate that the outer globular layer functions in attachment of the spermatophore to the female, while the middle layer is directly involved in spermatophore hardening and protection of spermatozoa during storage. The mechanism of spermatophoric reaction in the crayfish, that is, the changes responsible for the rupture of the hardened wall and the release of spermatozoa prior to fertilization, remains as yet unexplored, but it is possible that hydration or rapid decondensation of the inner matrix provides the force for the final expulsion of spermatozoa.

146

Chapter 8

Arachnida

8.1 Spermatophores and Sperm Transfer Mechanisms in Arachnids

The occurrence of spermatophores is widespread among arachnids. It has been observed and extensively studied in many Scorpiones (scorpions), Pseudoscorpiones (Chelonethi, Chernetes, false scorpions), Uropygi-Thelyphonida-Holopeltidia (whip scorpions), Amblypygi (amblypygids), Araneae (spiders), and Acari (mites and ticks). Differences in the structure of arachnid spermatophores are of such a high degree that several investigators have repeatedly drawn attention to the definite taxonomic value of spermatophore morphology (reviews by Cloudsley-Thompson 1976; Schaller 1971; Weygoldt 1974).

No less striking than the diversity in species characteristics of spermatophores are the differences in anatomy and function of the arachnid male genital system, which may include either paired or single testis (elongated, branched, coiled, either dorsally or ventrally located), a wide range of accessory structures, and a vas deferens, usually paired, but occasionally single, converging on the genital chamber. On the other hand, the histology of the arachnid testis seems to be fairly uniform, a most common feature being the presence of cysts containing germinal cells, but of variable number, which in some species can be very small; in *Hypoctonus formosus* (a whip scorpion), for example, there are only four gametes in each cyst (Warren 1939).

In many, though by no means all arachnids, the spermatophores, mostly stalk-like in general appearance, are transferred to the females indirectly: they are first deposited by the male on a substrate, and their contents subsequently picked up by the females. The prevailing theory based on phylogenetic considerations is that indirect sperm transfer is peculiar to those terrestrial creatures that left the sea without having had the chance of developing copulatory organs. In fact, however, amongst such animals, one finds within single orders some species without copulatory organs and others capable of using some kind of copulatory device. Many male Araneae, Solifugae, Ricinulei (Podogonata), Opiliones and, to some extent, also Acari, employ their legs or parts of external genitalia for transferring the contents of spermatophores directly to the females. Some pseudoscorpions use the first legs for this purpose, but ricinulids apparently prefer the third leg (Comstock 1940), and solifugids make use of the chelicae for lifting the "sperm balls" from the ground to the genital orifice of females (Heymons 1901).

8.2 Scorpiones

The amazingly complex copulatory behaviour and equally puzzling method of indirect spermatophore transfer in scorpions have long been the object of fascination by patient observers. The old literature abounds in vivid descriptions of the "mating parade" (Fabre's promenade à deux) in copulating scorpions and the ways in which the external genital organs or *les appendices pectiniformes* are used, „simplement aux deux individus à se maintenir dans la situation nécessaire, les lamelles des peignes s'enchevêtrant les unes dans les autres" (Blanchard 1853; Brongniart and Gauber 1891). These observations were followed by extensive studies on scorpions inhabiting various parts of the world. They provided more detailed accounts of the actual mating procedures and the methods employed by the males for depositing their spermatophores (Alexander 1956, 1959; Angermann 1957; Southcott 1955; Schaller 1971; Cloudsley-Thompson 1968, 1976).

Extrusion of the spermatophore by a male scorpion begins as soon as the mating parade is over. By pressing his genital opening against the ground, he attaches to it the stalk of the spermatophore and then pulls out the remaining part by raising his body. Next, he moves backwards and pulls the female towards the spermatophore. This enables the female to place her genital opening exactly over the sperm-containing portion of the spermatophore and draw its contents by sudden movement of the body. Neither male nor female appendages are involved in the actual injection of the sperm mass into the female genital atrium. The sperm-containing portion of the spermatophore itself is a complex structure composed of two separate halves which are mirror images of each other. Studies of the male genital system in *Euscorpius italicus* (Angermann 1955, 1957) and *Nebo hierochonticus* (Rosin and Shulov 1963) have shown that the paired sperm-container is formed by fusion of two hemispermatophores, produced earlier within special pockets of the male tract (the so-called paraxial organs; Schaller 1971). The actual shape of the hemispermatophores is variable. In North American scorpions, two basic types of spermatophore structure have been described, one flagelliform, the other lamelliform; the former is characteristic of Buthidae, the latter of several other families, including the Scorpionidae (Francke 1979). In South-West Africa (Namibia) an extensive survey of scorpions, which also included observations on the structure of their spermatophores, comprised 7 distinct genera, counting 56 species among them (Lamoral 1979).

The great diversity in the structure of spermatophores is appropriately matched by equally marked species differences in the morphology of scorpion spermatozoa, particularly in respect of the flagellar axonemes; in some instances the spermatozoa of even a single individual can differ structurally from each other (Alberti 1983).

8.3 Pseudoscorpiones

In general, the spermatophore of a pseudoscorpion is composed of three clearly discernible portions: a stalk which provides means for attachment; a droplet at the top of the stalk, consisting of fluid glandular secretion; and a tube-shaped

148

package of spermatozoa, enclosed within the droplet (Weygoldt 1966); but variants to this pattern are frequently encountered; a rather complex spermatophore with a lever-like mechanism is characteristic of *Withius subruber* (Weygoldt 1969 a). Encapsulation of spermatozoa by the glandular secretion probably occurs chiefly in the anterior region of the male genital atrium, while the posterior region is responsible for the formation of the stalk, including, in the case of *Chthonius ischnocheles*, the two lateral collars (Legg 1973).

The ways in which pseudoscorpions deposit and transfer the spermatophores are highly variable. While in some the stalked spermatophores are extruded in the absence of females, in others pairing is a regular feature. The aforementioned *Chthonius* male, for example, deposits spermatophores on the substratum in the absence of a female which has then to locate them herself (probably helped by the presence of a pheromone) and take up the contents. The male *Chelifer latreilli*, on the other hand, not only depends on female presence, but goes so far as to keep her genital orifice open (using the modified claws of his first legs for this purpose) to make certain that she can properly cover with her orifice the sperm mass located at the top of the spermatophore (Kew 1912). Moreover, before pairing has taken place, the *Chelifer* male performs a complicated dance in the presence of the more or less inactive female, and in its course proceeds to extrude the spermatophore. Somewhat similar events take place in *Dinocheirus tumidus* (Weygoldt 1966), *Withius subruber* (Weygoldt 1969 a), and *Hysterochelifer meridianus* (Boissin 1973). Sperm transfer in the latter is accomplished in several stages. First, the female places her genital orifice exactly over the sperm package. At this moment the male comes to her assistance. By holding her firmly, he guides the spermatozoa in the direction of the spermatheca, where they can remain in storage for several months.

Unlike morphology and function, the chemistry of spermatophores in pseudoscorpions (as indeed, in most of the other arachnids) still remains a largely unmapped area of research, a notable exception being the study of *Chthonius ischnocheles* in which glycine, alanine, and serine were shown to be present in higher concentration than any of the other amino acids (Hunt and Legg 1971).

8.4 Uropygi – Thelyphonida – Holopeltidia

Pairing behaviour and spermatophore deposition in whip scorpions, and Pedipalpi generally, appear to be even more diverse and complex than in either scorpions or pseudoscorpions, as shown by extensive research, which has also greatly contributed to better knowledge of the indirect sperm transfer mechanisms in general (Klingel 1963; Weygoldt 1978). Comparative studies on *Thelyphonellus, Mastigoproctus,* and *Typopeltis* provide a particularly illuminating example of this type of research (Weygoldt 1978).

During the characteristically short courtship, the male *Thelyphonellus amazonicus* proceeds first to seize one antenniform leg of the female, then turns until both partners face in the same direction and, as soon as the "mating parade" is over, deposits a spermatophore and pulls the female over it. The female picks up two sperm packages from the spermatophore and releases the male's opistho-soma. Now the male turns, embraces the female's opisthosoma from above and with his palpal chelae pushes the sperm packages into her gonopore. Thereafter, the sperm packages are emptied and a moment later, a second mating dance begins, culminating in the transfer of a second spermatophore. The sperm packages of *Thelyphonellus* and *Mastigoproctus* are elongate, and their tips are partly protected by a special structure; after having been pushed into the female's gonopore, only the blunt ends of the sperm packages protrude and these are manipulated by the male with the tips of the movable fingers of the palpal chelae. In *Typopeltis* the sperm packages are not elongated and they are pushed more effectively into the female's gonopore. This ensures a more complete evacuation of spermatozoa than in either *Thelyphonellus* or *Mastigoproctus*.

8.5 Amblypygi – Phrynichidea

Amblypygids exhibit certain special features of courting behaviour and spermatophore formation different from those of whip scorpions (Alexander 1962; Klingel 1963; Weygoldt 1969 b, 1972, 1977). Research on reproductive patterns in Tarantulidae has been particularly helpful in unravelling the complexities of male reproductive function in Amblypygi (Weygoldt et al. 1972). In *Tarantula marginemaculata,* the male genital system consists of paired testes (flat and oval, large diameter 200–500 μm), which contain cysts filled with spermatocytes or spermatids, and produce cork-screw-shaped spermatozoa that later become rolled up and encysted; two vasa deferentia; two vesiculae seminales which function as reservoirs for the spermatozoa; a pair of ventral glands which produce a secretion containing characteristic granular material; a pair of lateral glands forming a transparent secretion; two special reservoirs into which the secretions of the ventral and lateral glands are voided and there stored; and a genital chamber with the cone-shaped spermatophore organ, a complex structure formed by a pair of gonopods which splits first into two pairs, but later unites to form a hollow organ with a complicated system of cavities. No precursor of spermatophore is present in the spermatophore organ until deposition has started, when the spermatozoa are pressed out from the vesiculae seminales and when simultaneously (or shortly thereafter) the secretions begin to flow from the reservoirs and along the latero-ventral tubes to the two ventral openings. There the secreted materials are fixed to a substrate and the male then proceeds to mould the spermatophore by raising his body. Thereafter, the central cavity of the spermatophore organ is filled with secreted material acting as a matrix which presses the upper parts of the spermatophore. This is finally pulled out through a mid-dorsal opening. As soon as this has happened, the secretion reservoirs are filled again with new secretory material, but it takes some time before the seminal vesicles, that is the sperm reservoirs, are filled with spermatozoa, and that is probably the reason why a second spermatophore cannot be produced by the male immediately after the first one. In its final form (Fig. 42) the deposited spermatophore consists of an obli-

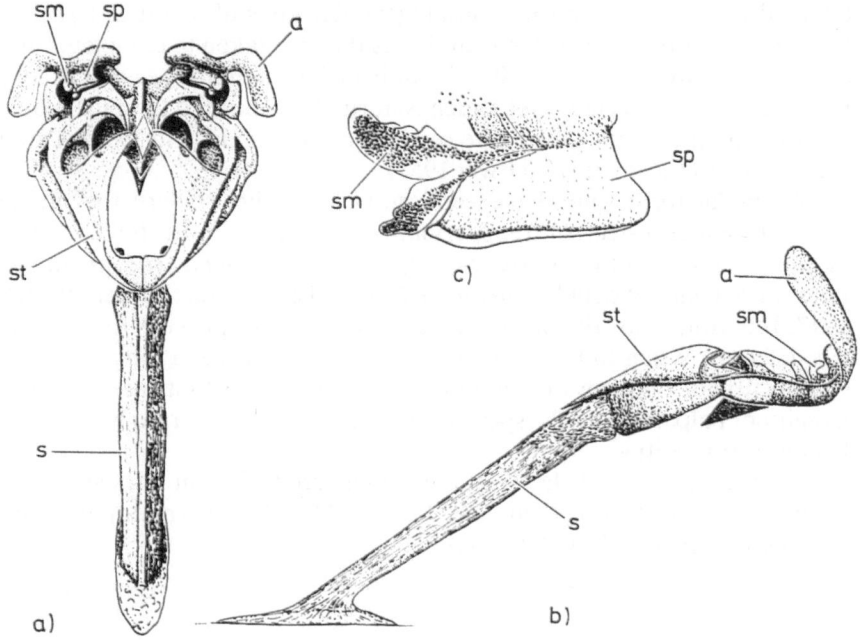

Fig. 42 a–c. Spermatophore of *Tarantula marginemaculata* (Chelicerata, Amblypygi). *a* Seen from above. *b* Seen sidewards; in order to pick up the sperm package, the female would approach from the left. *c* The base-plate, dissected with the sperm package carrier; *a* wing-like arms; *s* stalk; *sm* sperm package containing the spermatozoa; *sp* the base-plate; *st* sperm-package carrier. (For further details see Weygoldt 1969 b and Weygoldt et al. 1972)

quely inclined stalk, two sperm packages and a package-carrier terminating at the distal end in two wing-like arms to which, at their bases, the sperm packages are attached.

To remove the two sperm packages from the spermatophore, the female *Tarantula* employs her gonopods, each carrying a claw-like sclerite, pointing backwards and inwards. It is by means of these sclerites that the sperm packages are pulled out from the spermatophore and, at the same time, the spermatozoa squeezed out, to be then sucked into the seminal receptacles which are located just beneath the bases of the sclerites.

8.6 Araneae

In common with other arachnids, the spermatozoa of spiders belong to the encysted form. In most instances, fertilization is effected by transfer of free spermatozoa during mating, a drop of semen having been previously deposited by the male on his own web or that of the partner (Robinson 1982). The position adopted by the spiders during mating (the female is often much larger than the male) is extraordinarily variable, and so is the time taken to complete copulation; according to Montgomery (1903), 1 s only in *Epeira marmorea,* but as long as 36 h in *E. quadrata.* The fact that semen is normally deposited by spiders on the web gives them an obvious advantage over other arachnids with indirect transfer of spermatozoa (Alexander and Ewer 1957).

Telemidae form a notable exception. Unlike other spiders, they make use of spermatophores instead of free spermatozoa. Among recent studies on such spermatophores, those of the Pyrenean cave spider, *Telema tenella,* are of special interest on account of detailed histological and ultrastructural observations (Lopez 1977; Juberthie et al. 1981). In the male, the spermatophore is formed within the vas deferens. When fully assembled, it is in the shape of a long inverted trough equipped with two rows of digitations. Having been first inserted into his intromittent palpal organ, the spermatophore is then transferred by the male to the female during coitus.

Another member of the Telemidae lately studied is an African spider of the genus *Apneumonella* (Legendre and Lopez 1978). The spermatophore appears to be less complex than that of *Telema.*

8.7 Acari: Water Mites (Hydrachnidae); Terrestrial Mites (Trombiculidae, Trombidiformes, Oribatei, and Eriophyidae); Ticks (Ixodidae and Argasidae)

Both direct and indirect insertion of spermatophores are of common occurrence in Acari. The direct method is employed in particular by all species of the genera *Acarus* and *Lardoglyphus* (Griffith and Boczek 1977). The duration of copulation on which this mode of transfer depends can extend in some mites over exceedingly long periods of time. Thus, for example, in two species of phytoseiid mites commonly encountered in apple orchards in Europe, *Amblyseius potentillae* and *Typhlodromus pyri,* copulation or at least the period spent by the male and female in venter-to-venter position, can be as long as 3 and 8 h, respectively (Overmeer et al. 1982). During that time, the male phytoseiid mite produces a spermatophore surrounded by an outer wall (exospermatophore), and transfers its contents (endospermatophore) to the spermatheca of the female (Amano and Chant 1978). By interrupting the matings in either *A. potentillae* or *T. pyri* it has been possible to show that the volume of the endospermatophore transferred at a given time is in fact directly related to the length of that time, and that there also exists a direct relationship between duration of copulation and the number of eggs produced. In other mites copulation can be of much shorter duration. For example, in *Poecilochirus* (Gamasida), in which "the whole life of the males is characterized by mating and fighting", the total period of contact between the two sexes does not exceed 20 min and actual copulation during which spermatophores are transferred by the male chelicerae to the gonoporus of the female takes 30 s at the most (Korn 1982).

In mites and ticks alike, the indirect method of insemination is on the whole more frequently employed than direct insertion of spermatophores. In some instances, though indirect, the deposition of spermatophores occurs while the male and female are in close contact with each other, but there are other situations where deposition takes place with little, if any, regard for the presence of a female. To illustrate these differences and to provide at the same time more details about the structural peculiarities of acarid spermatophores, a series of examples is given below, taken from studies of the following species of aquatic and terrestrial mites, and of ixodid and argasid ticks.

Water mites
    Eylais discreta
    Hydrachna conjecta
    Hydrodroma despiciens
    Hydryphantes ruber
    Unionicola (Pentatax) intermedia. U. tricuspis

Terrestrial mites
  Trombiculidae
    Ascoschöngastia latyshevi
    Trombicula splendens

  Trombidiformes
    Abrolophus rubipes
    Bdella longicornis
    Calyptostoma velutinus
    Erythraeus regalis
    Eunicolina tuberculata
    Nicoletiella denticulata

  Oribatei
    Hermannia gibba
    Oppia nitens
    Pelokylla malabarica

  Eriophyidae
    Aculus cornutus
    Eriophyes sheldoni
    Phyllocoptruta oleivora

  Ticks
    Argas persicus
    Boophilus decoloratus, B. microplus
    Dermacentor variabilis
    Hyalomma excavatum
    Ixodes ricinus
    Ornithodoros gurneyi, O. moubata, O. savignyi, and *O. tholozani.*

Water Mites (Hydrachnidae). While in some water mites, such as, for example, *Eylais discreta,* sperm transfer is direct and takes place during pairing involving close bodily contact, in many others the spermatophores are deposited externally, either in the presence of a female, as in the genus *Unionicola,* or in her absence, as in *Hydrachna conjecta* (Böttger 1962, 1965; Davids and Belier 1980; Hevers 1978). Males of all six species of *Unionicola,* in which the structure, deposition, and transfer of spermatophores were studied by Hevers (1978), extruded spermatophores only when females of their own species were present. One of the six species was *Unionicola (Pentatax) intermedia,* which lives and reproduces in the mantle cavity of mussels. To deposit spermatophores, the male of this species seizes the spermatophores with his 4th pair of legs and merely touches the genital region of the female with the sperm from the spermatophore's head. The other five *Unionicola* species deposit their spermatophores on firm substrates, but in only one, *U. tricuspis,* has pairing been observed. Structurally, the spermatophores of the six species are quite distinct, but in each instance there is a sperm-filled head (29–99 μm long, 17–58 μm wide) and a stalk consisting of a strand of secretory material.

Hydryphantes, Hydrachna, and *Hydrodroma* provide further examples of certain unique features of reproductive patterns in hydrachnellids. Members of the genus *Hydryphantes* are among the commonest of the bright red mites in warm shallow waters of the Northern Hemisphere. *Hydryphantes ruber,* studied by Mitchell (1958), is an actively swimming mite which, however, unlike many other swimming forms, is at times a crawler in moss at or near the water line. The males deposit the spermatophores while crawling over the substrate with no regard to the presence of females. At the moment when a male lowers its body so that the distended anterior part of the gonopore can touch the substrate, the base of the spermatophore is extruded and attached. Then, as the body is elevated, a stalk (40–50 μm long) and the head of the spermatophore are extruded. Similarly, the males of *Hydrachna conjecta* have been observed by Davids and Belier (1980) to deposit their spermatophores irrespective of females being present, using plant material as substrate; under experimental conditions employed by these authors, the mean number of spermatophores deposited by single males was 121/24 h. By walking over the substrate with evaginated ovipositor, the female has to locate a spermatophore first, then squeeze it by means of the ovipositor, and finally, to pick up the content.

Hydrodroma despiciens is a conspicuous red water mite commonly found in lakes and ponds, which, as demonstrated by Wiles (1982), feeds almost entirely on chironomid egg masses. When presented with this kind of substratum, the males not only feed on it, but lay their spermatophores there. The widespread distribution of *H. despiciens* all over the world may well be due to the utilization of chironomid eggs which represent a relatively untapped food source available in most permanent waters. It also provides an interesting example of the use of food sources as a mating site.

Terrestrial Mites: Trombiculidae, Trombidiformes, Oribatei, and Eriophyidae. Equally as striking as in water mites are the reproductive patterns in the terrestrial mites, whose number is immense, distribution world-wide, and breeding sites present wherever decaying plant or animal remains are available (Hughes 1959). Many of the terrestrial mites, including those that are parasitic, employ spermatophores as a means of sperm transfer from males to females.

154

Trombiculid mites, in common with some of the water mites referred to above, deposit the spermatophores in places where the females themselves have to find them, as demonstrated convincingly by Lipovsky et al. (1957) in the study of reproductive behaviour in *Trombicula splendens*. Another trombiculid, *Ascoschöngastia latyshevi,* has been studied in detail as regards the structure of spermatophores (Sixl 1971). In this species the spermatophore, measuring 92–98 μm in total length, is equipped with an elastic stalk, 48–54 μm long and 8 μm wide at the base, which expands at the top into a fork-like structure forming a support for the egg-shaped sperm drop, 42 μm × 34–38 μm. Between 3 and 10 min are needed by the male to deposit each spermatophore on the substratum, and more time elapses before a female, even if artificially brought into contact with the spermatophore, decides to place her genital opening there.

Some Trombiculidae attack man and domestic animals and may act as vectors of rickettsial diseases. Of special interest in this respect is the difference in reproductive performance between Rickettsia-infected and -non-infected colonies of *Leptotrombidium* mites (Roberts et al. 1977). In these trombiculids, in which the uptake of spermatophore seems to be an absolute requisite for oviposition, the female: male ratio of progeny is normally about 2:1, but the colonies infected with *Rickettsia tsutsugamushi* produce, with a few exceptions, only females.

Trombidiform mites also exhibit some unusual properties as regards the structure and function of spermatophores (Schuster and Schuster 1969, 1977; Vistorin-Theis 1975; Vistorin 1978). In Rhagidiidae, the spermatophore consists of a thin stalk topped by either a single, coated sperm drop or several such droplets (Ehrnsberger 1977). In Nicoletiellidae (Labidostomidae), it is also composed mostly of a stalk and sperm drop (Schuster and Schuster 1969), but in at least one species, *Eunicolina tuberculata,* it carries in addition, a crown in the form of two wings, just above the sperm drop (Vistorin 1978). The stalk of this crowned spermatophore, which is about 370 μm long, is also equipped with a filamentous thread (Fig. 43). The deposition of spermatophores by the Nicoletiellidae takes place all the year round. The male *Nicoletiella denticulata* deposits several hundred spermatophores during the year (in one case 1269 have been counted within a 9-month period March–November), and the average distance separating the deposited spermatophores is 1.5 mm; in this particular species, unlike in some others, the male deposits the spermatophores only if a female is present. Another peculiarity of this species is that the male genital system includes only one testis, but two vasa deferentia which, however, unite nearer the ductus ejaculatorius, into a single vas deferens (Vistorin 1981).

Several other families of Trombidiformes have lately attracted special attention. One of them, the Erythraeidae (in particular, the species *Abrolophus rubipes* and *Erythraeus regalis*) has been studied in a detailed manner as regards secretory products of the male reproductive tract (Witte 1975; Witte and Storch 1973). The testis of the mites belonging to this family is composed of two distinct parts, a germ layer and a glandular portion. It produces three types of proteinaceous secretion in the glandular portion. Type I forms a thin envelope around the sperm cells, type II gives rise to the stalk of the spermatophore, and type III secretes (in *Abrolophus*) the crown which tops the sperm drop in the spermatophore. In the testes, the sperm cells and the droplets of secretory fluids float freely, but in the vasa deferentia the spermatozoa and the stalk- and crown secretions congregate

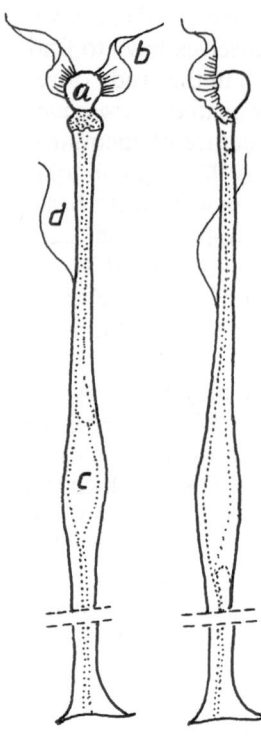

Fig. 43. Spermatophore of *Eunicolina tuberculata* (Acari, Trombidiformes). *Left* frontal view; *right* seen from the side. (*a*) Sperm drop containing the mass of spermatozoa; (*b*) crown in the form of two wings; (*c*) stalk (170 μm long) with (*d*) and whip-like appendage. (Vistorin 1978)

to form uniform complexes, the two secretions subsequently undergoing in the ejaculatory duct a hardening process, apparently dependent on the oxidation of sulphydryl groups to disulphide bonds. The sperm drop as such is formed in the ejaculatory duct. It is covered by a layer of lipid material which also serves to glue the different parts of the spermatophore together. Deposition of the spermatophore does not require the presence of a female, but a track which is set by the male probably contains a pheromone that is supposed to make it easier for the female to detect the spermatophores. In its natural environment, the male *Abrolophus* deposits numerous spermatophores on rocks in the sun within a short period of time; in culture jars up to 24 are deposited within 2 h.

The presence of two distinct parts, a germinal stratum and a glandular part, is also characteristic of the testis in the trombidiform families of Bdellidae and Calyptostomidae. In Bdellidae, such as *Bdella longicornis,* the secretions produced by the glandular testicular tissue form a coat around the spermatozoa in the lumen of the testis, and the encysted spermatozoa are then incorporated into spermatophores (Alberti and Storch 1976). In *Calyptostoma velutinus* (an orange-coloured mite, up to 3.5 mm long), the cavity of the testicular lobes serves as a reservoir for both spermatozoa and three types of secretion; the spermatophore (total length 530 μm) is composed of a thin straight stalk (290–370 μm long) and a large, round orange-coloured "sperm droplet" filled with numerous spermatozoa which are sphere-shaped and 1.3–1.9 μm in diameter; under laboratory conditions, a male commonly produces 2–3 spermatophores/48 h and continues to deposit them (externally) for several months (Theis and Schuster 1974; Vistorin-Theis 1975, 1977).

Oribatid mites, like some of the other mites referred to earlier, deposit spermatophores regardless of the presence of females. A female has to locate the spermatophores and insert them into her genital pouch. The stalked, in some instances nearly 1 mm long, spermatophores have been described in a number of oribatid mites (Fernandez 1981; Pauly 1956; Shereef 1972; Sengbusch 1958; Taberly 1957; Woodring 1970). Among the recently studied species are *Pelokylla malabarica* (Haq and Adolph 1981), *Oppia nitens* (Kümmel 1982), and *Hermannia gibba* (Waitzbauer 1983). Adult males of *Pelokylla* deposit the spermatophores within 10–15 days after emergence, to be picked up by the females later. Although not essential for the production of spermatophores, the presence of females certainly increases the rate of deposition; under these conditions, a single male turns out a mean maximum of 27.8 spermatophores/day. In common with that of other Oribatei, the spermatophore of *Oppia* is composed of a head and stalk. The spherical head, about 40 µm in diameter, consists of two distinct portions. Internally, at its base, it contains a sac-like sperm capsule consisting of two layers, in which are enclosed the spermatozoa together with some secretion. The rest of the head is filled with other viscous secretions which separate the sperm capsule completely from the external wall of the head. At least two other secretions contribute to the formation of the 80-µm-long stalk with an enlarged base which anchors the spermatophore to the substratum.

Eriophyid mites which are parasitic on plants and evoke the formation of a variety of galls on their hosts are also endowed with some special characteristic features of spermatophores. Glued to the substrate by its drop-like pad, a spermatophore deposited by *Eriophyes sheldoni* consists of a stalk with a basal stiff section and a distal flexible part ending in a head shaped like a convex lense, where the mass of spermatozoa is concentrated (Sternlicht and Griffith (1974). Similarly shaped are the spermatophores of other eriophyoids, such as *Phyllocoptruta oleivora* and *Aculus cornutus;* these are produced at a rate of 16–30/male/day; in *A. cornutus,* the spermatozoa remain viable within spermatophores on the leaf for 3 days; in the female for at least 4 days (Oldfield et al. 1970). Protogynes (summer-form females) of *A. cornutus* exhibit a characteristic behavioural pattern which includes visits to the spermatophores. Insemination involves the removal of spermatozoa from the sperm reservoir in the head of the spermatophore and their subsequent deposition in one of the two spermathecae. When the spermatheca has been filled with spermatozoa, it takes the shape of a sphere which is of nearly the same size as the sperm reservoir in the spermatophore. Newly emerged protogynes of *A. cornutus* inseminated from one fresh spermatophore containing about 50 spermatozoa produced predominantly females for several days, but thereafter only males (Oldfield 1973; Oldfield et al. 1972; Oldfield and Newell 1973).

Ticks: Ixodidae and Argasidae. Reproductive performance of ticks and the special character of their spermatozoa and spermatophores have been receiving a great deal of attention ever since the role of blood-sucking ticks as vectors of infectious diseases became fully recognized. Having observed the mating behaviour of *Ixo-*

157

des ricinus, Samson (1909) gained the impression that the male actually presses the freshly extruded spermatophore into the female genital opening, making thereby certain that it does not slide backwards. Two years later, Nuttall and Merriman (1911) published their observations on the sequence of events which accompany copulation in *Ornithodoros moubata.*

Spermatogenesis in ixodid ticks is of exceptional interest for several reasons, not least because in metastriate ixodids generally the testis of an adult male represents a large pool of spermatocytes, but contains no advanced products of spermiogenesis. Even while stored in the seminal vesicles, the gametes of ticks resemble more closely spermatids than spermatozoa and are, accordingly, commonly referred to as advanced spermatids or prospermia. Furthermore, only the early stage of spermatid transformation takes place in the male, while the final stages of maturation occur later in the female tract, resulting in the formation of spermatozoa, sometimes called *spermiophores,* a name which must not be confused with that of *spermaphores,* introduced by Duvernoy (1853) (Chap. 1) to describe spermatophores (Dumser and Oliver 1981; Oliver 1982; Suleiman and Brown 1978; Tuzet and Millot 1937). In the dog tick *Dermacentor variabilis* and other ixodids, there also exists a close relationship between reproductive activity and feeding habits, which, moreover, is linked to the function of spermatophores; the stimulus that the male dog tick provides for feeding of the females is closely dependent on the presence of spermatophores. Males unable to extrude spermatophores because of blocked genital aperture are incapable of rapid and complete feeding of the females; the stimulus for female rapid engorgement involves receiving the spermatophore and/or its contents (Pappas and Oliver 1972). Ixodid ticks of the Metastriata group are known to copulate on their hosts, the females completing engorgement after copulation. The ixodid female, which oviposits only once in her lifetime, theoretically needs only one supply of sperm; counts of two spermatophores of *Hyalomma excavatum* amounted to 80,000 and 120,000 spermatozoa, respectively, whereas the number of eggs commonly oviposited by a female does not exceed 10,000 (Feldman-Muhsam and Borut 1971).

Most unusual is also the copulatory behaviour of certain ixodid males towards females of different species or strains. Males of two distinct species, *Boophilus decoloratus* and *Boophilus microplus* mate readily with females of either species, as proved by counts of spermatophore capsules in the seminal receptacles, but the eggs produced as a result of cross matings are sterile. Furthermore, when the *B. microplus* females of a South-African strain are mated to *B. microplus* males of an Australian strain, the majority of them produce viable hybrid progeny, but a reciprocal cross results in less than 2% hatch of non-viable larvae (Spickett and Malan 1978).

As intriguing as in the ixodids are the events associated with sperm development and spermatophore production in the argasid genera *Ornithodoros* and *Argas.* Both the argasid and the ixodid ticks have been studied carefully and extensively by several investigators, Feldman-Muhsam and her associates in particular. The ultrastructure of the prospermium and its relation to sperm maturation and capacitation has been fully explored in *Ornithodoros gurneyi* and *O. tholozani* (Feldman-Muhsam and Filshie 1979), and a series of important publications from the same laboratory provided a great deal of fundamental information on

158

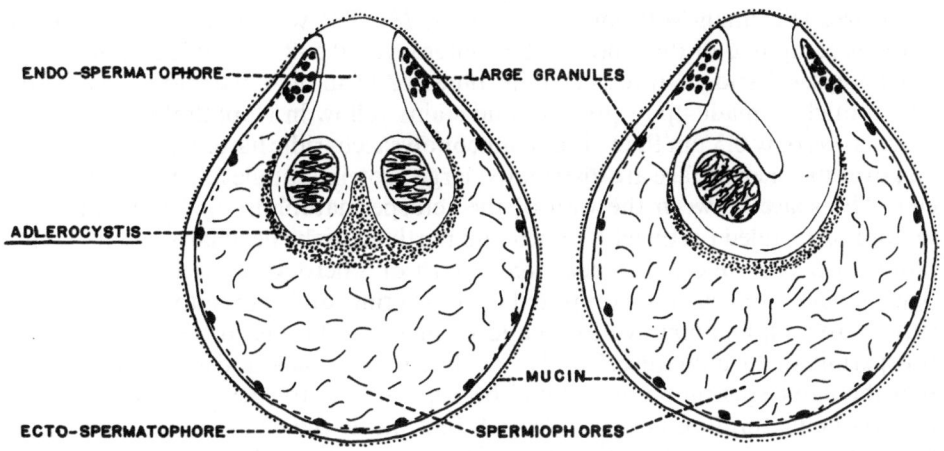

ENDO-SPERMATOPHORE

LARGE GRANULES

ADLEROCYSTIS

MUCIN

ECTO-SPERMATOPHORE

SPERMIOPHORES

Fig. 44. Schematic representation of two longitudinal sections, at an angle of 90°, through a fully formed spermatophore of the tick *Ornithodoros*. (Courtesy of Prof. B. Feldman-Muhsam)

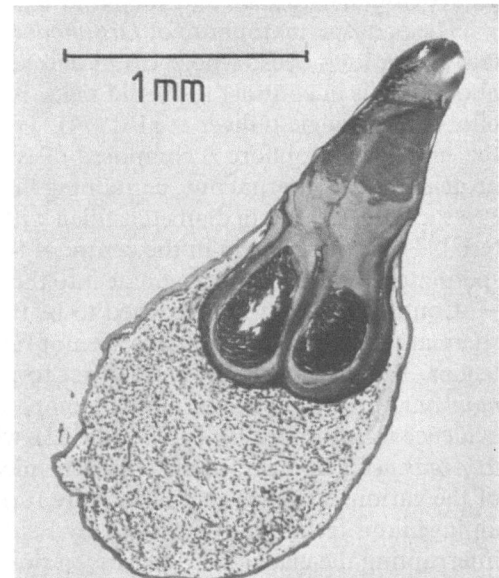

1 mm

Fig. 45. A longitudinal section of a spermatophore of *Ornithodoros savignyi,* at the initiation of evagination. The proteinous sacs of the bilobed spermatophore at the top; the ectospermatophore packed with spermatozoa forms the lower part. (Feldman-Muhsam et al. 1980)

the mechanisms of spermatophore formation in argasid and also in ixodid ticks (Feldman-Muhsam 1967; Feldman-Muhsam and Borut 1971, 1978, 1983; Feldman-Muhsam and Havivi 1963; Feldman-Muhsam et al. 1973, 1980).

In its final form, the argasid spermatophore consists of two containers, one inside the other: the outer one called ectospermatophore, and the inner one endospermatophore (Figs. 44 and 45). In Fig. 44 are shown schematically two longitudinal sections, at an angle of 90°, through a fully formed spermatophore of *Or-*

159

nithodoros (Feldman-Muhsam and Borut 1978). The wall of the ectospermatophore is composed of three layers. The outer layer, 0.5–2.2 µm thick, is of acidic-
mucopolysaccharide nature and stains blue in Alcian blue; the middle layer, 10–
20 µm thick, is made up of proteins and stains yellow in Naphthol yellow S; the
inner layer, only 0.5 µm thick, is of a mucopolysaccharidic nature (PAS-positive),
and over its inner surface are distributed large protein granules, 4–13 µm in diameter. The space between the wall of the ectospermatophore and the endospermatophore is filled partly with semen and partly with the peculiar yeast-like sym-
bionts of the sperm cells, named *Adlerocystis* or adlerocysts (after the parasitol-
ogist Adler). The semen is composed of prospermia and spermatophoric plasma
(PAS-positive); Feldman-Muhsam and Filshie (1979) prefer to restrict the name
prospermia to the immature, freshly ejaculated sperm cells, and are of the opinion
that the name spermiophores should be applied only to spermatozoa that have
undergone maturation (spermateleosis), first inside the ectospermatophore and
later, after release, in the oviducts. Adlerocysts grow, multiply and shelter in a
pair of accessory lobes of the male genital system, and they are transferred to the
female during copulation, that is by means of the spermatophore. These sym-
bionts are found not only in sexually mature males, but are also present in the
early spermatophores of the sexually immature males, still devoid of the sperm
cells (Feldman-Muhsam and Havivi 1963, 1967).

The endospermatophore of *Ornithodoros* has the form of a bilobed bottle fitted with a long neck, whose distal end seals the opening of the ectospermatophore; this is in contrast to ixodid ticks, in which the endospermatophore is not
bilobed, but single (Oliver et al. 1974). The membrane of the bottle-like part of
the endospermatophore is composed of two layers: an external one, made up of
protein, and an internal one, containing mucopolysaccharides. Each lobe encloses a sac, 300–400 µm in diameter, filled with some amorphous proteinaceous material. These sacs remain in the centre of the lobes until the time when the endospermatophore begins to evaginate into the female genital tract (Fig. 45).

Contrary to what was believed to be true at one time, it is now obvious that
spermatophores of argasid ticks are not fully formed at the time when extrusion
begins. Robinson (1942) was the first to provide evidence that in *Ornithodoros
moubata,* the process of assembly is completed outside the male body, and further
evidence was supplied by Tatchell (1962), whose study of spermatophore transfer
in *Argas persicus* included also histochemical observations. The final elucidation
of the various stages in spermatophore formation came with the introduction of
an ingenious technique, developed by Feldman-Muhsam and her associates, for
interrupting the ejaculatory process in ticks and observing the successive phases
independent of each other (Feldman-Muhsam 1967; Feldman-Muhsam and Borut 1978; Feldman-Muhsam et al. 1973). The procedure, in some ways allied to the
coitus-interruptus and split-ejaculate methods used in determining the sequence
of ejaculatory events in mammals (Mann and Lutwak-Mann 1981), is as follows.

At first, copulation is allowed to take its normal course: the male tick climbs
on the back of the female, then crawls underneath, folding his legs with hers, in-
troduces his capitulum into the female genital opening and next, 2–3 min later,
contracts the abdomen, thereby creating some space between his and her venters.
This contraction is the first clear sign that the production of a spermatophore is

160

about to begin, and it is at this point that copulation can already be interrupted by forcibly separating the two partners, yet without preventing in the least the spermatophore from developing further and finally emerging, fully formed, about 30–45 min later. Alternatively, copulation can be interrupted a little later at various points during the period of spermatophore formation, by following the sequence of events (under a stereoscopic microscope) and removing each individual component part of the spermatophore as soon as it has emerged from the male genital aperture. Under such circumstances, the first to be extruded by the male is the ectospermatophore, not all at once, however, but in three consecutive steps corresponding to its three layers. The first to appear is a droplet of transparent mucopolysaccharide-rich material, from which, in due course, the outer layer will be formed. The next to emerge is a second droplet, proteinaceous in nature, which is injected directly into the first one, and carries material for the formation of the intermediate layer. The third to appear is a cluster of large protein granules engulfed in mucopolysaccharide material needed for the formation of the inner layer, over the surface of which the granules will eventually spread. Almost immediately afterwards, the male ejaculates directly into the ectospermatophore the whole semen, that is, the mixture of prospermia and spermatophoric plasma, and this is closely followed by several thousands of the adlerocysts which will fill the most proximal part of the ectospermatophore. Only now, after the ectospermatophore has been formed, does the extrusion of the bilobed endospermatophore (with its two sacs filled with protein material) follow. The last to be secreted by the male is the distal part of the neck which connects the endospermatophore to the opening of the ectospermatophore, thus plugging the spermatophore, which now becomes ready for transfer to the female.

It is at this point that the male, using his chelicerae, catches the spermatophore at the neck and then deposits it straight on the genital aperture of the female, making in addition use of his saliva to prevent the spermatophore from sticking to the chelicerae; as the chelicerae touch the spermatophore, the saliva is spread neatly over the upper portion of the spermatophore, thus preventing it from becoming stuck to the integument of either the male or the female (Feldman-Muhsam et al. 1970).

8.8 Triggering of Spermatophoric Reaction in Ticks by Carbon Dioxide

A minute or two after the male tick has finally attached a spermatophore to the female aperature, the spermatophoric reaction is set in motion at the moment when the endospermatophore is evaginated into the female genital system, so that the whole semen can now be pushed out from the spermatophore into the female spermatophoroteca (also called uterus). The thus emptied and now collapsed wall of the ectospermatophore is the only part of the spermatophore that stays outside the female body, remaining attached to it for several more hours or days.

Like the spermatophoric reaction in the giant octopus (Chap. 3), that in the tick has been successfully reproduced under conditions in vitro. This has been accomplished by removing a spermatophore from the male, placing it in a humid chamber, and observing the changes that follow. The first to be affected is the tip of the spermatophore which begins to elongate, and as the neck of the endospermatophore is straightened, the tip of the external capsule is pushed outward and a channel is formed, into which the contents of the endospermatophore are drawn and evaginated. Evagination, as observed in vitro, is accompanied by two extraordinary bubbling activities, following upon each other in quick succession and affecting two distinct sites in the spermatophore (Feldman-Muhsam et al. 1973). The first to be affected is the narrow tip, which causes the initial elongation of the neck. The second occurs at the periphery of the spermatophore, following the neck's elongation. Mass-spectrographic analysis has shown that both bubbling processes are associated with the formation of CO_2, and it does appear that it is the pressure of CO_2 that triggers the evagination process and provides the force for the transfer of spermatozoa.

8.9 Occurrence of Spermine in Tick Spermatophores

When a fresh spermatophore of *Ornithodoros savignyi* is examined under the microscope, the proteinaceous material inside the two sacs located in the centre of the bilobed endospermatophore has a frothy appearance, but in 15–120 min this material gradually transforms into needle-like crystals; the formation of the crystals can be enhanced if the excised sacs are placed in a phosphate solution; the first needles then appear within 5 min. Using thin-layer, ion-exchange, and gas chromatography, combined with mass spectrometry, Feldman-Muhsam et al. (1980) were able to identify these crystals as spermine phosphate.

This is the first demonstration of spermine in arthropod semen. It comes more than 300 years after Van Leeuwenhoek's report to the Royal Society of London, describing for the first time motile spermatozoa in mammalian semen and pointing out that if human semen "had stood a little while, some three-sided bodies were seen in it, terminating at either end in a point" and looking "as bright and clear as if they had been crystals". It took two centuries before the crystals in human semen were identified as spermine phosphate, and another half a century before the structure of spermine itself was confirmed by synthesis. These and other developments in the biochemistry of spermine have been reviewed (Mann 1964; Mann and Lutwak-Mann 1981).

Chapter 9

Chaetognatha and Pogonophora

9.1 Exchange of Spermatophores in Hermaphroditic Chaetognatha

Chaetognata (arrow worms) are a hermaphroditic phylum. The male reproductive tract, consisting of paired testes, vasa deferentia and seminal vesicles, is located in the caudal region of the body. Within the seminal vesicles of mature organisms, the filiform spermatozoa are packed into aggregates which are of the same size and shape. as the seminal vesicles. Emission occurs by rupture of the vesicle wall. In the literature (reviewed by Reeve and Cosper 1975), the sperm aggregates of chaetognaths, such as *Sagitta* or *Spadella,* are invariably referred to as spermatophores (rather than spermatozeugmata). It seems that in some species at any rate, following discharge from the spermatophore, the spermatozoa can be used for both self-fertilization and cross-fertilization. Ghirardelli (1968), in his extensive study of spermatophores in the benthic genus *Spadella,* described cross-fertilization as a process in which a spermatophore from each individual is first placed on the neck of the partner and later, after the posterior portion of a spermatophore had dissolved, the spermatozoa pass into the seminal receptacles. From there they migrate to the ovary and fertilize the eggs which are still positioned in the gonad. In some chaetognaths, the seminal vesicles are equipped with special serrated structures which separate together with the spermatophore. Such structures, tubular in shape, are particularly prominent on spermatophores of *Eukrohnia,* and they are believed to act as a kind of copulatory device that facilitates the attachment of the spermatophore to the female genital orifice.

9.2 Special Features of Spermatophore Function in Pogonophora

Pogonophora (beard worms) constitute a phylum of marine invertebrates which in some ways resemble the present-day annelids, from which they differ, however, in certain respects, such as absence of the gut and production of characteristic individual chitinous tubes located on the sea floor, in which they live. Maurice Caullery (1914) was the first to identify these benthic creatures in material collected by the Netherlands *Siboga* Expedition around the Malayan Archipelago, but it has taken a long time before the formation and transfer of spermatophores have been found to constitute an outstanding feature of male reproductive function in *Siboglinum* (Ivanov 1957, 1963) (Fig. 46).

163

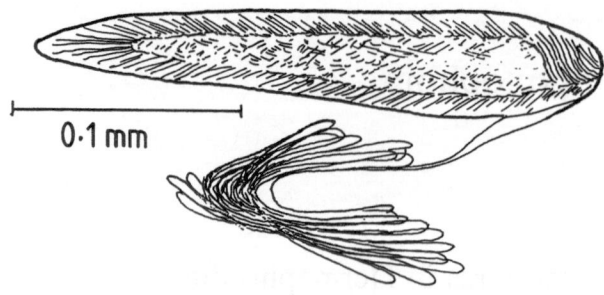

0·1 mm

Fig. 46. Spermatophore of
Siboglinum (Pogonophora).
(Redrawn from Ivanov 1957)

Siboglinum is the most widely distributed and best-studied genus (reviewed by Southward 1975a, b). Mature males can be easily identified by the large numbers of spermatophores within the transparent sperm duct. The gonads of the male are filled with floating masses of germ cells at various stages of development, but as the spermatozoa enter the thick-walled, glandular proximal portion of the sperm duct, they are bundled together and encased into thin, polysaccharide-rich envelopes, each bearing at one end a long, but tightly coiled filament. Depending on species, a pogonophore spermatophore can be spindle-, cigar- or leaf-shaped, and from about 40 μm to nearly 2 mm long (the length roughly correlated with the size of the species). From the proximal portion of the duct, the spermatophores ultimately pass into a distal portion of the sperm duct lined with ciliated cells, and here they are stored in their hundreds, all neatly arranged side by side, with the ends carrying the coiled filaments pointing forward.

From the genital aperture, the spermatophores are ejected one by one, with the filament-carrying portions emerging first, and they are picked up by the male with the tentacle. In most pogonophores, ejection probably occurs into seawater and involves no direct contact between the males and females, but some form of copulation cannot be ruled out in certain situations, particularly among the smaller species that exist in dense associations. Once in the water the filaments uncoil and spread out, and the spermatophores may then float for some time near the sea floor until the filaments which act as the floating and adhesive apparatus have become entangled with the tentacles of the females and drawn into the tubes. The covering membranes of spermatophores disintegrate after a few hours in seawater, and fertilization of eggs by the spermatozoa occurs inside the female's tube (Ivanov 1963; Southward and Southward 1963; Southward 1963, 1975a, b; Webb 1963).

The adhesive properties and strength of the filament are probably derived from its special tubular components, as indicated by the ultrastructural study of the spindle-shaped spermatophores in *Siboglinum ekmani* from the Skagerrak (Flügel 1977). In this species, the major structural components of the filament and of the spermatophore wall, are dense-core tubuli with an outside diameter of 150–170 Å, and several μm long, embedded in an amorphous ground substance and linked together in regularly spaced arrangements. The spermatophoral filament is equipped with numerous electron-dense threads, each thread containing 12–180 longitudinally oriented, dense-core tubuli. When a spermatophore is exposed to seawater, the regular arrangement of the dense-core tubuli is gradually lost and the spermatophore wall starts to disintegrate.

Chapter 10

Sporadic Occurrence of Spermatophores and Spermatozeugmata in Vertebrata

10.1 The Difficulty of Naming Correctly Various Forms of Sperm-Aggregates

Certain internally fertilizing fishes and urodele amphibians differ from other vertebrates by discharging semen in which the spermatozoa are not completely free, but form large aggregates, either naked or encapsulated. Only the latter kind resemble spermatophores proper, such as are found in invertebrates, and even in instances of this kind, the capsules are often underdeveloped. The naked aggregates on the other hand, are closely similar to the spermatozeugmata of invertebrates, those of insects in particular (Chap. 6), and have been known under this name ever since Philippi (1907, 1909) described them in certain fishes. They are mostly round or oblong in shape, occasionally quite large, and consist of tight masses of spermatozoa, usually with sperm-heads pointing to the periphery, and the tails converging centrally where they are firmly cemented together by some kind of gluey, sticky material. By and large, Philippi's proposal to call them spermatozeugmata, as Ballowitz (1890) called similar sperm-aggregates in insects, has been generally accepted, though even now they are occasionally referred to, inappropriately, as spermatophores.

As regards the encapsulated sperm-aggregates designated as spermatophores, such as are found in certain live-bearing fishes, not only are their capsules less developed than in molluscs or arthropods, but in some instances at any rate, they appear to be formed in the testes rather than in the excurrent parts of the male reproductive system. There is some evidence that such capsules can be derived directly from the testicular germinal cysts.

The difficulty that one occasionally experiences in defining properly sperm-aggregates or -conglomerates in vertebrates extends to Chordata in general. As an example may serve the situation encountered in *Saccoglossus (Dolichoglossus) otagoensis* (Hemichordata : Enteropneusta). In this acorn worm, which is found in considerable numbers on the coasts of New Zealand, Kirk (1938) reported that "the sperm of the male are freed from the gonad in rounded spermatophoric masses", and one of these he "found within a cocoon, but the sperm of the mass were not motile". It would require further study to conclude which of the several known forms of sperm-encompassing devices (Chap. 6), such as sperm-bundles, spermatodesm or spermatozeugma, most closely resembles the "spermatophoric masses" of the acorn worm.

10.2 Chondrichthyes

Sporadic occurrence of spermatophores or similarly functioning sperm packages has been reported in both major groups of cartilaginous fishes, that is, Elasmobranchii and Holocephali. The relatively small number of eggs produced by the selachians is correlated with the fact that following internal fertilization, in the majority of them the embryos develop and the young are hatched within the female's body. For transferring spermatozoa (whether pre-packaged or not) the males commonly use their pelvic claspers as intromittent organs.

A clasper, that is the prolonged caudal portion of the pelvic fin, shaped to form the stout rod-like copulatory organ, is rolled up in a scroll-like fashion, enclosing on its medial side a tube, the so-called clasper groove. It is along this tube that during copulation the spermatozoa pass from the urogenital papilla of the male to the oviducts of the female. In most such instances, the spermatozoa are found to be freely suspended in a fluid medium composed of accessory secretions, that is, seminal plasma, but in the shark, as we shall see later, the spermatozoa are pre-packaged prior to ejaculation.

10.2.1 5-Hydroxytryptamine in the Clasper Siphon of the Spiny Dogfish

The male accessory secretions which make up the seminal plasma of sharks and dogfishes originate partly in the seminal ducts and their ampullae (occasionally also called seminal vesicles) and partly in a pair of special accessory organs which void their secretion into the clasper groove. In sharks and dogfishes, these accessory organs usually take the form of muscular sacs located dorsolaterally from the two claspers, immediately under abdominal skin. They are called clasper siphons (Fig. 47). In the skate, on the other hand, being of more solid, glandular nature, they are usually referred to as clasper glands. Leigh-Sharpe (1920–1926), who studied in detail the comparative anatomy of secondary sexual characteristics of elasmobranch fishes, came to the conclusion that the clasper siphon of the Selachii acts as a "reservoir of sea-water" which is siphoned off and emptied at copulation by muscular contractions of the siphon wall, helping in this manner to pump the spermatozoa. Gilbert and Heath (1955), on the other hand, expressed the opinion that the siphons may actually have some secretory function and contribute definitive secretory products.

Our investigation (Mann 1960) has shown that the clasper siphon of the spiny dogfish, *Squalus acanthias,* secretes surprisingly large amounts of 5-hydroxytryptamine (serotonin). In mature males, as much as 6.25% of the total dry weight of the clasper siphon secretion is accounted for by this biogenic amine, but in sexually immature males the concentration is 200 times less. The presence of 5-hydroxytryptamine, an amine renowned for its strong muscle-contracting and oxycotic properties, suggested to us that this may be the substance that facilitates the passage of semen from the male to the female, by inducing muscular contractions. When, on flushing out the contents of the siphon sac with 10 ml water, 0.1–0.2 ml

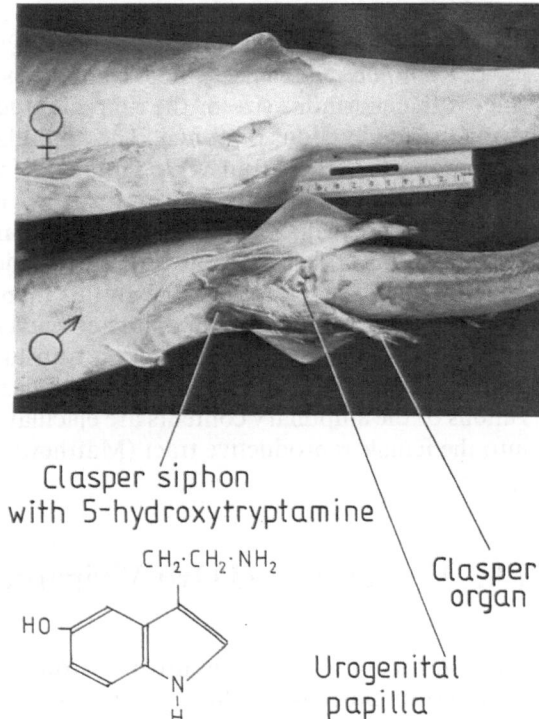

Clasper siphon
with 5-hydroxytryptamine

$CH_2 \cdot CH_2 \cdot NH_2$

Clasper
organ

Urogenital
papilla

Fig. 47. Pelvic fin region of the male and female spiny dogfish, *Squalus acanthias;* 5-hydroxytryptamine (serotonin) is secreted in the clasper siphon of the male. (Mann and Lutwak-Mann 1981)

of this dilute fluid was applied to the isolated uterine horns of the dogfish, powerful contractions occurred immediately, then persisting for hours (Mann and Prosser 1963). It may well be, however, that in addition to supplying the oxcytocic principle in the form of 5-hydroxytryptamine, the clasper siphon helps to propel the spermatozoa in the direction of the oviducts by pumping its secretion after it had been diluted with seawater (Gilbert and Heath 1972).

10.2.2 Spermatophores in *Callorhynchus antarcticus* and *Cetorhinus maximus*

Two studies, one of *Callorhynchus antarcticus* (Holocephali, rabbit-fishes) and the other of the basking shark *Cetorhinus maximus* (Euselachii) are briefly described below to illustrate how spermatozoa are pre-packaged in the male reproductive tract of cartilaginous fishes (Parker and Haswell 1897; Matthews 1950).

In *Callorhynchus antarcticus*, the aggregates of spermatozoa are formed in the highly coiled portion of the vas deferens (usually referred to as epididymis) and take the form of ovoidal capsules surrounded by a membrane which envelopes the sperm bundles imbedded in plentiful gelatinous material. The lower part of the vas deferens expands into a large cylindrical ampulla ("vesicula seminalis"), divided into a number of compartments which are filled with some greenish-coloured jelly. Through these compartments the spermatophores have to pass before making their way into the urogenital sinus (Parker and Haswell 1897).

In *Cetorhinus maximus,* a most characteristic feature of the male reproductive tract is the enormous size of the ampulla ductus deferentis, which is capable of storing several gallons of semen. The ampulla is divided into numerous pocket-like compartments by transverse septa, each with a perforation. At the point of entry into the ampulla, the spermatozoa are already in the form of small aggregates, but these, while passing through the compartments, are rotated by the cilia of the ampulla and at the same time surrounded by concentric layers of jelly-like material secreted by the ampulla. In this manner individual spermatophores are assembled, each up to 3 cm in diameter, each composed of an opaque core made up by the mass of spermatozoa and a translucent hyaline cortex. These spermatophores float freely in a large volume of clear fluid. During copulation, several gallons of the ampullary contents are ejaculated along the clasper groove straight into the female reproductive tract (Matthews 1950).

10.3 Viviparous and Ovo-Viviparous Teleosti

The majority of teleost fishes are oviparous, that is, their eggs are extruded prior to fertilization or are fertilized during the extrusion, but a minority of them are either viviparous or ovo-viviparous; in the latter case, the embryos are retained in the ovarian follicles until the time of birth. Internal fertilization in the live-bearing teleosts is brought about by spermatozoa that have been either directly ejaculated as semen into the female genital tract or released following deposition of sperm packages by the males. In some instances these packages look like spermatophores, that is to say, they contain the sperm mass firmly encased within some kind of external coat or capsule, but in a good many teleosts the external coating around the spermatozoa is not clearly discernible, and the sperm mass appears to be extruded as spermatozeugma.

10.3.1 Cyprinodontes (Microcyprini)

Several families of cyprinodont fishes are ovo-viviparous or viviparous. The study which led Philippi (1907, 1909) to bestow the name of spermatozeugmata on sperm aggregates of fishes, involved observations on copulation and sperm transfer in two ovo-viviparous Poeciliidae, namely, *Glaridichthys januarius* and *G. decem-maculatus*. He described the semen ejaculated by *G. januarius* as a hail of milk-white, elipsoid, rotating bodies (schrotschußartig, ein Hagel zahlreicher, milchweißer, drehrunder Gebilde), containing spermatozoa with heads pointing outwards. The spermatozeugmata were 122×73 μm in *G. januarius,* and 220×107 μm in *G. decem-maculatus*. From an examination of testicular sections, he gained the impression that the sperm aggregates were arranged similarly in the testis and in ejaculated semen.

In the years that have followed, spermatozeugmata have been described in the testes and semen of other poeciliid fishes, and material of this kind has been used successfully for artificial insemination of females, previously anaesthetized (Ginzburg 1968; Zander 1961). But the number of spermatozeugmata seems to vary considerably between species. The number that can be collected for artificial insemination on single occasions, has been reported to be 3000 from *Xiphophorus helleri*, 2000 from *Poecilia melanogaster*, but only 20 from *Heterandria formosa*. Likewise, the number of spermatozoa contained in a spermatozeugma differs between species; a guppy spermatozeugma contains 14,000–17,000 spermatozoa, but that of a swordtail 4000–5500.

A great deal of attention has also been given lately to testicular morphology of Poeciliidae, such as *Poecilia reticulata* (Billard and Fléchon 1969) and *P. latipinna* (Grier 1975). As a result of these studies it became clear that as spermatogenesis advances, the primary spermatocytes become surrounded by "Sertoli cells" at the periphery of the testis, and the germinal cysts thus formed move into the centre of the testis and towards the vasa efferentia. By the time spermiogenesis has reached the stage of spermatozoa, the spermatozeugmata are ready to enter these vasa. At this point they open, releasing the sperm mass from the embraces of the "Sertoli cells", thus enabling it to pass into the efferent duct system.

Similar sperm aggregates have been described in several other poeciliid teleosts, and also in other Cyprinodontes. In some instances, the sperm aggregates were shown to possess external coats of the kind that one would expect to find in spermatophores rather than in spermatozeugmata. In cases such as these, it is difficult to escape the impression that such simply coated spermatophores have evolved as structures representing an intermediate form between spermatophores proper, with their complex system of membranes, and the naked spermatozeugmata.

An early description of properly encapsulated spermatophores in cyprinodont teleosts was furnished by Kulkarni (1940) in his study of the family named by him Horaichthyidae (Figs. 48 and 49). A special feature of *Horaichthys* is the production by the mature testis of club-shaped bodies, about 0.6 mm long, surrounded by distinct hyaline capsules. About 250 of such bodies, in different stages of development, have been encountered in a male. Each (called spermatophore by the author), consists of two parts, one much broader than the other. The broad portion encloses "innumerable spermatozoa", while the thinner, which is sperm-free, contains at the tapering end a series of stiff, barb-like processes, some in the shape of conspicuous bifid spines. It is with the aid of these peculiar stiff processes that the freshly extruded spermatophore gets attached to the female's body, near her genital opening.

The way in which Kulkarni (1940) described the liberation of spermatozoa from the spermatophore of *Horaichthys* reminds one immediately of the spermatophoric reaction as observed in molluscs and arthropods. Having placed a spermatophore in a saline solution, this is what he observed:

"After a lapse of about 10 min a small rounded bulging appears between the basal portions of any of two bifid spines and begins to enlarge; this region of the spermatophore is presumably provided with somewhat weaker walls and some sort of osmotic pressure may be responsible for the bulging noted above. At a time the number of such protuberances may be even as many as three, each oc-

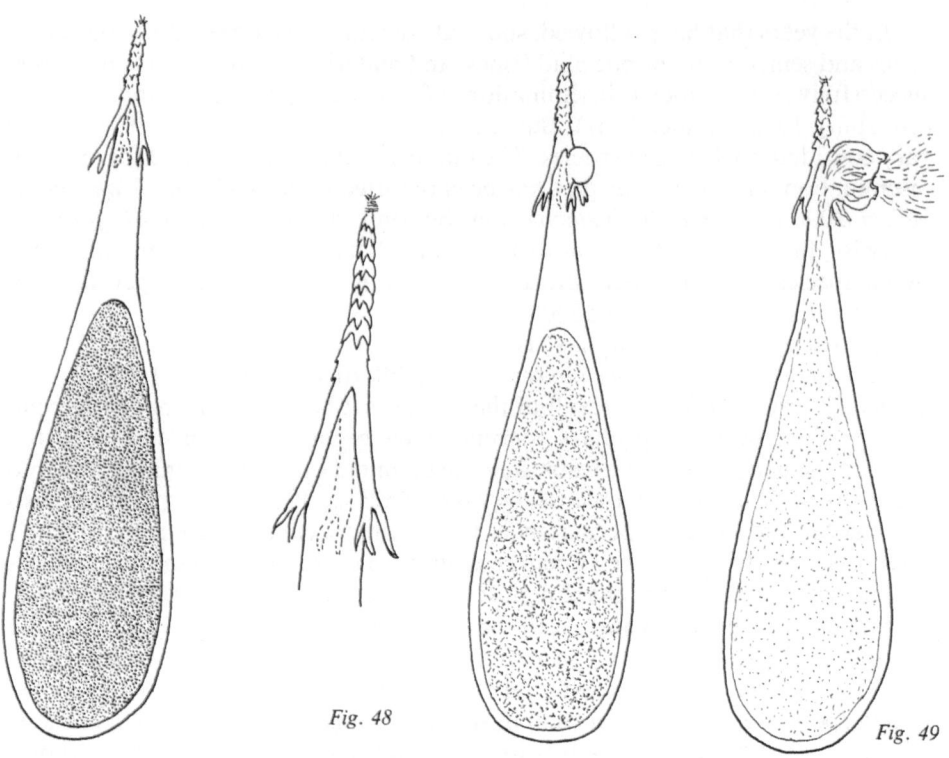

Fig. 48

Fig. 49

Fig. 48. Spermatophore of the cyprinodont fish *Horaichthys setnai*. *Left* a complete fresh spermatophore (× 140); *right* the tapering end-portion of the spermatophore with barb-like processes. (× 400). (Redrawn from Kulkarni 1940)

Fig. 49. Spermatozoa escaping from the spermatophore of *Horaichthys setnai*. *Left* an early stage with a bulge forming between the basal portions of the bifid spines; *right* a later stage with spermatozoa escaping from the ruptured distended wall. (Redrawn from Kulkarni 1940)

curring in the interspace between two adjoining spines, but in such cases only the biggest becomes enlarged while the others remain stationary.

When the protuberance is sufficiently large a semifluid substance from above the sperm mass slowly moves up and passes into it. This action is followed by the activation of a few adjoining sperms which slowly pass into the cavity. Other sperms do likewise with greater speed, and a large number of these concentrate into the bulging part and ultimately escape when the distended wall ruptures. By the time this opening is formed the movement of the sperms in the spermatophore is greatly accelerated and numbers of these are ejected at a time. Eventually the whole sperm mass inside the spermatophore is activated and there is a continuous movement up and down in the entire cavity of the spermatophore and innumerable sperms rush about in all directions to find an exit. After about an hour or so the whole of the spermatophore becomes empty".

10.3.2 Testicular Origin of Spermatophores in Percomorphi

Percomorphi, like Cyprinodontes, comprise an order of teleosts in which the occurrence of sperm packages in various shapes and size is well documented. As part

170

of the world-wide taxonomical survey undertaken by a Danish Deep Sea Expedition, Nielsen et al. (1968) made a morphological study of male gonads in 18 species of ophidioid fishes belonging to the families Brotulidae and Aphyonidae. Twelve of these were live-bearing and six oviparous. In 11 out of the 12 internally fertilizing fishes, the testes were shown to contain spermatophores, that is to say, encapsulated sperm masses; whenever females were examined, either spermatozoa or spermatophores (or both) were present in the ovaries. In none of the six oviparous ophidioid species could spermatophores be detected in the testes.

Another brotulid fish in which testicular spermatophores have been described is *Ogilbia cayorum*, commonly known as the key brotula (Suarez 1975). As the germinal cysts in the testis of this fish break down during spermiogenesis, some cells resembling lymphocytes leave the connective tissue of the lobules, enter the cysts and align along the sperm tail; it has been suggested that these cells may actually participate in the production of the capsule. The mature spermatophores are probably stored temporarily in the base of the lobules or in the sperm duct, before being extruded, but permanent spermatophore reservoirs are absent. The copulatory apparatus which enables the male to transfer the spermatophores includes a median penis surrounded by two pairs of pseudoclaspers.

Whatever the origin of the capsule in testicular spermatophore may be, the material composing such capsules is certainly of acellular nature. It consists of some strongly PAS-positive substance. In one of the Brotulidae examined by Nielsen et al. (1968), *Oligopus diagrammus,* the acellular capsules surrounding the oval spermatophores (80–170 μm long), were about 0.8 μm thick and made up of two clearly discernible membranes (both PAS-positive), separated by some faintly PAS-positive structureless material. In this species, the testicular spermatophores were contained within bag-like cavities, from which they passed first into the dorsal sperm ducts and later into a special reservoir. No reservoir of this kind could be found in another of the Brotulidae examined, namely, *Bellottia apoda.*

More recently, the mechanism of sperm encapsulation in the perciform testis has been subjected to a more detailed analysis by Gardiner (1978), who employed electron microscopy and autoradiography in his study of the shiner surfperch, *Cymatogaster aggregata* (Embiotocidae). The paired testes of *Cymatogaster* are suspended by a mesentery in the posterodorsal region of the coelom. The short sperm ducts leading from the medial surface of each testis fuse into a single duct which then continues to the urogenital papilla. The testis itself, as shown already by Wiebe (1968), is composed of radially arranged lobules with internal space divided by intracellular non-germinal cells (Sertoli cells) into cysts filled with the germinal cells, and in addition, it contains clusters of steroidogenic cells (Leydig cells). All germinal cells within any one cyst are at the same stage of development, their number varying from a few for late spermatogonia to about 600 for spermatozoa. In general, the cysts located in a lobule most centrally are also the most advanced, that is, they contain compact masses of spermatozoa. Progressively, the sperm mass assumes the oval shape of the final spermatophore which then passes from the cyst to the efferent ducts. Each of the several thousand spermatophores is about 60 μm long and 30 μm wide. On light-microscopic examination, no clearly discernible membranes can be seen around the spermatophores, and one may therefore be inclined to classify them as spermatozeugmata. However, electron microscopy of the *Cymatogaster* sperm packets has revealed the presence of external capsules. Obviously, the viability of making a distinction between true spermatophores and spermatozeugmata on the basis of light microscopic observations alone is highly questionable; the difference may well be mainly in the amount of the externally located material. Gardiner (1978) may be right in concluding from his electron microscopic observations on the sperm packages of the seaperch, that "for this reason I have referred to all such creatures as spermatophores". The results of his, and some of the earlier mentioned studies, also seem to be in accordance with the view that, in addition to functioning as a supporting element, the Sertoli cells of viviparous teleosts are directly involved in the formation of the spermatophore's extracellular matrix.

10.4 Amphibia

The majority of amphibians are terrestrial during much of their life, but return for breeding to the aquatic environment. In Anura, external fertilization follows after synchronous shedding of male and female gametes; internal fertilization is rare, but does occur in African live-bearing toads of the genus *Nectophrynoides* and the American tailed frog *Ascaphus truei* (in which the tail serves as intromittant organ). Urodela on the other hand, reproduce mainly by internal fertilization, following transfer of spermatozoa from the male to the female.

10.4.1 Spermatophores in Urodela

Except for Hynobiidae, Cryptobranchidae, and some Sirenidae, the urodeles extrude spermatozoa in the form of spermatophores. As they pass from the Wolffian ducts into the male cloaca, the spermatozoa become embedded in a matrix secreted by the pelvic glands and in this manner the cap of the spermatophore is formed, either simple in form or carrying in addition an external calcareous membrane, for example, in *Plethodon*. The stalk of the spermatophore, on which the cap is sitting, is made exclusively of cloacal secretory material which gives it its characteristic gelatinous appearance and enables it to become firmly attached to the substratum; it may take the shape of a simple tapering strand, but it may also be of a much more complex appearance (reviewed by Lofts 1974; Salthe and Mecham 1974). The structure of spermatophores has been described in a number of urodeles of the suborders Ambystomatoidea, Salamandroidea, and Proteida. By contrast, information on their biochemistry remains very scanty indeed, except for some histochemical observations relating to the localization of certain mucoproteins, amino acids, glycosaminoglycans (sulphated and non-sulphated) and sulphydryl groups (Benson 1964; Organ and Lowenthal 1963; Russell et al. 1981).

An early description of spermatophores in the axolotl (*Ambystoma*) (and their extrusion from the male cloaca) was given by Gasco (1881), who in his *Les*

Amours des Axolotls noted:

> „La masse gélatineuse qui en forme la base, s'élève a la hauteur d'un centimètre environ en se rétrécissant et prenant la forme d'un cône comprimé. À son sommet se trouve le peloton des spermatozoïdes qui tranche par sa blancheur sur tout la partie accessoire qui est très transparente».

In a later study of *Ambystoma punctatum*, Smith (1910) recognized clearly the existence of two distinct types of spermatophores, the simple type consisting of "an expanded hummocky base and a stout stalk, of very clear, transparent, gelatinous material, surmounted by a dome-shaped mass of snowy-white fluid" (average total height 6.2 mm), and the compound type in various shapes and sizes, the most common case being that of two simple spermatophores glued together.

Subsequently, largely as a result of electron microscopic studies, it was conclusively shown that within a "cap" the spermatozoa do not form a random mass, but are arranged in an orderly fashion. This has been demonstrated in a number of Ambystomatidae, Salamandridae, and Plethodontidae (Russell et al. 1981; Zalisko et al. 1984). An interesting example is provided by the microscopic organization of spermatozoa within the spermatophore cap of *Ambystoma texanum* (Fig. 50; Russell et al. 1981). In the mushroom-shaped spermatophore of this salamander, the spermatozoa are arranged within the cap in such a fashion that all the sperm-heads are near the periphery (area 1 a and b, Fig. 50), while the tails extend inwards and fill the central region (area 3, Fig. 50). Underlying the sperm heads and the proximal region of the tails is a membrane-like structure (area 2 a and 2 b of Fig. 50). In a fresh spermatophore observed on a slide under the microscope, groups of spermatozoa at the periphery of the cap have been seen to exhibit undulating movements. The stalk is composed of numerous globular bodies in close contact with each other (area 4 of Fig. 50), and it joins the cap in a region containing some tightly packed filamentous material which apparently serves as an adhesive which holds the cap and stalk together.

In other *Ambystoma* species, the spermatophores are somewhat differently shaped. While in *A. cingulatum* and *A. opacum* they are truncate and essentially quadrangular in shape, in *A. mabeei* and *A. jeffersonianum* they possess a wide base, tapered stalk and well-defined horns (Anderson and Williamson 1977). They also differ in size, those of *A. jeffersonianum* being about twice the size of the spermatophores in *A. laterale* (Uzzell 1969). An orderly arrangement of spermatozoa within spermatophores, similar to that in *Ambystoma,* is also typical of other urodeles such as newt (*Triturus*) and olm (*Proteus*). In the *Proteus anguinus* spermatophore (8–12 mm long), the spermatozoa (580 μm long) are interlaced, forming a strand several centimetres in length (Briegleb 1961).

The mechanism governing the release of spermatozoa from the deposited spermatophores has not been analysed in sufficient detail to warrant any firm statement regarding the nature of the spermatophore reaction in urodeles, but from the few observations on Salamandroidea, mainly plethodontids (Noble and Weber 1929), it would appear that phagocytosis, possibly in conjunction with activity of some mucolytic or proteolytic enzymes, plays a key role in destroying the spermatophore caps that had been lodged in the female's cloaca, thereby releasing the spermatozoa. In most salamanders that have been examined so far (Baylis 1939; Joly 1960), the spermatozoa were found to be stored by females for long periods; in *Salamandra salamandra,* for up to 2½ years.

173

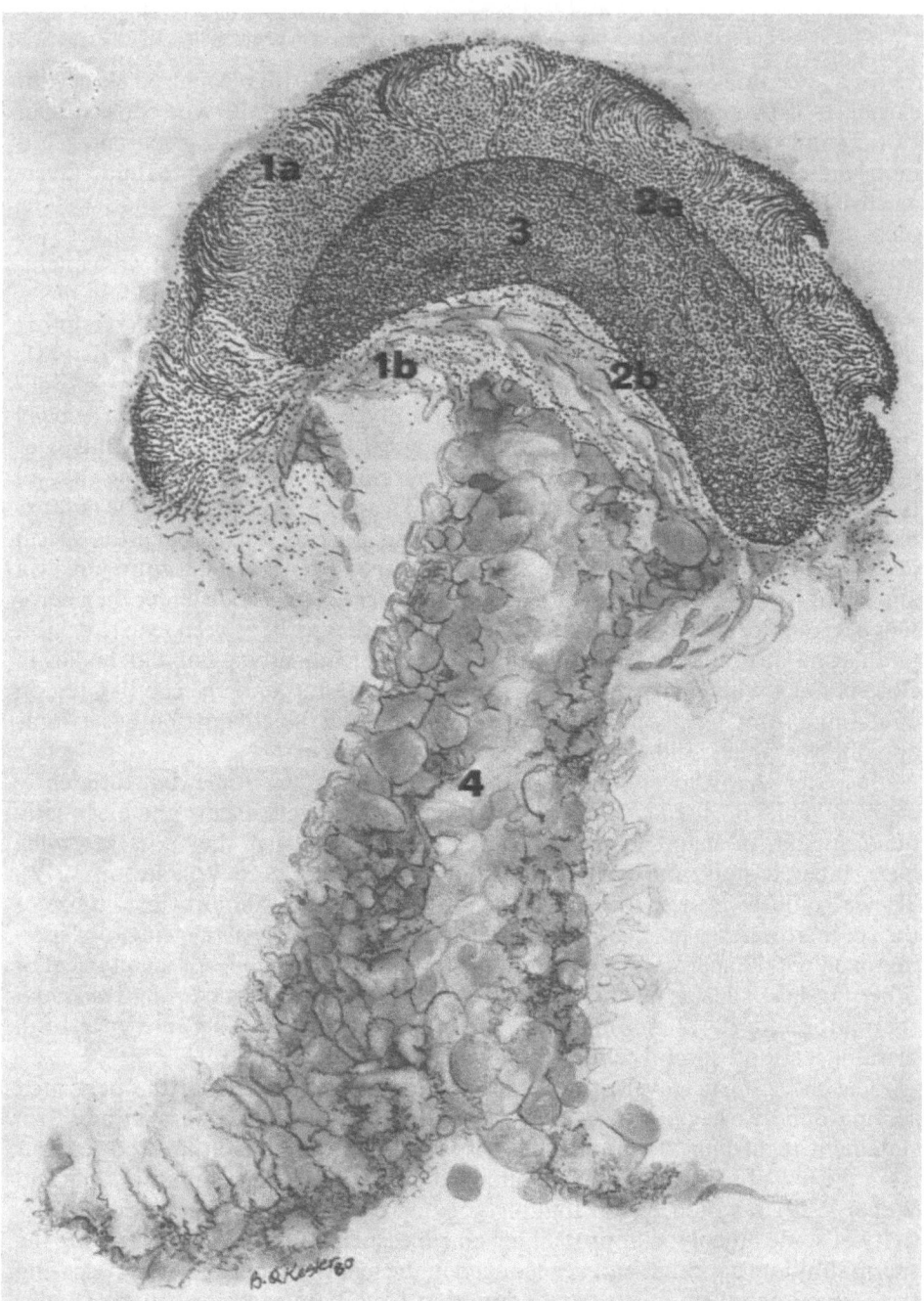

Fig. 50. Spermatophore of the salamander *Ambystoma texanum.* Drawing from longitudinal section stained with Toluidine blue. The stalk is composed of several apposed spheres of uniformly stained material (area *4*). Lightly stained material is present in the outer region of the cap (area *1a*) and those regions of the cap which make contact with the stalk (area *1b*); it is distributed around the heads and tails of spermatozoa. Underlying the surface and extending around the whole spermatophore is a membrane (areas *2a* and *2b;* PAS-positive) which separates the internal region of the cap (area *3*) containing the cytoplasmic droplets and the sperm tails, from the superficial zone (*1a* and *1b*). (Russell et al. 1981)

10.4.2 Courtship in Salamanders and its Effect on the Production of Spermatophores

On the subject of reproductive behaviour in urodeles there exists a voluminous literature which extends back to Gasco's account of *Les Amours des Axolotls* and Jordan's observations on male strategy directed towards the capture of females (Gasco 1881; Jordan 1891; Brandon 1970; Joly 1966; Organ and Organ 1968; Arnold 1977; Halliday 1977). Several new facts concerning courtship behaviour and its relation to spermatophore formation and deposition have been brought to light in recent years.

Arnold (1976) has shown that while males of *Ambystoma maculatum,* which court quickly and deposit many spermatophores, have a low success rate with each spermatophore, the males of *Plethodon jordani,* which court slowly and deposit mostly only a single spermatophore, have a high rate of success with their individual spermatophores. In the smooth newt, *Triturus vulgaris* (Halliday 1976), deposition of spermatophores occurs in response to a stimulus provided by the female when she touches the male's tail with her snout: the male then folds and lifts his tail, and deposits a spermatophore on the bottom of the pond. He then moves away, blocking the female's pathway in such a way that her cloaca is brought into a position right above the spot of spermatophore deposition. In experiments designated to determine the success rate of mating in the smooth newts, Halliday (1976) tested five males on 40 consecutive days, each time in a standard 5-min trial. In all the males there was a steady decline over the 40 days' period in the number of spermatophores produced. The study by Malacarne and Giacoma (1980) was concerned with mating behaviour of the crested newt, *Triturus cristatus carnifex.* It has shown that injections of testosterone into a castrated male had no reactivating effect on male sexual behaviour, and from their observations the authors concluded that male sexual behaviour is not solely dependent on the androgen. They found, however, that long-term treatment of the males with an anti-androgen (cyproterone) had a depressing effect on deposition of spermatophores. The red-spotted newt, *Notophthalmus viridescens,* is yet another species to be studied lately (Verrell 1982). Male sexual behaviour in this newt depends largely on the responsiveness of the female. If she is responsive when approached, then the male throws immediately his body into a series of serpentine undulations, called hula display (because of similarity to the Hawaian dance), at the end of which one or more spermatophores may be deposited. However, if she tries to run away, a struggle and prolonged period of amplexus ensues, to be followed only later by spermatophore-transfer behaviour. The probability of a successful spermatophore pick-up is much higher after the prolonged period of amplexus than at the end of a brief hula display.

Concluding Remarks

The bulk of secretory material necessary for the encapsulation of spermatozoa into a spermatophore is commonly derived from the male accessory organs of reproduction. Not surprisingly, the extraordinary diversity in size, shape, and structure of the spermatophores in different phyla is frequently a reflection of the equally spectacular variations in the anatomy and secretory performance of male reproductive tracts. Less obvious are the reasons why even within a given class or order of animals, only some species employ spermatophores, while others depend on liquid semen as the vehicle for spermatozoa. The most plausible explanation is that the development of methods for sperm transfer must have been influenced by the environment in which the animals breed, and that adaptation to habitat, rather than phylogeny, has played a decisive role. Reproduction in the giant octopus of the North Pacific, on which our attention has been focussed, provides an interesting example. During copulation in seawater, which may last 2 h, the sperm mass has to be pushed over the distance of 1 m separating the male and female genital orifices. The metre-long tubular spermatophore inside which the spermatozoa are conveyed offers an obvious advantage over liquid semen, which could hardly be hauled over such a long distance.

Apart from acting as a convenient transport vehicle, the spermatophore serves other purposes. Its gustatory and aphrodisiac attributes, the provision of an effective barrier to reinsemination, and stimulation of oogenesis and oviposition are all of great importance. Absolutely essential is its function as storage organ for spermatozoa, at least as effective as that of the epididymis for mammalian spermatozoa. Relevant in this connection may be the observation that the spermatophoric plasma which surrounds the sperm rope inside the spermatophore of the giant octopus bears close chemical similarity to mammalian epididymal plasma, reflected in the high concentration of mucoproteins, several glycosidases, and glycerylphosphorylcholine. Equally intriguing are the recent observations on the occurrence of spermine in the spermatophoric plasma of ticks and the prostaglandin-synthesizing potential of spermatophores in certain insects. To what extent spermatophores of other animals share biochemical properties with liquid semen is by no means clear at the present time. This is only one of the many gaps in our knowledge that remain to be filled in the future. If the present monograph can generate interest in this direction, it will have fulfilled its purpose.

References

Abele LG, Gilchrist S (1977) Homosexual rape and sexual selection in acanthocephalan worms. Science 197:81–83

Alberti G (1983) Fine structure of scorpion spermatozoa (*Buthus occitanus;* Buthidae, Scorpiones). J Morphol 177:205–212

Alberti G, Storch V (1976) Spermiocytogenese, Spermien und Spermatophore von Schnabelmilben (Bdellidae, Acari). Acta Zool (Stockh) 57:177–188

Alexander AJ (1956) Mating in scorpions. Nature 178:867–868

Alexander AJ (1959) Courtship and mating in the buthid scorpions. Proc Zool Soc Lond 133:145–169

Alexander AJ (1962) Courtship and mating in amblypygids. Proc Zool Soc Lond 138:379–383

Alexander AJ, Ewer DW (1957) On the origin of mating behavior in spiders. Amer Naturalist 91:311–317

Alexander RD (1964) The evolution of mating behaviour in arthropods. In: Highnam KC (ed) Insect reproduction. Roy Ent Soc Symposium Lond, no 2:78–94

Alexander RD, Otte D (1967) Cannibalism during copulation in the brown bush cricket, *Hapithus agitator* (Gryllidae). Flo Entomol 50:79–87

Amano H, Chant DA (1978) Mating behaviour and reproductive mechanisms of two species of predacious mites, *Phytoseiulus persimilis* Athias-Henriot and *Amblyseius andersoni* (Chant) (Acarina: Phytoseiidae). Acarologia 20:196–213

Anderson JD, Williamson GK (1977) Descriptions of the spermatophores of *Ambystoma cingulatum* and *Ambystoma mabeei* (Amphibia: Urodela). Herpetologica 33:253–256

Anderson JM (1950) A cytological study of the male accessory glands of the Japanese beetle, *Popillia japonica*. Biol Bull (Woods Hole) 99:49–64

Andrews EA (1910) Conjugation in the crayfish *Cambarus affinis*. J Exp Zool 9:235–264

Andrews E (1936) Spermatophores of the snail *Neritina reclivata*. J Morphol 60:191–209

Andrews E (1937) Certain reproductive organs in the Neritidae. J Morph 61:525–549

Angermann H (1955) Indirekte Spermatophorenübertragung bei *Euscorpius italicus* (Hbst.) (Scorpiones, Chactidae). Naturwiss 42:303

Angermann H (1957) Über Verhalten, Spermatophorenbildung und Sinnesphysiologie von *Euscorpius italicus* Hbst. und verwandten Arten (Scorpiones, Chactidae). Z Tierpsychol 14:276–302

Aristotle (ca 330 BC) Historia animalium. Transl. by D'Arcy Wentworth Thompson. In: Smith JA, Ross MA (eds) (1910) The works of Aristotle, vol 4. Clarendon Press, Oxford, p 541 b and p 544 a

Arnold JM (1962) Mating behaviour and social structure of *Loligo pealii*. Biol Bull (Woods Hole) 123:53–57

Arnold JM (1971) Cephalopods. In: Reverberi G (ed) Experimental embryology of marine and freshwater invertebrates. North-Holland, Amsterdam London, pp 265–311

Arnold JM, Williams-Arnold LD (1977) Cephalopoda: Decapoda. In: Giese AC, Pearse JS (eds) Reproduction of marine invertebrates, vol 4. Academic Press, New York San Francisco London, pp 243–290

Arnold SJ (1976) Sexual behavior, sexual interference, and sexual defence in the salamanders *Ambystoma maculatum, Ambystoma tigrinum,* and *Plethodon jordani*. Z Tierpsychol 42:247–300

Arnold SJ (1977) The evolution of courtship behaviour in New World salamanders with some comments on Old World salamandrids. In: Taylor DH, Gutman SI (eds) The reproductive biology of amphibians. Plenum Press, New York, pp 141–183

Austin CR, Lutwak-Mann C, Mann T (1964) Spermatophores and spermatozoa of the squid *Loligo pealii*. Proc R Soc Lond [Biol] 161:143–152

Ax P (1968) Das Fortpflanzungsverfahren von *Trilobodrilus* (Archiannelida). Mar Biol (Berl) 1:330–335

Ax P (1969) Populationsdynamik, Lebenszyklen und Fortpflanzungsbiologie der Mikrofauna des Meeressandes (Verh Dtsch Zool Ges Innsbruck 1968). Zool Anz Suppl 32:66–113

Ax P, Apelt G (1969) Organisation und Fortpflanzung von *Archaphanostoma agile* (Turbellaria, Acoela). (Verh Dtsch Zool Ges Innsbruck 1968). Zool Anz Suppl 32:339–343

Ax P, Borkott H (1969) Organisation und Fortpflanzung von *Macrostomum romanicum* (Turbellaria, Macrostomida) (Verh Dtsch Zool Ges Innsbruck 1968). Zool Anz Suppl 32:344–347

Baccetti B (ed) (1970) Comparative spermatology. Academia Nazionale dei Lincei, Rome. Academic Press, New York London, p 573

Badenhorst JH (1974) The morphology and histology of the male genital system of the squid *Loligo reynaudii* (d'Orbigny). Ann Univ Stellenbosch Ser A 49:1–36

Bairati A (1966) Filamentous structures in spermatic fluid of *Drosophila melanogaster* Meig. J Microsc (Paris) 5:265–268

Baker HR, Erséus C (1982) A new species of *Bacescuella* Hrabĕ (Oligochaeta, Tubíficidae) from the Pacific coast of Canada. Can J Zool 60:1951–1954

Baker TC, Nishida R, Roelofs WL (1981) Close-range attraction of female oriental fruit moth to herbal scent of male hairpencils. Science 214:1359–1361

Ballan-Dufrançais C (1968) Données morphologiques et histologiques sur les glandes annexes mâles et le spermatophore de *Blatella germanica*, au cours de la vie imaginale. Bull Soc Zool Fr 93:401–421

Ballowitz E (1890) Untersuchungen über die Struktur der Spermatozoen, zugleich ein Beitrag zur Lehre vom feineren Bau der kontraktilen Elemente. Die Spermatozoen der Insecten (1. Coleopteren). Z Wiss Zool 50:317–407

Ballowitz E (1895) Die Doppelspermatozoen der Dytisciden. Z Wiss Zool 60:458–499

Ballowitz E (1916) Spermiozeugmen bei Libellen. Biol Centralbl 36:209–216

Bareth C (1966) Études comparatives des spermatophores chez les Campodéidés. C R Acad Sci Paris 262:2055–2058

Barker JF, Davey KG (1982) Intraglandular synthesis of protein in the transparent accessory reproductive gland in the male *Rhodnius prolixus*. Insect Biochem 12:157–160

Battaglia B (1953) Il significato della presenza di polisaccharidi negli spermatozoi atipici dei Gastropodi Prosobranchi. Ric Sci Suppl 23:125–129

Bauer RT (1976) Mating behaviour and spermatophore transfer in the shrimp *Heptacarpus pictus* (Stimpson) (Decapoda: Caridea: Hippolytidae). J Nat Hist 10:415–440

Bawa SR, Marwaha RK (1975) The sperm bundles of honeybee *Apis cerana indica* Fabr. Experientia 31:684–686

Baylis HA (1939) Delayed reproduction in the spotted salamander. Proc Zool Soc Lond 109:243–246

Beauregard H (1890) Les insectes vésicants. Baillière, Paris, p 544

Beeman RD (1977) Gastropoda: Opisthobranchia. In: Giese AC, Pearse JC (eds) Reproduction of marine invertebrates, vol 4. Academic Press, New York San Francisco London, pp 115–179

Bell PD (1980) Multimodal communication by the black-horned tree cricket, *Oecanthus nigricornis* (Walker) (Orthoptera: Gryllidae). Can J Zool 58:1861–1868

Belonoschkin B (1929a) Die Geschlechtswege von Oktopus vulgaris und ihre Bedeutung für die Bewegung der Spermatozoen. Z Zellforsch Mikrosk Anat 9:643–662

Belonoschkin B (1929b) Das Verhalten der Spermatozoen zwischen Begattung und Befruchtung bei Oktopus vulgaris. Z Zellforsch Mikrosk Anat 9:750–753

Bennike GAB (1943) Contribution to the ecology and biology of the Danish fresh-water leeches (Hirudinea). Fol Limmol Scand 2:1–109

Benson DG (1964) The histochemistry of the spermatophore of *Ambystoma maculatum*. Am Zool 4:287

Benz G (1969) Influence of mating, insemination, and other factors on oögenesis and oviposition in the moth *Zeiraphera diniana*. J Insect Physiol 15:55–71

Berry AJ (1977) Gastropoda: Pulmonata. In: Giese AC, Pearse JD (eds) Reproduction in marine invertebrates, vol 4. Academic Press, New York San Francisco London, pp 181–226

Berry AJ, Lim R, Sase Kumar A (1973) Reproductive systems and breeding conditions in *Nerita birmanica* (Archaeogastropoda: Neritacea) from Malayan mangrove swamps. J Zool Proc Zool Soc Lond 170:189–200

Betsch-Pinot M-C (1974) Description du spermatophore d'*Isotoma viridis* Bourlet, 1839 (Isotomidae) et comparaison des spermatophores connus dans chaque groupe de Collemboles. Rev Ecol Biol Sol 11:541–552

Betsch-Pinot M-C (1977) Les parades sexuelles primitives chez les Collemboles Symphypléones. Rev Ecol Biol Sol 14:15–20

Billard R, Fléchon JE (1975) Spermatogonies et spermatocytes flagéllés chez *Poecilia reticulata* (Téléostéens, cyprinodontiforme). Ann Biol Anim Bioch Biophys 9:281–286

Black PN, Landers MH, Happ GM (1982) Cytodifferentiation in the accessory glands of *Tenebrio molitor*. 8. Crossed immunoelectrophoretic analysis of terminal differentiation in the postecdysial tubular accessory glands. Dev Biol 94:106–115

Blades PI (1977) Mating behavior of *Centropages typicus* (Copepoda: Calanoida). Mar Biol (Berl) 40:57–64

Blades PI, Youngbluth MJ (1979) Mating behavior of *Labidocera aestiva* (Copepoda: Calanoida). Mar Biol (Berl) 51:339–356

Blades PI, Youngbluth MJ (1981) Ultrastructure of the male reproductive system and spermatophore formation in *Labidocera aestiva* (Crustacea: Copepoda). Zoomorphology (Berl) 99:1–22

Blanchard E (1853) L'organisation du règne animal; cf Brongniart and Gaubert (1891)

Blancquaert J-P (1981) Mating behaviour in some Sminthurididae (Collembola) with reference to the systematics of Symphypleona. Pedobiologia 22:1–4

Blancquaert T (1925) L'origine et la formation des spermatophores chez les céphalopodes décapodes. Cellule Rec Cytol Histol 36:315–356

Blunck H (1912) Das Geschlechtsleben des *Dytiscus marginalis* L. 1 Teil. Die Begattung. Z Wiss Zool 102:169–248

Blunck H (1913) Kleine Beiträge zur Kenntnis des Geschlechtsleben und der Metamorphose der Dytisciden. 2 Teil. *Acilius sulcatus* L. Zool Anz 41:586–597

Bode W (1975) Der Ovipositor und die weiblichen Geschlechtswege der Thripiden (Thysanoptera, Terebrantia). Z Morph Tiere 81:1–53

Boggs CL (1981) Selection pressures affecting male nutrient investment at mating in heliconiine butterflies. Evolution 35:931–940

Boggs CL, Gilbert LE (1979) Male contribution to egg production in butterflies: Evidence for transfer of nutrients at mating. Science 206:83–84

Boissin L (1973) Biologie sexuelle du Pseudoscorpion *Hysterochelifer meridianus* (L. Koch): Accouplement et description du spermatophore. Bull Soc Zool Fr 98:521–528

Boldyrev BT (1915) Contributions à l'étude de la structure des spermatophores et des particularitès de la copulation chez Locustodea and Gryllodea. Horae Soc Entomol Ross 41:1–245

Boldyrev BT (1927) Einige Daten über die Spermatophoren-Befruchtung bei den Insekten. Rev Russe Entomol 21:133–136

Boldyrev BT (1929) Spermatophore fertilization in the migratory locust (*Locusta migratoria* L.). Izv Prikl Entomol Leningrad 4:189–218

Böttger K (1962) Zur Biologie und Ethologie der einheimischen Wassermilben *Arrenurus* (*Megaluracarus*) *globator* (Müll.) 1776, *Piona nodata nodata* (Müll.) 1776, und *Eylais infundibulifera meridionalis* (Thon) 1899. Zool Jahrb (Abt System) 89:501–584

Böttger K (1965) Ökologie und Fortpflanzungsbiologie von *Arrenurus valdiviensis* K.O. Viets 1964. Z Morph Oekol Tiere 55:115–141

Boudou-Saltet P (1972) Les Dolichopodes (Orth. Rhaph.) de Grèce: VII. Nouvelles espèces du Péleponnèse. Bull Soc Hist Nat Toulouse 108:420–425

Boudou-Saltet P (1980) Les Dolichopodes (Orth. Rhaph.) de Grèce. IX. Une espèce nouvelle en Eubée: *D. makrykapa*. Biol Gallo-Hell 9:123–134

Boudou-Saltet P, Capolongo D (1975) Le spermatophore chez quelques espèces de Dolichopoda (Orth. Rhaph.) de Grèce et d'Italie. Biol Gallo-Hell 6:3–8

Brandes G (1901) Die Begattung der Hirudineen. Abhdl Naturf Ges Halle 22:373–392

Brandon RA (1970) Courtship, spermatophores, and eggs of the Mexican Achoque, *Ambystoma* (*Bathysiredon*) *dumerili* (Duges). Zool J Linnean Soc 49:247–254

Breucker H (1970) Die Struktur des samenableitenden Gangsystems bei *Geophilus linearis* Koch (Chilopoda). Z Zellforsch 108:225–242

Breure ASH, Eskens AAC (1977) Observations on the formation of spermatophores in a bulimulid land snail, *Drymaeus canaliculatus* (Pfeiffer, 1845) (Mollusca, Gastropoda, Pulmonata). Neth J Zool 27:271–276

Briegleb W (1961) Die Spermatophore des Grottenolms. Zool Anz 166:87–91

Brinck P (1960) Die Entwicklung der Spermaübertragung der Odonaten. Verhandl XI Intern Kongr Entomol (Wien 1960) 1:715–718

Brock J (1879) Über die Geschlechtsorgane der Cephalopoden. Z Wiss Zool 32:1–116

Brock J (1882) Zur Anatomie und Systematik der Cephalopoden. Z Wiss Zool 36:543–606

Brongniart C, Gaubert (1891) Fonctions de l'organe pectiniforme des Scorpions. CR Acad Sci Paris 113:1062–1064

Brooks DE, Lutwak-Mann C, Mann T, Martin AW (1971) Motility and energy-rich phosphorus compounds in spermatozoa of *Octopus dofleini martini*. Proc R Soc Lond [Biol] 178:151–160

Brooks DE, Hamilton DW, Mallek AH (1974a) Carnitine and glycerylphosphorylcholine in the reproductive tract of the male rat. J Reprod Fert 36:141–160

Brooks DE, Mann T, Martin AW (1974b) The occurrence of carnitine and glycerylphosphorylcholine in the octopus spermatophore. Proc R Soc Lond [Biol] 186:79–82

Brooks DE, Tate ME, Mann T, Martin AW (1981) Phosphoglycopeptide, a major constituent of the spermatophoric plasma of the octopus (*Octopus dofleini martini*). J Reprod Fert 63:515–521

Brooks WK, Cowles RP (1905) *Phoronis architecta*. Its life history, anatomy, and breeding habits. Mem Nat Acad Sci Washington 10:69–111

Brousse-Gaury P, Goudey-Perriere F (1983) Spermatophore et vitellogenèse chez *Blabera fusca* Br. (Dictyoptère, Blaberidae). C R Acad Sci Paris, Ser III, 296:659–664

Brower JH (1979) Radiosensitivity of adults of the almond moth. J Econ Entomol 72:43–47

Brown GG (1970) Some comparative aspects of selected crustacean spermatozoa and crustacean phylogeny. In: Baccetti B (ed) Comparative spermatology. Academic Nazionale dei Lincei, Rome, and Academic Press, New York London, pp 183–204

Brown GG, Humphreys WJ (1971) Sperm-egg interactions of *Limulus polyphemus* with scanning electron microscopy. J Cell Biol 51:904–907

Brown KS (1981) The biology of *Heliconius* and related genera. Ann Rev Entomol 26:247–256

Brumpt E (1900) Reproduction des Hirudinées. Mém Soc Zool France 13:286–430

Bulnheim H-P (1962a) Untersuchungen zum Spermatozoendiomorphismus von *Opalia crenimarginata* (Gastropoda, Prosobranchia). Z Zellforsch Mikr Anat 56:300–343

Bulnheim H-P (1962b) Elektronmikroskopische Untersuchungen zur Feinstruktur der atypischen und typischen Spermatozoen von *Opalia crenimarginata* (Gastropoda, Prosobranchia). Z Zellforsch Mikr Anat 56:371–386

Callahan PS, Cascio T (1963) Histology of the reproductive tracts and transmission of sperm in the corn earworm, *Heliothis zea*. Ann Entomol Soc Am 56:553–556

Callahan PS, Chapin JB (1960) Morphology of the reproductive systems and mating in two representative members of the family Noctuidae, *Pseudaletia unipuncta* and *Peridroma margaritosa*, with comparison to *Heliothis zea*. Ann Entom Soc Am 53:763–782

Campion DG (1971) Chemosterilization of the red bollworm *Diparopsis castanea* Hmps. (Lep., Noctuidae): effects of certain s-triazines and a carbamate insecticide. Bull Entomol Res 61:351–355

Cantacuzène A-M (1968) Recherches morphologiques et physiologiques sur les glandes annexes mâles des orthoptères. III. Modes d'association des spermatozoïdes d'Orthoptères. Z Zellforsch 90:113–126

Cantelo WW (1973) Dye markers for moth of the tobacco hornworm. Environ Entomol 2:393–396

Carayon J (1959) Insémination par «spermalège» et cordon conducteur des spermatozoïdes chez *Stricticimex brevispinosus* Usinger (Heteroptera, Cimicidae). Rev Zool Bot Africaine 60:81–104

Carlberg U (1981) Spermatophores of *Baculum extradentatum* and other Phasmida. Entom Mon Mag 117:125–127

Carlberg U (1984) Sizes of eggs and oviposition females of *Extatosoma tiaratum* (MacLeay) (Insecta: Phasmida). Zool Anz (Jena) 212:226–228

Caullery M (1914) Sur les Siboglinidae, type nouveau d'invertébrés recuelli par l'expédition du *Siboga*. CR Acad Sci Paris 158:2014–2017

Chao W-Y (1981) The distribution, translocation, and function of semen in the female reproductive systems of the army moth (*Leucania separata*). Acta Entomol Sin 24:135–141; cf Biol Abs 73(3), Abstract no 17879 (1982)

Cerruti A (1908) Richerche sulla anatomia e sulla biologia del *Microspio mecznikowianus* Clpd, con speciale reguardo ai nefridi. Rend Accad Sci Fis Mat Napoli (Ser 2) 13:1–35

Chevaillier P (1962) Sur la genèse et la constitution histochimique du spermatodesme des Cercopidae (Homopt. Auch.). CR Acad Sci Paris 254:1148–1149

180

Chevaillier P (1963) Le spermatodesme des Cercopidae (Homoptères Auchénorhynches). Bull Biol Fr Belg 97:553–571

Chevaillier P, Maillet PL (1965) Sur quelques aspects du métabolisme des «cellules-satellites» du testicule des Homoptères Auchénorhynches, et en particulier de *Cicada orni* L. CR Acad Sci Paris 260:1255–1258

Cholodkovsky N (1880) Ueber die Hoden der Schmetterlinge. Zool Anz 3:115–117

Cholodkovsky N (1884) Ueber die Hoden der Lepidopteren. Zool Anz 7:564–568

Cholodkovsky N (1913) Über die Spermatodosen der Locustiden. Zool Anz 41:615–619

Chopard L (1934) Sur la présence d'un spermatophore chez certains Insectes Orthoptères de la famille des Phasmides. CR Acad Sci Paris 199:806–807

Chow S, Ogasawara Y, Taki Y (1982) Male reproductive system and fertilization of the palaemonid shrimp *Macrobrachium rosenbergii*. Bull Jpn Soc Sci Fish 48:177–184

Claparède E, Mecznikow E (1869) Beiträge zur Kenntnis der Entwicklungsgeschichte der Chaetopoden. Z Wiss Zool 19:163–205

Clark JT (1974) A conspicuous spermatophore in the phasmid *Extatosoma tiaratum* Macleay. Entomol Mon Mag 110:81–82

Clark WC (1981) Sperm transfer mechanisms: some correlates and consequences. NZJ Zool 8:49–66

Clarke C, Johnston G, Johnston B (1983) All-female broods of *Hypolimnas bolina:* A survey of West Fiji after 60 years. Biol J Linn Soc 19:221–236

Clay T (1968) Contributions towards a revision of Myrsidea. III. Bull Br Mus Nat Hist (Ent) 21:203–243

Clay T (1971) A new species of *Austrogoniodes* (Phthiraptera: Philopteridae) from a duck (Anseriformes). J Aust Entomol Soc 10:293–298

Cloudsley-Thompson JL (1968) Spiders, scorpions, centipedes, and mites (2nd edn). Pergamon Press, Oxford, p 819

Cloudsley-Thompson JL (1976) Evolutionary trends in the mating of Arthropoda. Meadowfield, Shildon England, p 85

Comstock JH (1940) The spider book. Doubleday, Doran & Co, New York. Republished, 1948, Comstock Publ Co Ithaca NY, p 729

Conchie J, Mann T (1957) Glycosidases in mammalian sperm and seminal plasma. Nature 179:1190–1191

Cooney JD, Gehrs CW (1980) Effects of varying food concentration on reproduction in *Diaptomus clavipes*. Am Midl Nat 104:63–69

Cotton B (1938) Mollusca, Part I: The spermatophore of *Rossia australis* Berry. Proc Roy Soc Victoria 50:338–340

Cressey RF (1972) Revision of the genus *Alebion* (Copepoda: Caligoida). Smithson Contrib Zool 123:1–29

Cummings BF (1916) Studies on the Anoplura and Mallophaga. Proc Zool Soc Lond 1916, pp 643–673

Cuvier G (1805) [XIV] Leçons d'anatomie comparée (5 volumes). Crochard et Fantin, Paris. Tome V: Les organes de la géneration et ceux des sécretion excrémentitielles ou des excrétions; Quatrieme section: Organes de la géneration dans les animaux sans vertèbres (pp 164–195); this section deals with reproductive organs in cephalopods, gastropods, crustaceans, insects, and other invertebrates

Cuvier G (1817) Mémoires pour servir à l'histoire et à l'anatomie des Mollusques (1 volume), Deterville, Paris. This volume consists of several Mémoires, the second of which is the Mémoire sur les Céphalopodes et sur leur anatomie (54 pages); it is in 3 sections: Sur le Poulpe (pp 1–41), De la Seiche (pp 43–50), and Des Calmars (pp 50–54)

Cuvier G (1829) Mémoire sur un ver parasite d'un nouveau genre (*Hectocotylus octopodis*). Ann Sci Nat 18:147–156

Da Cruz-Landim C, Ferreira·A (1977) Spermatophore formation in *Conocephalus saltator* (Saussure) (Orthoptera: Conocephalidae). Int J Insect Morph Embryol 6:97–104

Dailey PJ, Happ GM (1983) Cytodifferentiation in the accessory glands of *Tenebrio molitor*. XI. Transitional cell types during establishment of pattern. J Morph 178:139–154

Dall P (1979) Ecology and production of the leeches *Erpobdella octoculata* and *Erpobdella testacea* in Lake Esrom, Denmark. Arch Hydrobiol Suppl 57:188–220

Dallai R (1975) Ultrastructural and polarized light microscope studies on spermatophores of *Dicyrtoma ornata* (Insecta, Collembola). J Ultrastruct Res 50:355–361

181

Damas D (1968) Origine et structure du spermatophore de *Glossiphonia complanata* (L) (Hirudinée Rhynchobdelle). Arch Zool Exp Gen 109:79–85

Davey KG (1959) Spermatophore production in *Rhodnius prolixus*. QJ Micr Sci 100:221–230

Davey KG (1960a) A pharmacologically active agent in the reproductive system of insects. Can J Zool 38:39–45

Davey KG (1960b) The evolution of spermatophores in insects. Proc R Ent Soc London A35:107–113

Davey KG (1965) Reproduction in the insects. Oliver & Boyd, Edinburgh London, p 98

David K (1936) Beiträge zur Anatomie und Lebensgeschichte von *Osmylus chrysops* L. Z Morph Ökol Tiere 31:151–206

Davids C, Belier R (1980) Spermatophores and sperm transfer in the water mite, *Hydrachna conjecta:* Descent of water mites from terrestrial forms. Acarologia (Paris) 21:84–90

Davies L (1965) On spermatophores in Simuliidae (Diptera). Proc R Ent Soc Lond (A) 40:30–34

Dawson RMC, Mann T, White IG (1957) Glycerylphosphorylcholine and phosphorylcholine in semen, and their relation to choline. Biochem J 65:627–634

Dazo BC (1965) The morphology and natural history of *Pleurocera acuta* and *Gonobiasis livescens* (Gastropoda: Cerithiacea: Pleuroceridae). Malacologia 3:1–80

Delle Chiaje S (1822–1830) Memorie sulla storia et notomia degli animali senza vertebre del regno di Napoli. Fratelli Fernandes/Societa Tipographica, Napoli. *Memoria II su' Cefalopodi* forms part (pp 39–61) of vol 4, published in 1829; *Scoloce bilobato* and *Monostoma del polpo:* pp 53–54; the statement „Hoi il più fondato sospetto che le famose anguille di Needham, di cui parla Cuvier in un modo bastantemente preciso, avendo l'apparenza di filamenti bianchi, siano gli entozoi in discussione": p 54

Denys-Monfort P (1802[X]) Histoire naturelle, générale et particuliere des Mollusques, animaux sans vertèbres et a sang blanc. Dufart, Paris, vol 1, p 234 and 288

Destephano DB, Brady UE (1977) Prostaglandin and prostaglandin synthetase in the cricket, *Acheta domesticus*. J Insect Physiol 23:905–912

Devine MC (1975) Copulatory plugs in snakes: enforced chastity. Science 187:844–845

Drecktrah HG (1978) Morphology of the internal reproductive system of the adult female army cut-worm, *Euxoa auxilliaris*. Ann Entomol Soc Am 71:923–927

Drecktrah HG, Brindley TA (1967) Morphology of the internal reproductive systems of the European corn borer. Iowa State J Sci 41:467–480

Drew GA (1911) Sexual activities of the squid, *Loligo pealii* (Les.). I. Copulation, egg-laying and fertilization. J Morph 22:327–360

Drew GA (1919a) Sexual activities of the squid, *Loligo pealii* (Les.). II. The spermatophore, its structure, ejaculation, and formation. J Morph 32:379–435

Drew GA (1919b) The structure and ejaculation of the spermatophores of *Octopus americana*. Carnegie Inst Washington. Papers from the Tortugas Laboratory 13:35–47

Du Bois AM, Geigy R (1935) Beiträge zur Oekologie, Fortpflanzungsbiologie und Metamorphose von *Sialis lutaria* L. (Studien am Sempachersee). Rev Suisse Zool 42:169–248

Dudenhausen EE, Talbot P (1983) An ultrastructural comparison of soft and hardened spermatophores from the crayfish *Pacifastacus leniusculus* Dana. Can J Zool 61:182–194

Dumser JB (1980) The regulation of spermatogenesis in insects. Ann Rev Entomol 25:341–369

Dumser JB, Oliver JH Jr (1981) Kinetics of spermatogenesis in the adult tick, *Dermacentor variabilis*. J Insect Physiol 27:681–688

Duncan TK (1983) Sexual dimorphism and reproductive behaviour in *Almyracuma proximoculi* (Crustacea: Cumacea): The effect of habitat. Biol Bull (Woods Hole) 165:370–378

Dunham PJ (1978) Sex pheromones in crustacea. Biol Rev 53:555–583

Durward RD, Vessey E, O'Dor RK, Amaratunga T (1980) Reproduction in the squid, *Illex illecebrosus:* First observations in captivity and implications for the life cycle. Int Comm Northwest Atl Fish Sel Pap (6):7–14

Duvernoy (Monsieur) (1853) Fragments sur les organes de génération de divers animaux. Quatrième Fragment: De spermaphores dans la Sèpiole de Rondelet et dans le Calmar subulé, et des organes qui les produisent dans ces deux espèces et dans plusieurs autre Céphalopodes; de leur composition par ces organes, et de leur décomposition dans l'eau et dans les organes sexuels des femelles. Mem Acad Sci Inst Fr 23:215–281

Dyster FD (1858) Notes on *Phoronis hippocrepia*. Trans Linn Soc Lond (Zool) 22:251–256

Edwards (Milne) H (1840) see Milne Edwards H (1840)

182

Edwards (Milne) H (1842) see Milne Edwards H (1842)

Ehrlich AH, Ehrlich PR (1978) Reproductive strategies in the butterflies: I. Mating frequency, plugging, and egg number. J Kansas Entomol Soc 51:666–697

Ehrnsberger R (1977) Fortpflanzungsverhalten der Rhagidiidae (Acarina, Trombidiformes). Acarologia (Paris) 19:67–73

Eisig H (1887) Die Capitelliden des Golfes von Neapel. Fauna Fl Golfes Neapel 16:1–906

Elliott JM (1973) The life cycle and production of the leech *Erpobdella octoculata* (L.) (Hirudinea: Erpobdellidae) in a Lake District stream. J Anim Ecol 42:435–448

Engelmann F (1970) The physiology of insect reproduction. Pergamon Press, Oxford New York Toronto Sydney Braunschweig, p 307

Erséus C (1978) Two new species of the little-known genus *Bacescuella* Hrabě (Oligochaeta, Tubificidae) from the North Atlantic. Zool Scr 7:263–268

Erséus C (1980) Taxonomic studies on the marine genera *Aktedrilus* and *Bacescuella* (Oligochaeta, Tubificidae), with descriptions of 7 new species. Zool Scr 9:97–112

Ewen AB, Pickford R (1975) Morphology of the male sex organs, spermatophore formation, and insemination in the clear-winged grasshopper, *Camnula pellucida* (Scudder). Acrida 4:195–203

Fahrenbach WH (1962) The biology of a harpacticoid copepod. Cellule 58:301–376

Fahrenbach WH (1973) Spermiogenesis in the horseshoe crab, *Limulus polyphemus*. J Morph 140:31–52

Fain-Maurel M-A (1970) Le spermatozoïde des isopodes. In: Baccetti B (ed) Comparative spermatology. Academia Nazionale dei Lincei, Rome, and Academic Press, New York London, pp 221–236

Falk-Petersen S, Hopkins CCE (1981) Ecological investigations on the zooplankton community of Balsfjorden, northern Norway: Population dynamics of the euphausiids, *Thysanoessa inermis, Thysanoessa raschii,* and *Meganyctiphanes norvegica,* in 1976 and 1977. J Plankton Res 3:177–192

Fallon AM, Wyatt GR (1975) Cyclic guanosine 3',5'-monophosphate: High levels in the male accessory gland of *Acheta domesticus* and related crickets. Biochim Biophys Acta 411:173–185

Farfante IP (1975) Spermatophores and thelyca of the American white shrimps, genus *Penaeus* subgenus *Litopenaeus*. US Natl Mar Fish Serv Fish Bull 73:463–486

Feldman-Muhsam B (1967) Spermatophore formation and sperm transfer in *Ornithodoros* ticks. Science 156:1252–1253

Feldman-Muhsam B, Borut S (1971) Copulation in ixodid ticks. J Parasitol 57:630–634

Feldman-Muhsam B, Borut S (1978) Further observations on the spermatophore formation in argasid ticks. J Insect Physiol 24:693–697

Feldman-Muhsam B, Borut S (1983) On the spermatophore of ixodid ticks. J Insect Physiol 29:449–457

Feldman-Muhsam B, Filshie BK (1979) The ultrastructure of the prospermium of *Ornithodoros* ticks and its relation to sperm maturation and capacitation. In: Fawcett DW, Bedford JM (eds) The spermatozoon. Urban & Schwarzenberg, Baltimore, pp 355–369

Feldman-Muhsam B, Havivi Y (1963) On *Adlerocystis* n. gen. (*Phycomycetes*) a symbiote of *Ornithodoros* ticks. Parasitology 53:183–188

Feldman-Muhsam B, Havivi Y (1967) Juvenile sterility in male ticks of *Ornithodoros tholozani*. Nature 213:422–423

Feldman-Muhsam B, Borut S, Saliternik-Givant S (1970) Salivary secretion of the male tick during copulation. J Insect Physiol 16:1945–1949

Feldman-Muhsam B, Borut S, Saliternik-Givant S, Eden C (1973) On the evacuation of sperm from the spermatophore of the tick, *Ornithodoros savignyi*. J Insect Physiol 19:951–962

Feldman-Muhsam B, Bachrach U, Ben-Joseph M (1980) The presence of spermine in the spermatophore of a soft tick (*Ornithodoros savignyi*). J Insect Physiol 26:407–414

Fernandez NA (1981) Argentine fauna of oribatid mites: 3. Spermatophore of *Epilohmannia maurii*. Acarologia (Paris) 22:239–242

Ferrari F (1978) Spermatophore placement in the copepod *Euchaeta norvegica* Boeck 1972 from Deepwater Dumpsite 106. Proc Biol Soc Wash 91:509–521

Ferrari FD, Bowman TE (1980) Pelagic copepods of the family Oithonidae (Cyclopoida) from the east coasts of Central and South America. Smithson Contrib Zool no 312:1–27

Ferro DN, Akre RD (1975) Reproductive morphology and mechanics of mating of the codling moth, *Laspeyresia pomonella*. Ann Entomol Soc Am 68:417–424

Fields WG (1965) The structure, development, food relations, reproduction, and life history of the squid *Loligo opalescens* Berry. Calif Fish and Game 131:1–108

Fields WG, Thompson KA (1976) Ultrastructure and functional morphology of spermatozoa of *Rossia pacifica* (Cephalopoda, Decapoda). Can J Zool 54:908–932

Finlayson LH (1949) The life-history and anatomy of *Lepinotus patruelis* Pearman (Psocoptera: Atropidae). Proc Zool Soc Lond 119:301–323

Fiori G (1954) Morfologia addominale, anatomia ed istologia degli apparati genitali di *Pimelia angulata confalonierii* Grid. (Coleoptera: Tenebrionidae) e formazione dello spermatoforo. Boll Ist Entomol Bologna 20:377–422

Fisher TW (1959) Occurrence of spermatophores in certain species of *Chilocorus* (Coleoptera: Coccinellidae). Pan-Pacif Ent 35:205–208

Flanders SE (1939) Environmental control of sex in hymenopterous insects. Ann Entomol Soc Am 32:11–26

Flanders SE (1945) The role of the spermatophore in the mass propagation of *Macrocentrus ancylivorus*. J Econ Entomol 38:323–327

Fleminger A (1979) *Labidocera* (Copepoda: Calanoida): New and poorly known Caribbean species with a key to species in the western Atlantic. Bull Mar Sci 29:170–190

Flügel H (1977) Ultrastructure of the spermatophores of *Siboglinum ekmani* Jägersten (Pogonophora). Nature 269:800–801

Folliot R, Maillet P-L (1970) Ultrastructure de la spermiogenèse et du spermatozoïde de divers insectes homoptères. In: Baccetti B (ed) Comparative spermatology. Academia Nazionale dei Lincei, Rome. Academic Press, New York London, pp 289–300

Foot K (1898) The cocoons and eggs of *Allolobophora foetida*. J Morph 14:481–505

Fort G (1937) Le spermatophore des Céphalopodes. Etude du spermatophore d'*Eledone cirrhosa* (Lamarck 1799). Bull Biol Fr Belg 71:357–373

Fort G (1941) Le spermatophore des Céphalopodes. Etude du spermatophore d'*Eledone moschata* (Lamarck 1799). Bull Biol Fr Belg 75:249–256

Francke OF (1979) Spermatophores of some North American scorpions (Arachnida, Scorpiones). J Arachnol 7:19–32

Franzén Å (1956) On spermiogenesis, morphology of the spermatozoa, and biology of reproduction among invertebrates. Zool Bidr, Uppsala 31:355–482

Franzén Å (1967) Spermiogenesis and spermatozoa of the Cephalopoda. Ark Zool (Stockh) 19:323–334

Franzén Å (1970) Phylogenetic aspects of the morphology of spermatozoa and spermiogenesis. In: Baccetti B (ed) Comparative spermatology. Academia Nazionale dei Lincei, Rome and Academic Press, New York London, pp 29–46

Franzén Å, Ahlfors K (1980) Ultrastructure of spermatids and spermatozoa in *Phoronis*, Phylum Phoronida. J Submicr Cytol 12:585–597

Frenk E, Happ GM (1976) Spermatophore of the mealworm beetle: Immunochemical characteristics suggest affinities with male accessory gland. J Insect Physiol 22:891–895

Fretter V (1953) The transference of sperm from male to female prosobranchs, with reference, also, to the pyramidellids. Proc Linn Soc Lond 164:217–224

Fretter V, Graham A (1962) British prosobranch molluscs, their functional anatomy and ecology. Ray Society Publ 114. Ray Society, London, p 755

Fretter V, Graham A (1964) Reproduction. In: Wilbur KM, Yonge CM (eds) Physiology of Mollusca, vol 1. Academic Press, New York, pp 127–164

Frick K, Wilson RF (1982) Some factors influencing the fecundity and flight potential of *Bactra verutana*. Environ Entomol 11:181–186

Friedel T, Gillott C (1977) Contribution of male-produced proteins to vitellogenesis in *Melanoplus sanguinipes*. J Insect Physiol 23:145–151

Froesch D, Marthy H-J (1975) The structure and function of the oviducal gland in octopods (Cephalopoda). Proc R Soc Lond [Biol] 188:95–101

Gabbutt PD (1954) Notes on the mating behaviour of *Nemobius sylvestris* (Bosc) (Orth., Gryllidae). Brit J Anim Behav 2:84–88

Gadzama NM, Happ GM (1974) The structure and evacuation of the spermatophore of *Tenebrio molitor* L. (Coleoptera: Tenebrionidae). Tissue & Cell 6:95–108

Gardiner DM (1978) The origin and fate of spermatophores in the viviparous teleost *Cymatogaster aggregata* (Perciformes: Embiotocidae). J Morph 155:157–172

184

Gasco F (1881) Les amours des axolotls. Zool Anz 4:313–316, 328–334

Gascoigne T (1956) Feeding and reproduction in the Limapontiidae. Trans R Soc Edinburgh 63:129–151

Gates GE (1979) Contributions to a revision of the earthworm family Lumbricidae: XXIII. The genus *Dendrodrilus* Omodeo, 1956 in North America. Megadrilogica 3:151–162

Gauld DT (1957) Copulation in calanoid copepoda. Nature 180:510

Gauld DT (1966) The swimming and feeding of planktonic copepoda. In: Barnes H (ed) Some contemporary studies in marine science. Allen and Unwin, London, pp 313–334

Gavrilov K (1955) Über die uniparentale Vermehrung von *Paranadrilus*. Zool Anz 155:302–306

George JA, Howard MG (1968) Insemination without spermatophores in the oriental fruit moth, *Grapholitha molesta* (Lepidoptera: Tortricidae). Can Entomol 100:190–192

Gerber GH (1970) Evolution of the methods of spermatophore formation in pterygotan insects. Can Entomol 102:358–362

Gerber GH, Church NS (1976) The reproductive cycles of male and female *Lytta nuttalli* (Coleoptera: Meloidae). Can Entomol 108:1125–1136

Gerber GH, Church NS, Rempel JG (1971) The structure, formation, histochemistry, fate, and functions of the spermatophore of *Lytta nuttalli* Say (Coleoptera: Meloidae). Can J Zool 49:1595–1610

Gerhardt U (1913) Copulation und Spermatophoren von Grylliden und Locustiden. I. Zool Jahrb Syst 35:415–532

Gerhardt U (1914) Copulation und Spermatophoren von Grylliden und Locustiden. II. Zool Jahrb Syst 37:1–64

Gerhardt U (1921) Neue Studien über Copulation und Spermatophoren von Grylliden und Locustiden. Acta Zool (Stockh) 2:293–327

Gharagozlou-van Ginneken ID (1978) Sécrétion et organisation de la paroi stratifiée du spermatophore chez quelques Copépodes: Ultrastructure et cytochimie. Cytobiologie 18:231–243

Ghirardelli E (1968) Some aspects of the biology of the chaetognaths. Adv Mar Biol 6:271–375

Ghiselin MT (1963) On the functional and comparative anatomy of *Runcina setoensis* Baba, an opisthobranch gastropod. Publ Seto Mar Biol Lab 11:219–228

Ghiselin MT (1966) Reproductive function and the phylogeny of opisthobranch gastropods. Malacologia 3:327–378

Giglioli MEC, Mason GF (1966) The mating plug in anopheline mosquitoes. Proc R Ent Soc Lond (A) 41:123–129

Gilbert LE (1976) Postmating female odor in *Heliconius* butterflies: a male-contributed antiaphrodisiac? Science 193:419–420

Gilbert PW, Heath GW (1955) The functional anatomy of the claspers and siphon sacs in the spiny dogfish (*Squalus acanthias*) and smooth dogfish (*Mustelus canis*). Anat Rec 121:433

Gilbert PW, Heath GW (1972) The clasper-siphon sac mechanism in *Squalus acanthias* and *Mustelus canis*. Comp Biochem Physiol 42A:97–119

Ginzburg AS (1968) Fertilization in fishes and the problem of polyspermy (Oplodotvoranie u ryb i problema polispermii). Academy of Sciences of the USSR, Moscow. English translation (Detlaf TA ed). Israel Program for Scientific Translations, Jerusalem 1972, p 366

Godlewski E (1910–1914) Physiologie der Zeugung. In: Winterstein H (ed) Handbuch der vergleichenden Physiologie, vol 3, part 2. Gustav Fischer Verlag, Jena, S. 457–775

Goodrich ES (1930) On a new hermaphrodite syllid. QJ Micr Sci 73:651–666

Gourbault N, Renaud-Mornant J (1982) Un nouveau mode de fécondation, par spermatophore, chez les Nématodes. CR Acad Sci Paris 294 (Ser III):285–287

Gourbault N, Renaud-Mornant J (1983) Système reproducteur d'un Nématode marin à fécondation par spermatophore. Revue Nematol 6:51–56

Greenfield MD, Coffelt JA (1983) Reproductive behaviour of the lesser waxmoth, *Achroia grisella* (Pyralidae: Galleriinae): signalling, pair formation, male interaction, and mate guarding. Behaviour 8:287–316

Greenwood JG (1972) The male reproductive systems and spermatophore formation in *Pagurus noveazealandiae* (Dana) (Anomura: Paguridae). J Nat Hist 6:561–574

Gregory GE (1965a) The formation and fate of the spermatophore in the African migratory locust, *Locusta migratoria migratorioides* Reiche and Fairmaire. Trans R Ent Soc Lond 117:33–66

Gregory GE (1965b) On the initiation of spermatophore formation in the African migratory locust, *Locusta migratoria migratorioides* Reiche and Fairmaire. J Exp Biol 42:423–435

185

Greve W (1974) Planktonic spermatophores found in a culture device with spionid polychaetes. Helg Wiss Meeresunters 26:370–374

Grier HJ (1975) Aspects of germinal cyst and sperm development in *Poecilia latipinna* (Teleostei: Poeciliidae). J Morph 146:229–250

Griffin LE (1902) The anatomy of *Nautilus pompilius*. Mem Nat Acad Sci 8:103–197

Griffith DA, Boczek J (1977) Spermatophores of some acaroid mites (Astigamata: Acarina). Int J Insect Morphol Embryol 6:231–238

Grimes MJ, Happ GM (1980) Fine structure of the bean-shaped accessory gland in the male pupa of *Tenebrio molitor* (Coleoptera: Tenebrionidae). Int J Insect Morphol Embryol 9:281–296

Gruber A (1879) Beiträge zur Kenntnis der Generationsorgane der freilebenden Copepoden. Z Wiss Zool 32:407–442

Gundevia HS, Ramamurty PS (1977) The male accessory glands and spermatophore in *Hydrophilus olivaceus* (Polyphaga-Coleoptera). Z Mikrosk-Anat Forsch (Leipz) 91:475–492

Gupta AP (1965) The digestive and reproductive systems of the Meloidae (Coleoptera) and their significance in the classification of the family. Ann Entomol Soc Am 58:442–474

Gupta AP (1967) Further studies on the internal anatomy of the Meloidae. III. The digestive and reproductive systems as bases for tribal designation of *Pseudomeloe miniaceomaculata* (Blanchard) (Coleoptera: Meloidae). JNY Entomol Soc 75:93–99

Gupta BL (1964) Cytological studies of the male germ cells in some freshwater ostracods and copepods. Ph D dissertation, University of Cambridge

Gupta BL (1968) Aspects of cell motility in the non-flagellate spermatozoa of fresh water ostracods. In: Aspects of cell motility. 22nd Symp Soc Exp Biol, pp 117–129

Gupta PD (1946) On the structure and formation of spermatophore in the cockroach, *Periplaneta americana* (Linn). Ind J Entom 8:79–84

Gurney AB (1947) A new species of *Pristoceuthopilus* from Oregon and remarks on certain special glands of Orthoptera. J Wash Acad Sci 37:430–435

Gwynne DT (1981) Sexual difference theory: Mormon crickets show reversal in mate choice. Science 213:779–780

Gwynne DT (1982) Mate selection by female katydids (Orthoptera: Tettigoniidae, *Conocephalus nigropleurum*). Anim Behav 30:734–738

Gwynne DT (1983) Male nutritional investment and the evolution of sexual differences in Tettigoniidae and other Orthoptera. In: Gwynne DT, Morris G (eds) Orthopteran mating systems: Sexual competition in a diverse group of insects. Westview Press, Boulder Colorado, pp 337–366

Gwynne DL (1984) Courtship feeding increases female reproductive success in bushcrickets. Nature 307:361–363

Gwynne DL, Bowen BJ, Codd CG (1984) The function of the katydid spermatophore and its role in fecundity and insemination (Orthoptera: Tettigoniidae). Aust J Zool 32:15–23

Haacker U (1968) Sperma-Transport beim Kugeltausendfüßler (*Sphaerotherium*). Naturwiss 55:89

Haacker U (1969) Spermaübertragung von *Glomeris*. Naturwiss 56:467

Haines LC (1981) Changes in colour of a secretion in the reproductive tract of adult males of *Spodoptera littoralis* (Lepidoptera: Noctuidae) with age and mated status. Bull Entol Res 71:591–598

Halffter G, Lopez Y (1977) Development of the ovary and mating behavior in *Phanaeus*. Ann Entomol Soc Am 70:203–213

Hallbèrg E (1984) The spermathecal complex in *Ips typographus* (L.): Differentiation of the spermathecal gland related to age and reproductive state. J Insect Physiol 30:197–202

Halliday TR (1976) The libidinous newt: An analysis of variations in the sexual behaviour of the male smooth newt, *Triturus vulgaris*. Anim Behav 24:386–414

Halliday TR (1977) The courtship of European newts. An evolutionary perspective. In: Taylor DH, Gutman SI (eds) The reproductive biology of amphibians. Plenum Press, New York, pp 185–232

Hammer RM (1978) Scanning electron microscope study of the spermatophore of *Acartia tonsa* (Copepoda: Calanoida). Trans Am Microsc Soc 97:386–389

Hamon M (1939a) Les constituánts chimiques des enveloppes des spermatophores de Céphalopodes. CR Acad Sci Paris 208:387–389

Hamon M (1939b) Charactérisation de quelques acides α-aminés entrant dans la constitution des spermatophores des Céphalopodes. CR Acad Sci Paris 208:835–837

Hanson D, Mann T, Martin AW (1973) Mechanism of the spermatophoric reaction in the giant octopus of the north Pacific, *Octopus dofleini martini*. J Exp Biol 58:711–723

186

Hanson J, Randall JT, Bayley ST (1952) The microstructure of the spermatozoa of the snail *Viviparus*. Exp Cell Res 3:65–78

Happ GM, Happ CM (1975) Fine structure of the spermatheca of the mealworm beetle (*Tenebrio molitor* L.). Cell Tissue Res 162:253–269

Haq MA, Adolph C (1981) Spermatophore deposition and transfer in *Pelokylla malabarica* (Acari: Oribatei). Entomon 6:135–142

Harmer SF (1889) Notes on the anatomy of Dinophilus. J Mar Biol Assoc UK 1:119–143

Hartmann R (1970) Experimentelle und histologische Untersuchungen der Spermatophorenbildung bei der Feldheuschrecke *Gomphocarus rufus* L. (Orthoptera, Acrididae). Z Morph Tiere 68:140–176

Hartmann R, Loher W (1974) Control of sexual behavior pattern ‚secondary defense' in the female grasshopper *Chorthippus curtipennis*. J Insect Physiol 20:1713-1728

Hartree EF, Mann T (1959) Plasmalogen in ram semen and its role in sperm metabolism. Biochem J 71:423–434

Hartree EF, Mann T (1961) Phospholipids in ram semen: metabolism of plasmalogen and fatty acids. Biochem J 80:464–476

Hatanaka H (1979) Spawning season of common octopus (*Octopus vulgaris*) off the northwest coast of Africa. Bull Jpn Soc Sci Fish 45:805–810 (In Japanese with English summary)

Haven N (1977a) Cephalopoda: Nautiloidea. In: Giese AC, Pearse JS (eds) Reproduction of marine invertebrates, vol 4. Academic Press, New York San Francisco London, pp 227–241

Haven N (1977b) The reproductive biology of *Nautilus pompilius* in the Philippines. Mar Biol (Berl) 42:177–184

Hayward TL (1981) Mating and depth distribution of an oceanic copepod. Limnol Oceanogr 26:374–377

Heberer G (1926) Zur Kenntnis der männlichen Generationsorgane der Cyclopiden. Zool Anz Suppl 2:141–148

Heberer G (1932) Untersuchungen über Bau und Funktion der Genitalorgane der Copepoden. I. Der männliche Genitalapparat der calanoiden Copepoden. Zeit Mikr-Anat Forsch 31:250–424

Heberer G (1937) Weitere Ergebnisse über Bildung und Bau der Spermatophoren und Spermatophorenkoppelapparate bei calanoiden Copepoden. Verh Dtsch Zool Ges 39:86–93

Heberer G (1955a) Physiologischer Spermiendualismus bei harpaticoiden Copepoden. Naturwiss 42:261–263

Heberer G (1955b) Die Spermatophoren der Corycaeidae. Biol Zbl 74:9–10, 555–565

Heller J, Chojnacki T, Piechowska M (1960) On pyrophosphate in the hawk moth *Celerio euphorbiae*. Acta Biochim Pol 7:187–192

Hendricks DE, Graham HM, Fernandez AT (1970) Mating of female tobacco budworms and bollworms collected from light traps. J Econ Entomol 63:1228–1231

Henneberry TJ, Shorey HH, Kishaba AN (1966) Mating frequency and copulatory aberrations, response to female sex pheromone, and longevity of male cabbage loopers treated with Tepa. J Econ Entomol 59:573–576

Herrick FH (1911) Natural history of the American lobster. Bull US Bur Fish 29:149–408

Hertling H (1930) Über ein Hedylide von Helgoland und Bemerkungen zur Systematik der Hedyliden. Wiss Meeresunters, Abt. Helgoland 18:1–10

Hevers J (1978) Zur Sexualbiologie der Gattung *Unionicola* (Hydrachnellae, Acari). Zool Jahrb Syst 105:33–64

Hewer HR (1934) Studies in *Zygaena* (Lepidoptera). Part II. The mechanism of copulation and the passage of sperm in the female. Proc Zool Soc Lond 1934, 513–527

Heymons R (1901) Biologische Beobachtungen an asiatischen Solifugen. Abhandl K Preuss Akad Wiss Berlin, 1901, 1–65

Hinsch G, Walker MH (1974) The vas deferens of the spider crab, *Libinia emarginata*. J Morph 143:1–19

Hinton HE (1964) Sperm transfer in insects and the evolution of haemocoelic insemination. In: Highnam KC (ed) Insect reproduction. Symp R Entomol Soc Lond, no 2, pp 95–107

Hirai K, Shorey HH, Gaston LK (1978) Competition among courting male moths: the male-to-male inhibitory pheromone. Science 202:644–645

Hoffmann KH, Behrens W (1982) Free ecdysteroids in adult male crickets, *Gryllus bimaculatus*. Physiol Entomol 7:269–279

Hohorst W (1936) Die Begattungsbiologie der Grille *Oecanthus pellucens Scopoli*. Z Morph Oekol Tiere 32:227–275

Holt GG, North DT (1970) Effects of gamma irradiation on the mechanisms of transfer in *Trichoplusia ni*. J Insect Physiol 16:2211–2222

Hopkins CCE (1978) The male genital system, and spermatophore production and function in *Euchaeta norvegica* Boeck (Copepoda: Calanoida). J Exp Mar Biol Ecol 35:197–232

Hopkins CCE (1982) The breeding biology of *Euchaeta norvegica* (Copepoda: Calanoida) in Loch Etive, Scotland, UK: Assessment of breeding intensity in terms of seasonal cycles in the sex ratio, spermatophore attachment and egg-sac production. J Exp Mar Biol Ecol 60:91–102

Hopkins CCE, Machin D (1977) Patterns of spermatophore distribution and placement in *Euchaeta norvegica* (Copepoda: Calanoida). J Mar Biol Assoc UK 57:113–131

Horstmann E (1968) Die Spermatozoen von *Geophilus linearis* Koch (Chilopoda). Z Zellforsch 89:410–429

Horstmann E (1970) The acrosome complex of the diplopode *Spirostreptus* sp. In: Baccetti B (ed) Comparative spermatology. Academia Nazionale dei Lincei; Rome. Academic Press, New York London, pp 255–262

Horstmann E, Breucker H (1969a) Spermatozoen und Spermiohistogenese von *Graphidostreptus* spec. (Myriapoda, Diplopoda). I. Die reifen Spermatozoen. Z Zellforsch 96:505–520

Horstmann E, Breucker H (1969b) Spermatozoen und Spermiohistogenese von *Spirostreptus* spec. (Myriapoda, Diplopoda). II. Die Spermiohistogenese. Z Zellforsch 99:153–184

Houbrick R (1980) Observations of the anatomy and life history of *Modulus modulus* (Prosobranchia: Modulidae). Malacologia 20:117–142

Houbrick R (1981) Anatomy and systematics of *Gourmya gourmyi* (Prosobranchia: Cerithiidae), a tethyan relict from the soutwest Pacific. Nautilus 95:2–11

Howell JF, Hutt RB, Hill WB (1978) Codling moth: Mating behavior in the laboratory. Ann Entomol Soc Am 71:891–896

Hoyle WE (1885) Diagnoses of new species of Cephalopoda collected by HMS 'Challenger'. Part 1. The Octopoda. Ann Mag Nat Hist (5) 15:222–236

Hoyle WE (1886) Cephalopoda. In: Report of the scientific results of the voyage of HMS Challenger during the years 1873–1876. Sect V (Zoology), vol 16 (I):1–245. The whole report, comprising 50 volumes, originally published in London, 1880–1895, has been reprinted in 1965, by the Johnson Reprint Corporation, New York London

Hoyle WE (1907) Presidential address to the zoological section (D) of the British Association for the Advancement of Science, Leicester 1907. Br Assoc Adv Sci Rep 77:520–539

Hryniewiecka-Szyfter Z, Redziniak E (1976) The localization of the lysosomal enzymes in the bursa copulatrix of the snail *Helix pomatia* L. Bull Soc Amis Sci Lett Poznan (Poland), Ser D Sci Biol 16:125–134

Hubendick B (1947) Phylogenie und Tiergeographie der Siphonariidae. Zur Kenntnis der Phylogenie in der Ordnung Basommatophora und des Ursprungs der Pulmonatengruppe. Zool Bidr Uppsala 24:1–216

Hudson CT (1883) On *Asplanchna Ebbesbornii* nov. sp. JR Micr Soc 3:621–628

Hudson CT (1891) The president's address on some doubtful points in the natural history of Rotifera. J Roy Micr Soc (1891):6–18

Hughes TE (1959) Mites or Acari. University of London, Athlone Press, London, p 225

Huignard J (1969) Action stimulatrice du spermatophore sur l'ovogenese chez *Acanthoscelides obtectus* Say (Insecte Coleoptère). CR Acad Sci Ser (D) Sci Natur (Paris) 268:2938–2940

Huignard J (1974) Influence de la copulation sur la fonction reproductrice femelles chez *Acanthoscelides obtectus* (Coleoptère, Bruchidae). Ann Sci Nat Zool Biol Anim 16:361–434

Huignard J (1983) Transfer and fate of male secretions deposited in the spermatophore of females of *Acanthoscelides obtectus* Say (Coleoptera Bruchidae). J Insect Physiol 29:55–63

Huignard J, Lamy M (1972) Etude de l'evolution des proteines du spermatophore apres l'accouplement chez *Acanthoscelides obtectus* Say (Insecte Coleoptère). CR Acad Sci Ser (D) Sci Nat 275:1067–1070

Huignard J, Quesneau-Thierry A, Barbier M (1977) Isolement, action biologique et evolution des substances paragoniales contenues dans le spermatophore d'*Acanthoscelides obtectus* (Coleoptère). J Insect Physiol 23:351–357

Hunt S, Legg G (1971) Characterization of the structural protein component in the spermatophore of the pseudoscorpion *Chthonius ischnocheles* (Hermann). Comp Biochem Physiol Biochem 40B:475–479

Ihering H (1892) Morphologie und Systematik des Genitalapparates von *Helix*. Z Wiss Zool 54:425–520

Iriki S (1941) The two sperm types in the silkworm and their function. Zool Mag 53:123–124

Itaya PW (1979) Electron microscopic investigation of the formation of spermatophores of *Armadillidium vulgare*. Cell Tissue Res 196:95–102

Itgi NB, Biradar VK, Mathad SB (1982) Primary stimulus for oviposition and egg production in the house cricket *Gryllodes sigillatus*. Z Angew Entomol 94:35–41

Ivanov AV (1957) Neue Pogonophoren aus dem nord-westlichen Teil des Stillen Oceans. Zool Jahrb Syst 85:431–500

Ivanov AV (1963) Pogonophora. English edition of Ivanov's (1960) ‚Pogonophores'. Akad Nauk USSR, translated from the Russian and edited by DB Carlisle; the spermatophores of *Siboglinum fedotoui* and *Lamellisabella johanssoni* are illustrated in Fig. 65, on p 97. Academic Press, London, p 479

Iwanowa SA (1925) Zur Frage über die Spermatophorbefruchtung bei den Acridodea (*Locusta migratoria* L.). Zool Anz 65:75–86

Jaana H (1982) The ultrastructure of the epithelial lining of the male genital tract and its role in spermatozeugma formation in *Tubifex hattai* Nomura (Annelida, Oligochaeta). Zool Anz 209:159–176

Jacoby CA, Youngbluth MJ (1983) Mating behavior of three species of *Pseudodiaptomus* (Copepoda: Calanoida). Mar Biol (Berl) 76:77–86

Jägersten G (1934a) Studien über den histologischen Bau der männlichen Geschlechtsorgane und die Ausbildung des Spermiums bei *Myzostomum*. Zool Bidr Uppsala 15:1–22

Jägersten G (1934b) Ueber den Befruchtungsmechanismus der Myzostomiden. Arch Zool (Stockh) 30B:1–4

Jägersten G (1939) Über die Morphologie und Physiologie des Geschlechtsapparats und den Kopulationsmechanismus der Myzostomiden. Zool Bidr Uppsala 18:163–242

Jägersten G (1944) Über den Bau des Kopulationsapparates und den Kopulationsmechanismus bei *Dinophilus*. Zool Bidr Uppsala 22:61–86

Jägersten G (1952) Studies on the morphology, larval development, and biology of *Protodrilus*. Zool Bidr Uppsala 29:427–511

Jamieson BGM (1978) *Rhyacodrilus arthingtonae*, a new species of freshwater oligochaetae (Tubificidae) from North Stradbroke Island, Queensland (Australia). Proc RS Queensl 89:34–44

Jenkins BW (1983) Redescription and relationship of *Siphonaria zelandica* Quoy and Gaimard to *S. australis* Quoy and Gaimard with a description of *S. propria* sp. nov. (Mollusca: Pulmonata: Siphonariidae). J Malac Soc Aust 6:1–35

Jewell D (1931) Reproduction in *Goniobasis*. Nautilus 44:115–119

Joly J (1960) La conservation des spermatozoides et les particularités histophysiologiques du réceptacle séminale chez la salamandre *Salamandra salamandra taeniata*. CR Acad Sci Paris 250:2269–2271

Joly J (1966) Sur l'éthologie sexuelle de *Salamandra salamandra*. Z Tierpsychol 23:8–27

Joosse ENG, Brugman FA, Veld CJ (1973) The effects of constants and fluctuating temperatures on the production of spermatophores and eggs in populations of *Orchesella cincta* (Linne), (Collembola, Entomobryidae). Neth J Zool 23:488–502

Jordan EO (1891) The spermatophores of *Diemyctylus*. J Morph 5:263–270

Jouin C (1968) Sexualité et biologie de la reproduction chez *Mesonerilla* Remane et *Meganerilla* Boaden (Archiannélides Nerillidae). Cah Biol Mar 9:31–52

Jouin C (1970) Recherches sur les Protodrilidae (Archiannélides). I. Etude morphologique et systématique du genre *Protodrilus*. Cah Biol Mar 11:367–434

Juberthie C, Lopez A, Kovoor J (1981) Spermiogenesis and spermatophore in *Telema tenella* (Araneae: Telemidae); an ultrastructural study. Int J Invertebr Reprod 3:181–192

uberthie-Jupeau L (1956) Existence de spermatophores chez les Symphyles. CR Acad Sci Paris 243:1164–1166

uberthie-Jupeau L (1963) Recherches sur la réproduction et la mue chez les Symphyles. Arch Zool Exp Gen 102:1–172

189

Kato K (1952) On the development of Myzostome. Sci Rep Saitama Univ B1:1–16

Katona SK (1973) Evidence for sex pheromones in planktonic copepoda. Limnology and Oceanography 81:574–583

Kaulenas MS (1976) Regional specialization for export protein synthesis in the male cricket accessory gland. J Exp Zool 195:81–86

Kearn GC (1970) The production, transfer, and assimilation of spermatophores by *Entobdella soleae*, a monogenean skin parasite of the common sole. Parasitology 60:301–311

Kew HW (1912) On the pairing of the pseudoscorpions. Proc Zool Soc Lond 25:376–390

Khalifa A (1949 a) The mechanism of insemination and the mode of action of the spermatophore in *Gryllus domesticus*. QJ Micr Sci 90:281–292

Khalifa A (1949 b) Spermatophore production in Trichoptera and some other insects. Trans R Ent Soc Lond 100:449–479

Khalifa A (1950 a) Spermatophore production in *Galleria mellonella* L. (Lepidoptera). Proc R Ent Soc Lond (A) 25:33–42

Khalifa A (1950 b) Spermatophore production in *Blattella germanica* L. (Orthoptera,. Blattidae). Proc R Ent Soc (A) 25:53–61

Kirk H (1938) Notes on the breeding habits and early development of *Dolichoglossus otagoensis* Benham. Trans Proc R Soc NZ 68:49–50

Klaver E (1975) Some aspects of the reproductive biology of *Bourletiella (Cassagnaudiella) pruinosa* (Tullberg 1871) (Collembola: Sminthuridae). Bull Zool Mus Univ Amsterdam 4:178–186

Klingel H (1957 j) Indirekte Spermatophorenübertragung beim Skolopender (*Scolopendra cingulata* Latreille). Naturwiss 44:338

Klingel H (1959) Indirekte Spermatophorenübertragung bei Geophiliden (Hundertfüsser, Chilopoda). Naturwiss 46:632–633

Klingel H (1960) Vergleichende Verhaltensbiologie der Chilopoden *Scutigera coleoptrata* („Spinnenassel") und *Scolopendra cingulata* Latreille (Skolopender). Z Tierpsychol 17:11–30

Klingel H (1963) Paarungsverhalten bei Pedipalpen (*Telyphonus caudatus* L., Holopeltidia, Uropygi und *Sarax sarawakensis* Simon, Charontidae, Amblypygi). Verh Dtsch Zool Ges 1962, Zool Anz Suppl 26:452–459

Kohnert R, Storch V (1984) Zur Ultrastruktur der eupyrenen Spermien und Spermiogenese von *Neritina communis* (Quoy and Gaimard) (Mollusca, Gastropoda). Zool Anz (Jena) 212:85–94

Kokwaro ED, Odhiambo TR (1981) Spermatophore of the tsetse *Glossina morsitans morsitans:* An ultrastructural study. Insect Sci Application 1:185–190

Kokwaro ED, Odhiambo TR, Murithi JK (1981) Ultrastructural and histochemical study of the spermatheca of the tsetse *Glossina morsitans morsitans* Westwood. Insect Sci Application 2:135–143

Kölliker A (1841) Beiträge zur Kenntnis der Geschlechtsverhältnisse und der Samenflüssigkeit wirbelloser Thiere, nebst einem Versuch über das Wesen und die Bedeutung der sogenannten Samenthiere. W. Logier, Berlin, S. 88

Kölliker A (1845) Some observations upon the structure of two new species of Hectocotyle parasitic upon *Tremoctopus violaceus* Delle Chiaje, and *Argonauta argo* L., with an exposition of the hypothesis that those Hectocotylae are the males of the Cephalopoda on which they are found. Proc Linn Soc Lond 1:237–238. The same title (and virtually the same text) appears in a "memoir" communicated the same year (1845), by The Lord Bishop of Norwich, President, to the Linnean Society, on behalf of Prof. Kölliker (Ann Mag Nat Hist 16:414–415)

Kölliker A (1848) Beiträge zur Kenntniss niederer Tiere. Z Wiss Zool 1:1–37

Kölliker A (1853) Nachwort (Postscript to the paper by H. Müller). Z Wiss Zool 4:35

Kollmann J (1876) Die Cephalopoden in der zoologischen Station des Dr. Dohrn. Z Wiss Zool 26:1–23

Kooda-Cisco MJ, Talbot P (1982) A structural analysis of the freshly extruded spermatophore from the lobster, *Homarus americanus*. J Morph 172:193–207

Kooda-Cisco MJ, Talbot P (1983) A technique for electrically stimulating extrusion of spermatophores from the lobster, *Homarus americanus*. Aquaculture 30:221–228

Korn W (1982) Zur Fortpflanzung von *Poecilochirus carabi* G. u. R. Canestrini 1882 (syn. *P. necrophori* Vitzt.) und *P. austroasiaticus* Vitzhum 1930 (Gamasina, Eugamasidae). Spixiana 5:261–288

Kulkarni CV (1940) On the systemic position, structural modifications, bionomics, and development of a remarkable new family of cyprinodont fishes from the province of Bombay. Rec Indian Mus 42:379–423

Kümmel G (1982) Zur Ultrastruktur von Spermatophoren der Hornmilbe *Oppia nitens* (Acari: Sarcoptiformes: Oribatei). Entomol Gen 7:301–311

Labine PA (1964) Population biology of the butterfly *Euphydryas editha*. I. Barriers to multiple insemination. Evolution 18:335–336

Laidlaw HH (1944) Artificial insemination of the queen bee (*Apis mellifera*), morphological basis and results. J Morph 74:429–465

Lamoral BH (1979) The scorpions of Namibia (Arachnida: Scorpionida). Ann Natal Mus 23:497–784

Landa V (1960) Origin, development, and function of the spermatophore in cockchafer (*Melolontha melolontha* L.). Acta Soc Entomol Čechosloveniae 57:297–316

Landa V (1961) Use of an artificial spermatophore in the study of activation of spermatozoa and development of spermatophores in the cockchafer. Nature 190:935–936

Lang A (1881) Der Bau von *Gunda segmentata* und die Verwandschaft der Plathelminthen mit Coelenteraten und Hirudineen. Mitteilungen Zool St Neapel 3, S. 222–224

Lang A (1884) Die Polycladen (Seeplanarien) des Golfes von Neapel und der angrenzenden Meeresabschnitte. Fauna und Flora des Golfes von Neapel 11:1–688

Lange K (1948) Monographie der Harpacticiden (2 vol, 1682 pp). Nordiska Bokhandeln, Stockholm Lund

Lankester ER (1871) On the structure and origin of the spermatophores, or sperm ropes, in two species of Tubifex. QJ Micr Sci 11:180–187

Lanzieri PD, Rezende HEB, Inada T (1972) Caracterizacao histoquimica da capsula de espermatoforo de *Protoglyptus lopesi* Rezende, Lanzieri + Inada, 1972 (Mollusca, Gastropoda, Pulmonata, Bulimulidae) [Histochemical characterization of the capsule of the spermatophore of *Protoglyptus lopesi* Rezende, Lanzieri + Inada 1972 (Mollusca, Pulmonata, Bolimulidae)]. Rev Bras Biol 32:389–390

Lasserre P (1975) Clitellata. In: Giese AC, Pearse JS (eds) Reproduction of marine invertebrates, vol 3 (Annelids and Echurians). Academic Press, New York San Francisco London, pp 215–275

Latzel R (1884) Die Myriopoden der österreich-ungarischen Monarchie, 2. Hälfte. A. Hölder, Wien, S. 48

Laubier-Bonichon A, Mangold K (1975) La maturation sexuelle chez les males d'*Octopus vulgaris* (Cephalopoda: Octopoda), en relation avec le reflexe photo-sexuel. Mar Biol (Berl) 29:45–52

Laviale ML (1964) Présence de spermatophores chez *Stylopauropus pedunculatus* (Lubb) (Pauropode, Myriapode). CR Acad Sci Paris 259:652–654

Leahy MG (1973) Oviposition of *Schistocerca gregaria* (Forskal) (Orthoptera: Acrididae) mated with males unable to transfer spermatophores. J Entomol (A) 48:79–84

Lee CM (1972) Structure and function of the spermatophore and its coupling device in the *Centropagidae* (Copepoda: Calanoida). Bull Mar Ecol 8:1–20

Legendre R, Lopez A (1978) Présence d'un spermatophore dans le genre *Apneumonella* (Araneae, Telemidae): Valeur systématique et problèmes de biologie sexuelle. Bull Soc Zool Fr 103:35–42

Legg G (1973) Spermatophore formation in the pseudoscorpion *Chthonius ischnocheles* (Chthoniidae). J Zool Proc Zool Soc Lond 170:367–394

Leigh-Sharpe WH (1920–1926) The comparative morphology of the secondary sexual characters of elasmobranch fishes. J Morph, I (1920) 34:245–265; II (1921) 35:359–380; III, IV, and V (1922) 36:191–243; VI and VII (1924) 39:553–577; VIII, IX, X and XI (1926) 42:307–358

Leopold RA (1976) The role of male accessory glands in insect reproduction. Ann Rev Entomol 21:199–221

Leppla NC, Carlysle TC, Guy RH (1975) Reproductive systems and the mechanics of copulation in *Plecia nearctica* Hardy (Diptera: Bibionidae). Int J Insect Morphol Embryol 4:299–305

Lespès C (1855a) Mémoire sur les spermatophores de grillons. Ann Sci Nat (4) Zool 3:366–377

Lespès C (1855b) Deuxieme note sur les spermatophores du Gryllus sylvestris. Ann Sci Mat (4) Zool 4:244–249

Lew A, Ball HJ (1980) Effect of copulation time on spermatozoan transfer of *Diabrotica virgifera* (Coleoptera, Chrysomelidae). Ann Entomol Soc Am 73:360–361

Leydig F (1849) Zur Anatomie von *Piscicola geometrica* mit theilweiser Vergleichung anderer einheimischer Hirudineen. Z Wiss Zool 1:103–134

Lind H (1973) The functional significance of the spermatophore and the fate of spermatozoa in the genital tract of *Helix pomatia* (Gastropoda: Stylommatophora). J Zool Proc Zool Soc Lond 169:39–64

191

Linley JR (1975) Termination of copulation and associated behaviour in *Culicoides melleus* (Coq.) (Diptera, Ceratopogonidae). Bull Entomol Res 65:143–156

Linley JR (1981 a) Ejaculation and spermatophore formation in *Culicoides melleus* (Coq.) (Diptera: Ceratopogonidae). Can J Zool 59:332–346

Linley JR (1981 b) Emptying of the spermatophore and spermathecal filling in *Culicoides melleus* (Coq.) (Diptera: Ceratopogonidae). Can J Zool 59:347–356

Linley JR, Adams C (1972) A study of the mating behaviour of *Culicoides melleus* (Coquillett) (Diptera: Ceratopogonidae). Trans R Ent Soc Lond 124:81–121

Linley JR, Hinds MJ (1975) Quantity of the male ejaculate influenced by female unreceptivity in the fly, *Culicoides melleus*. J Insect Physiol 21:281–285

Linley JR, Simmons KR (1981) Sperm motility and spermathecal filling in lower Diptera. Int J Invertebr Reprod 4:137–146

Linley JR, Simmons KR (1983) Quantitative aspects of sperm transfer in *Simulium decorum* (Diptera: Simuliidae). J Insect Physiol 29:581–584

Lipcius RN, Edwards ML, Herrnkind WF, Waterman SA (1983) In situ mating behaviour of the spiny lobster *Panulirus argus*. J Crustacean Biol 3:217–222

Lipovsky LJ, Byers GW, Kardos EH (1957) Spermatophores – the mode of insemination of chiggers (Acarina: Trombiculidae). J Parasit 43:256–262

Llewellyn J (1983) Sperm transfer in the monogenean gill parasite *Gastrocotyle trachuri*. Proc R Soc Lond [Biol] 219:439–446

Lofts B (1974) Reproduction. In: Lofts B (ed) Physiology of the Amphibia, vol 2. Academic Press, New York, pp 107–218

Loher W, Huber F (1966) Nervous and endocrine control of sexual behaviour in a grasshopper (*Gomphocerus rufus*, L., Acridinae). Symp Soc Exp Biol 20:381–400

Loher W, Ganjian I, Kubo I, Stanley-Samuelson D, Tobe SS (1981) Prostaglandins: Their role in egg laying of the cricket *Teleogryllus commodus*. Proc Natl Acad Sci USA 78:7835–7838

Longo FJ, Anderson E (1970) Structural and cytochemical features of the sperm of the cephalopod *Octopus bimaculatus*. J Ultrastr Res 32:94–106

Lopez A (1977) Sur un nouveau mode de reproduction chez les araignées: Existence de spermatophores chez *Telema tenella* Simon, 1882 (Telemidae). Bull Soc Zool Fr 102:261–266

Lowndes AG (1935) The sperms of fresh-water ostracods. Proc Zool Soc Lond 1:35–48

Lupu D (1974) La révision des répresentants de la famille des Arionidae (Gastropoda-Pulmonata) de Roumanie. Travaux du Muséum d'Histoire naturelle Grigore Antipa, Bucarest 15:31–44

Lupu D (1979) *Tanzaniella* gen. n. (Gastropoda, Neritidae) from the coast of Indian Ocean (Tanzania). Travaux du Muséum d'Histoire naturelle Grigore Antipa, Bucarest 20:15–19

Lupu D (1980) Étude morpho-anatomique comparée concernant la systématique de la famille des Clausiliidae (Gastropoda, Pulmonata) de Roumanie. Travaux du Muséum d'Histoire naturelle Grigore Antipa, Bucarest 22:341–357

Macdonald S, Llewellyn J (1980) Reproduction in *Acanthocotyle greeni*, new species (Monogenea) from the skin of *Raia* spp. at Plymouth, England, UK. J Mar Biol Ass UK 60:81–88

MacFarlane JH, Tsao CH (1974) The neural control of spermatophore formation and sperm transfer in the silkworm, *Bombyx mori* L. Ann Entomol Soc Am 67:759–761

Maillet PL (1959) Sur la reproduction des Homoptères Auchénorhynques. CR Acad Sci Paris 249:1945–1947

Malacarne G, Giacoma C (1980) Influence of testosterone on mating behaviour in the male crested newt (*Triturus cristatus carnifex*). Boll Zool 47:107–112

Malek SRA, Bawab FM (1971) Tanning in the spermatophore of a crustacean (*Penaeus trisulcatus*). Experientia 27:1098

Malek SRA, Bawab FM (1974 a) The formation of the spermatophore in *Penaeus kerathurus* (Forskal, 1775) (Decapoda, Penaeidae): I. The initial formation of a sperm mass. Crustaceana (Leiden) 26:273–285

Malek SRA, Bawab FM (1974 b) The formation of the spermatophore in *Penaeus kerathurus* (Forskal, 1775) (Decapoda, Penaeidae): II. The deposition of the main layers of the body and of the wing. Crustaceana (Leiden) 27:72–83

Mann KH (1953) The segmentation of leeches. Biol Rev 28:1–15

Mann KH (1962) Leeches (Hirudinea). Their structure, physiology, ecology, and embryology. Pergamon, Oxford London New York Paris, p 201

Mann T (1945) Studies on the metabolism of semen. 1. General aspects, occurrence, and distribution of cytochrome, certain enzymes, and coenzymes. Biochem J 39:451–458

Mann T (1946) Studies on the metabolism of semen. 3. Fructose as a normal constituent of seminal plasma. Site of formation and function of fructose in semen. Biochem J 40:481–491

Mann T (1948) Fructose content and fructolysis in semen. Practical application in the evaluation of semen quality. J Agric Sci 48:323–331

Mann T (1960) Serotonin (5-hydroxytryptamine) in the male reproductive tract of the spiny dogfish. Nature 188:941–942

Mann T (1963) 5-Hydroxytryptamine in the spermatophoric sac of the octopus. Nature 199:1066–1067

Mann T (1964) The biochemistry of semen and of the male reproductive tract. Methuen, London, p 493

Mann T (1967) Appraisal of endocrine testicular activity by chemical analysis of semen and male accessory secretions. Ciba Found Coll Endocr (Endocrinology of the testis) 16:233–244

Mann T (1975) Spermatophores. In: Duckett JG, Racey PA (eds) The biology of the male gamete. Biol J Linn Soc (Suppl 1) 7:417–422

Mann T, Lutwak-Mann C (1981) Male reproductive function and semen. Themes and trends in physiology, biochemistry, and investigative andrology. Springer-Verlag, Berlin Heidelberg New York, p 495

Mann T, Lutwak-Mann C (1982) Passage of chemicals into human and animal semen: mechanisms and significance. Critical Reviews in Toxicology (CRC Press) 11:1–14

Mann T, Lutwak-Mann C (1983) Adverse effects of chemicals on male reproductive function in mammals. In: Vouk VB, Sheehan PJ (eds) Methods for assessing the effects of chemicals on reproductive functions. SCOPE Publication no 20, Willey, Chichester New York Brisbane Toronto Singapore, pp 135–147

Mann T, Prosser LC (1963) Uterine response to 5-hydroxytryptamine in the clasper-siphon secretion of the spiny dogfish, *Squalus acanthias*. Biol Bull (Woods Hole) 125:384–385

Mann T, Martin AW, Thiersch JB (1966) Spermatophores and spermatophoric reaction in the giant octopus of the North Pacific, *Octopus dofleini martini*. Nature 211:1279–1282

Mann T, Martin AW, Thiersch JB (1969) Film in colour on ,Mating and spermatophoric reaction in the Giant Octopus of the North Pacific,' made in the Department of Zoology, University of Washington, Seattle; presented by T. Mann to the Royal Society of London at the meeting on 5 February 1970

Mann T, Martin AW, Thiersch JB (1970) Male reproductive tract, spermatophores, and spermatophoric reaction in the giant octopus of the North Pacific, *Octopus dofleini martini*. Proc R Soc Lond [Biol] 175:31–61

Mann T, Karagiannidis A, Martin AW (1973) Glycosidases in the spermatophores of the giant octopus, *Octopus dofleini martini*. Comp Biochem Physiol 44A:1377–1386

Mann T, Martin AW, Thiersch JB, Lutwak-Mann C, Brooks DE, Jones R (1974) D(−)-Lactic acid and D(−)-lactate dehydrogenase in octopus spermatozoa. Science 185:453–454

Mann T, Martin AW, Lutwak-Mann C, Thiersch JB (1977) Glycogenolysis in octopus spermatozoa. J Exp Mar Biol Ecol 27:155–159

Mann T, Jones R, Sherins R, Dufau M (1981 a) Observations on cyclic nucleotides in human semen. J Androl 5:243–248

Mann T, Martin AW, Thiersch JB (1981 b) Changes in the spermatophoric plasma during spermatophore development and during the spermatophoric reaction in the giant octopus of the North Pacific, *Octopus dofleini martini*. Mar Biol (Berl) 63:121–127

Manton SM (1938) Studies on Onychophora. IV. The passage of spermatozoa into the ovary in *Peripatopsis* and the early development of the ova. Phil Trans R Soc Lond [Biol] 228:421–441

Marchand W (1907) Studien über Cephalopoden. I. Der männliche Leitungsapparat der Dibranchiaten. Z Wiss Zool 86:311–415

Marchand W (1913) Studien über Cephalopoden. II. Über die Spermatophoren. Zoologica (Stuttgart) 26:171–200

Marcus E (1953) The Brazilian sand-opisthobranchia. Univ Sao Paolo, Fac Filos, Cienc Let Bol (Zool) 18:165–203

Marcus E, Marcus E (1960) On *Siphonaria hispida*. Univ Sao Paolo, Fac Filos, Cienc Let, Bol (Zool) 23:107–140

Markow TA, Ankney PF (1984) *Drosophila* males contribute to oogenesis in a multiple mating species. Science 224:302–303

Marquis NR, Fritz IB (1965) Effects of testosterone on the distribution of carnitine, acetylcarnitine, and carnitine acetyltransferase in tissues of the reproductive system of the male rat. J Biol Chem 240:2197–2200

Martin AW (1961) The carbohydrate metabolism of the molluscs. In: Martin AW (ed) Comparative physiology of carbohydrate metabolism in heterothermic animals. Univ Washington Press, Seattle, pp 35–64

Martin AW (1983) Excretion. In: Saleuddin ASM, Wilbur KM (eds) The Mollusca, vol 5, Physiology, Part 2. Academic Press, New York London, pp 353–405

Martin AW, Thiersch JB, Dott HM, Harrison RAP, Mann T (1970) Spermatozoa of the giant octopus of the North Pacific *Octopus dofleini martini*. Proc R Soc Lond [Biol] 175:63–68

Martin AW, Lutwak-Mann C, McIntosh JEA, Mann T (1973) Zinc in the spermatozoa of the giant octopus, *Octopus dofleini martini*. Comp Biochem Physiol 45A:227–233

Martin AW, Jones R, Mann T (1976) D(–)Lactic acid formation and D(–)lactate dehydrogenase in octopus spermatozoa. Proc R Soc Lond [Biol] 193:235–243

Martin AW, Catala-Stucki I, Ward PD (1978) The growth rate and reproductive behavior of *Nautilus macromphalus*. N Jb Geol Paläont Abh 156:207–225

Matthews DC (1956) The origin of the spermatophoric mass of sand crab, *Hippa pacifica*. QJ Microsc Sci 97:257–268

Matthews LH (1950) Reproduction in the basking shark *Cetorhinus maximus* (Gunner). Phil Trans R Soc Lond [Biol] 234:247–316

Maxwell WL (1974) Spermiogenesis of *Eledone cirrhosa* Lamarck (Cephalopoda, Octopoda). Proc R Soc Lond [Biol] 186:181–190

Maxwell WL (1975) Spermiogenesis of *Eusepia officinalis* (L.), *Loligo forbesi* (Steenstrup), and *Alloteuthis subulata* (L.) (Cephalopoda, Decopoda). Proc R Soc Lond [Biol] 191:527–535

Medem FG von (1945) Untersuchungen über die Ei- und Spermawirkstoffe bei marinen Mollusken. Zool Jahrb Abt Anat Ontog Tiere 61:1–44

Meisenheimer J (1907) Biologie, Morphologie und Physiologie des Begattungsvorgangs und der Eiablage von *Helix pomatia*. Zool Jahrb Abt Syst Geog Biol 25:461–502

Meisenheimer J (1921) Geschlecht und Geschlechter im Tierreiche. I. Die natürlichen Beziehungen. Fischer, Jena, S. 344–347

Meves F (1903) Über oligopyrene und apyrene Spermien und ihre Entstehung nach Beobachtungen an *Palludina* und *Pygaera*. Arch Mikr Anat 61:1–84

Meyer WT (1911) Die Spermatophore von Polypus (Octopus) vulgaris. Zool Anz 37:404–405

Mickoleit G (1974) Über die Spermatophore von *Boreus westwoodi* Hagen (Insecta, Mecoptera). Z Morph Tiere 77:271–284

Mika G (1959) Über das Paarungsverhalten der Wanderheuschrecke *Locusta migratoria* R. und F. und deren Abhängigkeit vom Zustand der inneren Geschlechtsorgane. Zool Beitr, Berl 4:153–203

Milne Edwards H (1840) Observations sur les Spermatophores des Mollusques céphalopodes, et sur la structure des Carinaires, des Dendrophyllies, etc; extraites d'une lettre de M. Milne Edwards, datée de Nice le 28 avril 1840. Ann Sci Nat (2) 13 (Zool):193–197. Same year, Milne Edwards's coworker W. Peters published a short communication „Über den Bau der Needhamschen Körper" in: Arch Anat Physiol Wiss Med Jhrg 1840:98–100

Milne Edwards H (1842) Observations sur la structure et les fonctiones de quelques Zoophytes, Mollusques et Crustacées des côtes de France. IV. Sur les spermatophores des Céphalopodes. Ann Sci Nat (2) 18 (Zool):331–350

Miskimen GW, Rodriguez NL, Nazario ML (1983) Reproductive morphology and sperm transport facilitation and regulation in the female sugarcane borer, *Diatraea saccharalis* (F.) (Lepidoptera: Crambidae). Ann Entomol Soc Am 76:248–252

Mitchell R (1958) Sperm transfer in the water-mite *Hydryphantes ruber* Geer. Amer Midland Naturalist 60:156–158

Montgomery TH (1903) Studies on the habits of spiders, particularly those of the mating period. Proc Acad N Sc Philadelphia 55:59–149

Morton JE, Younge CM (1964) Classification and structure of the Mollusca. In: Wilbur KM, Yonge CM (eds) Physiology of Mollusca, vol 1. Academic Press, New York, pp 1–58

Mouchet S (1931) Spermatophores des crustacés décapodes, anomures et brachyoures et castration paraitaire chez quelques pagures. Ann Sta Océanogr Salammbô 6:1–203

Mozolowski W, Mann T, Lutwak-Mann C (1931) Über den Ammoniakgehalt und die Ammoniakbildung im Muskel und deren Zusammenhang mit Funktions- und Zustandsänderung. IX. Mitteilung: Die Stellung der Ammoniakbildung in der Reihenfolge der chemischen Vorgänge im tätigen Muskel. Biochem Z 231:290–305

Mukerji S, Sarma PS (1951) Spermatheca in suckling louse. Nature 168:612

Müller F (Fridericus Mueller) (1844) De Hirudinibus circa Berolinum hucusque observatis (Dissertatio). Typis Fratrum Schlesinger, Berlin

Müller GW (1894) Die Ostracoden des Golfes von Neapel und der angrenzenden Meeresabschnitte. Fauna und Flora des Golfes von Neapel. Monogr 21, Berlin

Müller H (1853) Ueber das Männchen von *Argonauta argo* und die Hectocotylen. Z Wiss Zool 4:1–35

Mullins DE, Keil CB (1980) Paternal investment of urates in cockroaches. Nature 283:567–569

Musgrave AD (1937) The histology of the male and female reproductive organs of *Ephestia kuhniella* Zeller (Lepidoptera). 1. The young imagines. Proc Zool Soc Lond 107:337–364

Nabors RA, Pless CD (1981) Inherited sterility induced by gamma radiation in a laboratory population of the European corn borer (*Ostrinia nubilalis*). J Econ Entomol 74:701–702

Nagano Z (1957) Observations on the breeding habits in a fresh-water leech, *Herpobdella lineata* O.F. Müller. J Fac Sci Hokkaido Univ, Ser 6, 13:192–196

Navon A, Levinson HZ (1976) Oral application of D-glucoascorbic acid to adult *Spodoptera littoralis* (Boisduval) (Lepidoptera, Noctuidae) inducing sterility by spermatophore malformation. Bull Entomol Res 66:437–442

Navon A, Marcus R (1982) D-Isoascorbic acid fed to *Spodoptera littoralis* moths, induce sterility due to spermatophore malformation. J Insect Physiol 28:823–828

Navon A, Heinzel JNW, Mulligan K, Mullen JA, Sugumaran M, Lipke H (1983) The effect of D-isoascorbic acid on spermatophore composition in *Spodoptera littoralis*. Insect Biochem 13:247–250

Needham John Tubervill (1745) New microscopical discoveries. Printed for F Needham, over-against Gray's Inn in Holborn, London, pp 47–49 and 56–57

Nemoto T, Brinton E, Kamada K (1977) Reproduction and growth of deep sea *Thysanopoda* euphausiids. Bull Plankton Soc Jpn 24:36–42

Nielsen ET (1959) Copulation of *Glyptotendipes* (*Phytotendipes*) *paripes* Edwards. Nature 184:1252–1253

Nielsen JG, Jespersen Å, Munk O (1968) Spermatophores in Ophidioidea (Pisces, Percomorphi). Galathea Reports (Danish Science Press, Copenhagen) 9:239–254

Nixon DA (1966) The meso-inositol concentration in the blood and tissue of some members of the Cephalopoda. Publ Staz Zool Napoli 35:105–114

Noble GK, Weber JA (1929) The spermatophores of *Desmognathus* and other plethodontid salamanders. Amer Mus Novit no 351:1–15

Noguchi H (1981) Mating frequency, fecundity, and egg hatchability of the smaller tea tortrix moth, *Adoxophyes* sp. (Lepidoptera: Tortricidae). Jpn J Appl Entomol Zool 25:259–264

Noordink JPW (1970) Autoradiography: A sensitive method in dispersal studies with *Adoxophyes orana* (Lepidoptera: Tortricidae). Entomol Exp Appl 13:448–454

Norris MJ (1932) Contributions towards the study of insect fertility. I. The structure and operation of the reproductive organs of the genera *Ephestia* and *Plodia* (Lepidoptera, Phycitidae). Proc Zool Soc Lond 1932, 595–611

Norris MJ (1933) Contributions towards the study of insect fertility. II. Experiments on the factors influencing fertility in *Ephestia kühniella* Z. (Lepidoptera, Phycitidae). Proc Zool Soc Lond 1933, 903–934

Norris MJ (1934) Contributions towards the study of insect fertility. III. Adult nutrition, fecundity, and longevity in the genus *Ephestia* (Lepidoptera, Phycitidae). Proc Zool Soc Lond 1934, 333–360

Nuttall GHF, Merriman G (1911) The process of copulation in *Ornithodorus moubata*. Parasitology 4:39–44

Obara Y, Tateda H, Kuwabara M (1975) Mating behavior of the cabbage white butterfly, *Pieris rapae crucivora* Boisduval: V. Copulatory stimuli inducing changes of female response patterns. Zool Mag (Tokyo) 84:71–76

195

Odhiambo TR (1969a) The architecture of the accessory reproductive glands of the male desert locust. I. Types of glands and their secretions. Tissue & Cell 1:155–182

Odhiambo TR (1969b) The architecture of the accessory reproductive glands of the desert locust. IV. Fine structure of the glandular epithelium. Phil Trans R Soc Lond [Biol] 256:85–114

Odhiambo TR, Kokwaro ED, Sequeira LM (1983) Histochemical and ultrastructural studies of the male accessory reproductive gland and spermatophore of the tsetse, *Glossina morsitans morsitans*. Insect Sci Application 4:227–236

Okelo O (1979) Mechanisms of sperm release from the *receptaculum seminis* of *Schistocerca vaga* (Orthoptera: Acrididae). Int J Invertebr Reprod 1:121–132

Okuda S (1946) Studies on the development of Annelida Polychaeta. J Fac Sci Hokkaido Univ (Ser 6) 9:115–219

Oldfield GN (1973) Sperm storage in female Eriophyoidea (Acarina). Ann Entomol Soc Am 66:1089–1092

Oldfied GN, Newell IM (1973) The role of the spermatophore in the reproductive biology of protogynes of *Aculus cornutus* (Acarina: Eriophyidae). Ann Entomol Soc Am 66:160–163

Oldfield GN Hobza RF, Wilson NS (1970) Discovery and characterization of spermatophores in the Eriophyoidea (Acari). Ann Entomol Soc Am 63:520–526

Oldfield GN, Newell IM, Reed RK (1972) Insemination of protogynes of *Aculus cornutus* from spermatophores and description of the sperm cell. Ann Entomol Soc Am 65:1080–1084

Oliver JH Jr (1982) Tick reproduction: sperm development and cytogenetics. In: Obenchain FD, Galun R (eds) Physiology of ticks, vol 1. Pergamon Press, Oxford New York Toronto Sydney Paris Frankfurt, pp 224–275

Oliver JH Jr, Al-Ahmadi Z, Osburn RL (1974) Reproduction in ticks (Acari: Ixodoidea). 3. Copulation in *Dermacentor occidentalis* Marx and *Haemaphysalis leporispalustris* (Packard) (Ixodidae). J Parasit 60:499–506

Omura S (1938) Structure and function of the female genital system of Bombyx mori with special reference to mechanism of fertilization. J Fac Agric Hokkaido Univ 40:111–128

Orelli M (1962) Die Übertragung der Spermatophore von *Octopus vulgaris* und *Eledone* (Cephalopoda). Rev Suisse de Zool 69:193–202

Organ JA, Lowenthal LA (1963) Comparative studies of macroscopic and microscopic features of spermatophores of some plethodontid salamanders. Copeia 1963:659–669

Organ JA, Organ DJ (1968) Courtship behaviour of the red salamander, *Pseudotriton ruber*. Copeia pp 217–223

Outram I (1970) Morphology and histology of the reproductive system of the female spruce budworm, *Choristoneura fumiferana* (Lepidoptera: Tortricidae). Can Entomol 103:32–34

Overmeer WPJ, Doodeman M, van Zon AQ (1982) Copulation and egg production in *Amblyseius potentillae* and *Typhlodromus pyri* (Acari, Phytoseiidae). Z Angew Entomol 93:1–11

Pantel J, de Sinéty R (1906) Les cellules la lignée mâle chez le *Notonecta glauca* L. Cellule 23:87–303

Pappas PJ, Oliver JH (1972) Reproduction in ticks (Acari: Ixodoidea): II. Analysis of the stimulus for rapid and complete feeding of female *Dermacentor variabilis* (Say). J Med Entomol 9:47–50

Parker GA (1970) Sperm competition and its evolutionary consequences in the insects. Biol Rev 45:525–567

Parker GA (1978) Evolution of competitive mate searching. Ann Rev Entomol 23:173–196

Parker TJ, Haswell WA (1897) A textbook of zoology. Vol. 2, Macmillan, London pp 179–181

Pauly F (1956) Zur Biologie einiger Belbiden (Oribatei, Moosmilben) und zur Funktion ihrer pseudostigmatischen Organe. Zool Jb (Syst) 84:275–328

Perotti ME (1971) Microtubules as components of *Drosophila* male paragonia secretion. An electron microscope study, with enzymatic tests. J Submicr Cytol 3:255–282

Perrier R, Fischer H (1914) Sur l'existence des spermatophores chez quelques opisthobranches. CR Acad Sci Paris 158:1366–1369

Petersen W (1907) Über die Spermatophoren der Schmetterlinge. Z Wiss Zool 88:117–130

Petersen W (1929) Über die Sphragis und das Spermatophragma der Tagfaltergattung *Parnassius*. Dtsch Entomol Z Jhr 1928:407–413

Peterson RP (1959) The anatomy and histology of the reproductive systems of *Octopus bimaculoides*. J Morph 104:61–88

Philippe R (1972) Les appareils genitaux male et femele chez *Chrysopa perla* (Neuroptera): Etude anatomique, histologique et fonctionelle. Ann Soc Entomol Fr 8:693–705

196

Philippi (Dr) (1839) Notiz, die sogenannten Samenmaschinen des Octopus betreffend. Arch Anat Phys Wiss Med Jhrg 1839:302–310

Philippi E (1907) „Spermatophoren" bei Fischen. Verh Dtsch Zool Ges 17:105–108

Philippi E (1909) Fortpflanzungsgeschichte der viviparen Teleosteer *Glaridichthys januarius* and *G. decem-maculatus* in ihrem Einfluss auf Lebensweise, makroskopische und mikroskopische Anatomie. Zool Jhrb, Abt Anat Ontog 27:1–94

Phillips DM (1974) Spermiogenesis. Academic Press, New York London, p 68

Phillips RA, Watson H (1930) *Milax gracilis* (Leydig) in the British Isles. J Conchol 19:65–93

Pickford GE (1946) *Vampyroteuthis infernalis* Chun. An archaic dibranchiate Cephalopod. 1. Natural history and distribution. Dana Rep 29:1–40

Pickford GE (1949) *Vampyroteuthis infernalis* Chun. An archaic dibranchiate Cephalopod. 2. External anatomy. Dana Rep 32:1–132

Pickford GE (1964) *Octopus dofleini* (Wülker). Bull Bingham Oceanogr Col (Yale Univ) 19(1):1–70

Pickford R, Gillott C (1971) Insemination in the migratory grasshopper *Melanoplus sanguinipes*. Can J Zool 49:1583–1588

Pickford R, Padgham DE (1973) Spermatophore formation and sperm transfer in the desert locust, *Schistocerca gregaria* (Orthoptera: Acrididae). Can Entomol 105:613–618

Plate LH (1885) Beiträge zur Naturgeschichte der Rotatorien. Inaug Diss Univ Jena, Gustav Fischer, Jena, S. 31

Plate L (1887) Über einige ectoparasitische Rotatorien des Golfes von Neapel. Mitteilungen aus der Zoologischen Station zu Neapel 7:234–263

Poinsot-Balaguer N (1976) Contribution a l'étude des spermatophores de quatre espèces de Collemboles. Bull Mus Natl Hist Nat Zool 290, 413:1235–1240

Pollock JN (1970) Sperm transfer by spermatophores in *Glossina austeni* Newstead. Nature (Lond) 225:1063–1064

Pollock JN (1972) The evolution of sperm transfer mechanism in Diptera. J Entomol (A) 47:29–35

Pomerantzev BJ (1932) Beiträge zur Morphologie und Anatomie der Genitalien von *Culicoides* (Diptera: Nematocera). Mag Parasit Leningr 3:183–214

Pritchard AW, Huston MJ, Martin AW (1963) Effects of *in vitro* anoxia on metabolism of octopus heart tissue. Proc Soc Exp Biol Med 112:27–29

Proshold FI, LaChance LE, Richard RD (1975) Sperm production and transfer by *Heliothis virescens*, *H. subflexa*, and the sterile hybrid males. Ann Entomol Soc Am 68:31–34

Przibram H (1907) Die Lebensgeschichte der Gottesanbeterinnen (Fang-Heuschrecken). Z Wiss Insektenbiol 3:117–123, 147–153

Purchon RD (1968) The biology of the Mollusca. Pergamon, Oxford, p 560

Racovitza E-G (1894 a) Notes de biologie. 1. Accouplement et fécondation chez l'Octopus vulgaris Lam. Arch Zool Exp Gen 2:23–49

Racovitza E-G (1894 b) Notes de biologie. 3. Moeurs et reproduction de la *Rossia macrosoma* (D. Ch.). Arch Zool Exp Gen 2:491–539

Racovitza E-G (1894 c) Sur l'accouplement de quelques Céphalopodes *Sepiola Rondeletii* (Leach), *Rossia macrosoma* (D. Ch.) et *Octopus vulgaris* (Lam.). CR Acad Sci Paris 118:722–724

Rahn R (1971) Evolution des potentialities d'accouplement des males et des femelles d'*Acrolepia assectella* Z. (Lepidoptera, Plutellidae). Ann Zool Econ Anim 3:337–345

Raymont JEG, Krishnaswamy S, Woodhouse MA, Griffin RL (1974) Studies on the fine structure of Copepoda: Observations on *Calanus finmarchicus* (Gunnerus). Proc R Soc Lond [Biol] 185:409–424

Reeve MR, Cosper TC (1975) Chaetognatha. In: Giese AC, Pearse JS (eds) Reproduction of marine invertebrates, vol 2. Academic Press, New York San Francisco London, pp 157–184

Regen J (1910) Kastration und ihre Folgeerscheinungen bei *Gryllus campestris* L. Zool Anz 35:427–432

Regen J (1924) Anatomisch-physiologische Untersuchungen über die Spermatophore von *Liogryllus campestris* L. Sitzgsber Akad Wien, Math-Naturwiss Kl, Abt 1, 133:347–359

Reger JF (1970) Some aspects of the fine structure of filiform spermatozoa (Ostracod, *Cypridopsis* sp.) lacking tubule sub-structure. In: Baccetti B (ed) Comparative spermatology. Academia Nazionale dei Lincei, Rome, and Academic Press, New York London, pp 237–245

Reger JF, Cooper DP (1968) Studies on the fine structure of spermatids and spermatozoa from the millipede *Polydesmus* sp. J Ultrastr Res 23:60–70

197

Reger JF, Fain-Maurel M-A (1973) A comparative study on the origin, distribution, and fine structure of extracellular tubules in the male reproductive system of species of isopods, amphipods, schizopods, copepods, and Cumacea. J Ultrastr Res 44:235–252

Reid JD (1964) The reproduction of the sacoglossan opisthobranch *Elysia maoria*. Proc Zool Soc Lond 143:365–393

Renner M, Kremer E (1980) Das Paarungsverhalten der Feldheuschrecke *Chrysochraon dispar* Germ. in Abhängigkeit vom Eiablagerhythmus. Spixiana (München) 3:25–32

Ribaga C (1897) Sopra un organo particolare delle cimici dei letti. Riv Patol Veg 5:223–226

Rice SA (1978) Spermatophores and sperm transfer in spionid polychaetes. Trans Am Micr Soc 97:160–170

Rice SA (1980) Ultrastructure of the male nephridium and its role in spermatophore formation in spionid polychaetes (Annelida). Zoomorphologie 95:181–194

Rice SA (1981) Spermatogenesis and sperm ultrastructure in three species of *Polydora* and in *Streblospio benedicti* (Polychaeta: Spionidae). Zoomorphology 97:1–16

Rice SA, Simon JL (1980) Intraspecific variation in the pollution indicator polychaete *Polydora ligni* (Spionidae). Ophelia 19:79–115

Richards OW (1927) Sexual selection and allied problems in the insects. Biol Rev 2:298–360

Richards SL (1970) Spawning and reproductive morphology of *Scolelepis squamata* (Spionidae: Polychaeta). Can J Zool 48:1369–1379

Richards TJ (1952) *Nemobius sylvestris* in S E Devon. Entomologist 85:83–87, 108–111, 136–141, 161–166

Richards TJ (1953) *Nemobius sylvestris* (F.) (Orthopt., Gryllidae): A correction and some further notes. Entomologist 86:133–134

Riddiford LM, Ashenhurst JB (1973) The switchover from virgin to mated behavior in female Cecropia moth: the role of the bursa copulatrix. Biol Bull (Woods Hole) 144:162–171

Riemann JG (1970) Metamorphosis of sperm of the cabbage looper, *Trichoplusia ni,* during passage from the testes to the female spermatheca. In: Baccetti B (ed) Comparative spermatology. Academia Nazionale dei Lincei, Rome, and Academic Press, New York London, pp 321–331

Riemann JG, Thorson BJ (1979) Foliate and granule-secreting cells in the ejaculatory duct (simplex) of the Mediterranean flour moth. J Ultrastr Res 66:1–10

Roberts LW, Rapmund G, Cadigan FC (1977) Sex ratio in *Rickettsia tsutsugamushi* – infected and noninfected colonies of *Leptotrombidium* (Acari: Trombiculidae). J Med Entomol 14:89–92

Robertson A, Gibbs AJ (1973) Spermatogenesis and fertilization in *Philaenus spumarius* Fallen. J Trop Med Hyg 40:257–262

Robinson AS (1974) Gamma radiation and insemination in the codling moth, *Laspeyresia pomonella* (Lepidoptera: Olethreutidae). Entomol Exp Appl 17:425–432

Robinson GG (1942) The mechanism of insemination in the argasid tick, *Ornithodorus moubata* Murray. Parasitology 34:195–198

Robinson MH (1982) Courtship and mating behavior in spiders. Ann Rev Entomol 27:1–20

Robison WG Jr (1966) Microtubules in relation to the motility of a sperm syncytium in an armored scale insect. J Cell Biol 29:251–266

Robison WG Jr (1968) Microtubular patterns in motility of sperm and sperm bundles of coccid insects. J Cell Biol 39:159a

Robison WG Jr (1970) Unusual arrangements of microtubules in relation to mechanisms of sperm movement. In: Baccetti B (ed) Comparative spermatology. Academia Nazionale dei Lincei, Rome; Academic Press, New York London, pp 311–320

Robison WG Jr (1972) Microtubular patterns in spermatozoa of coccid insects in relation to bending. J Cell Biol 52:66–83

Robison WG Jr (1977) Ultrastructure of Coccoidea sperm. Proc Symp: Recent advances in the study of scale insects. Virginia Polytechnic Institute and State University, 1976; Res Div Bull 127:35–50

Robson GC (1929) A monograph of the recent Cephalopoda, based on the collections of the British Museum (Natural History). Part I. Octopodinae, Brit Mus (Nat Hist) London, p 236

Robson GC (1932) A monograph of the recent Cephalopoda, based on the collections of the British Museum (Natural History). Part II. The Octopoda (Excluding the Octopidinae), Brit Mus (Nat Hist) London, p 359

Roosen-Runge EC (1977) The process of spermatogenesis in animals. University Press, Cambridge, p 214

Rosin R, Shulov A (1963) Studies on the scorpion *Nebo hierochonticus*. Proc Zool Soc Lond 140:547–575

Ross J (1971) Microtubules and the development of the corkscrew region of a mealybug sperm bundle. Tissue Cell 3:35–56

Ross J, Robison WG Jr (1969) Unusual microtubular patterns and three-dimensional movement of mealybug sperm and sperm bundles. J Cell Biol 40:426–445

Roth LM, Dateo GP Jr (1964) Uric acid in the reproductive system of males of the cockroach, *Blatella germanica*. Science 146:782–784

Roth LM, Dateo GP Jr (1965) Uric acid storage and excretion by accessory glands of male cockroaches. J Insect Physiol 11:1023–1029

Rothschild Lord (1965) A classification of living animals. Longmans, London, p 134

Rothschild Lord, Mann T (1950) Carbohydrate and adenosine triphosphate in sea-urchin semen. Nature 166:781

Russell LD, Brandon RA, Zalisko EJ, Martan J (1981) Spermatophores of the salamander *Ambystoma texanum*. Tissue Cell 13:609–621

Rutowski RL, Long CE, Marshall LD, Vetter RS (1981) Courtship solicitation by *Colias* females (Lepidoptera: Pieridae). Am Midl Nat 105:334–340

Sakaluk SK (1984) Male crickets feed females to ensure complete sperm transfer. Science 223:609–610

Sakaluk SK, Cade WH (1980) Female mating frequency and progeny production in singly and doubly mated house crickets (*Acheta domesticus*) and field crickets (*Gryllus integer*). Can J Zool 58:404–411

Saltet P (1969) Le spermatophore chez un Orthoptère *Rhaphidophoridae* (Dolichopoda Linderi Duf.). Ann Spéléologie 24:533–539

Salthe SN, Mecham JS (1974) Reproductive and courtship patterns. In: Lofts B (ed) Physiology of the amphibia, vol 2. Academic Press, New York, pp 309–521

Samson K (1909) Zur Anatomie und Biologie von Ixodes ricinus L. Z Wiss Zool 93:185–236

Sandifer P, Lynn J (1980) Artificial insemination of Caridean shrimp. In: Clark WH, Adams TS (eds) Advances in invertebrate reproduction. Elsevier North-Holland, Amsterdam, pp 271–288

Sasaki M (1929) A monograph of the dibranchiate cephalopods of the Japanese and adjacent waters. J Fac Agric Hokkaido Imp Univ, Sapporo 20 (Suppl):1–357

Sawyer RT, Chamberlain NA (1972) A new species of marine leech (Annelida: Hirudinea) from South Carolina, parasitic on the Atlantic menhaden, *Brevoortia tyrannus*. Biol Bull (Woods Hole) 142:470–479

Schal C, Bell WJ (1982) Ecological correlates of paternal investment of urates in a tropical cockroach. Science 218:170–173

Schaller F (1953) Untersuchungen zur Fortpflanzungsbiologie arthropleoner Collembolen. Z Morph Oekol Tiere 41:265–277.

Schaller F (1965) Mating behaviour of lower terrestrial arthropods from the phylogenetic point of view. Proc 12th Intern Congr Entomol, London 1964, 297–298

Schaller F (1971) Indirect sperm transfer by soil arthropods. Ann Rev Entomol 16:407–440

Schaller F (1979) Significance of sperm transfer and formation of spermatophores in arthropod phylogeny. In: Gupta AP (ed) Arthropod phylogeny. Van Nostrand Reinhold Co, New York London, pp 587–608

Schliwa W, Schaller F (1963) Die Paarbildung des Springschwanzes *Podura aquatica* (Apterygota, Collembola). Naturwiss 50:698

Schmekel L (1971) Histologie und Feinstruktur der Genitalorgane von Nudibranchien (Gastropoda, Euthyneura). 2. Z Morph Tiere 69:115–183

Schneider A (1883) Das Ei und seine Befruchtung. JU Kern's Verlag, Breslau, S. 57–67

Schömann K (1956) Zur Biologie von *Polyxenus lagurus* (L. 1758). Zool Jb 84:195–256

Schöne H (1961) Complex behavior. In: Waterman TH (ed) The physiology of Crustacea, vol 2. Academic Press, New York London, pp 465–520

Schroeder PC, Hermans CO (1975) Annelida: Polychaeta. In: Giese AC, Pearse JS (eds) Reproduction of marine invertebrates, vol 3 (Annelids and Echiurans). Academic Press, New York San Francisco London, pp 1–213

Schuster R, Hasenhütl K (1983) Die Spermatophore der Eurypauropodiden (Myriapoda, Pauropoda). Zool Anz 211:187–196

Schuster R, Schuster IJ (1969) Gestielte Spermatophoren bei Labidostomiden (Acari, Trombidiformes). Naturwiss 56:145

Schuster R, Schuster IJ (1977) Ernährungs- und fortpflanzungsbiologische Studien an der Milbenfamilie Nanorchestidae (Acari, Trombidiformes). Zool Anz 199:89–94

Schweder T (1979) Patterns of spermatophore distribution in *Euchaeta norvegica:* A further analysis. J Mar Biol Assoc UK 59:369–372

Scudder GGE (1971) Comparative morphology of insect genitalia. Ann Rev Entomol 16:379–406

Sedgwick A (1885) The development of Peripatus Capensis. QJ Micr Sci 25:449–468

Sengbusch H (1958) Zuchtversuche mit Oribatiden. Naturwiss 45:498–499

Shereef GM (1972) Observations on oribatid mites in laboratory cultures. Acarologia 14:281–291

Siebold CT v (1836a) Über die Spermatozoen der Crustaceen, Insecten, Gasteropoden und einiger anderer wirbellosen Tiere. Arch Anat Physiol Wiss Med Jhrg 1836:13–53

Siebold CT v (1836b) Fernere Beobachtungen über die Spermatozoen der wirbellosen Tiere. Arch Anat Physiol Wiss Med Jhrg 1836:232–255

Siebold CT v (1839) Über das Begattungsgeschäft der Cyclops castor. Neueste Schriften Naturf Ges Danzig, Bd 3, Heft 2: Beiträge zur Naturgeschichte der wirbellosen Tiere, S. 36–50

Siebold CT v (1845) Über die Spermatozoiden der Locustinen. Nova Acta Acad Caes Leopoldino-Carolinae 21:249–274

Siebold CT v (1848) Lehrbuch der vergleichenden Anatomie der wirbellosen Tiere. Veit & Co, Berlin, S. 679; the early observations on spermatophores are described on pp 408–411 (cephalopods), 494–498 (crustaceans), and 636–637 (insects)

Simroth H (1900) Über Selbstbefruchtung bei Lungenschnecken. Verh Dtsch Zool Ges 10:143–147

Sixl W (1971) Fortpflanzungsbeziehungen bei *Ascoschöngastia latyshevi* (Schluger, 1955) – Trombiculidae, Acari. Rev Suisse Zool 78:815–820

Smith BG (1910) The structure of the spermatophores of *Ambystoma punctatum*. Biol Bull (Woods Hole) 18:204–211

Söderström A (1920) Studien über die Polychaeten Familie Spionidae. Inaug Diss Uppsala, Almquist and Wiksells, S. 286

Sokolow AJ (1926) Zur Frage der Spermatophorbefruchtung bei der Wanderheuschrecke (*Locusta migratoria* L.). Das Weibchen. Z Wiss Zool 127:608–618

Southcott RV (1955) Some observations on the biology, including mating and other behaviour, of the Australian scorpion *Urodacus abruptus* Pocock. Trans R Soc Aust 78:145–154

Southward AJ, Southward EC (1963) Notes on the biology of some Pogonophora. J Mar Biol Ass UK 43:57–64

Southward EC (1963) Pogonophora. Oceanogr Mar Biol Ann Rev 1:405–428

Southward EC (1975a) Fine structure and phylogeny of the Pogonophora. Symp Zool Soc Lond 36:235–251

Southward EC (1975b) Pogonophora. In: Giese AC, Pearse JS (eds) Reproduction of marine invertebrates, vol 2. Academic Press, New York San Francisco London, pp 129–156

Spickett AM, Malan JR (1978) Genetic incompatibility between *Boophilus decoloratus* (Koch, 1844) and *Boophilus microplus* (Canestrini, 1888) and hybrid sterility of Australian and South African *Boophilus microplus* (Acarina: Ixodidae). Onderstepoort J Vet Res 45:149–154

Srivastava US, Srivastava BP (1957) Notes on the spermatophore formation and transference of sperms in the female reproductive organs of *Leucinodes orbonalis* Guen. (Lepidoptera: Pyraustidae). Zool Anz 158:258–267

Stanley-Samuelson DW, Loher W (1983) Arachidonic and other longchain polyunsaturated fatty acids in spermatophores and spermathecae of *Teleogryllus commodus:* significance in prostaglandin-mediated reproductive behaviour. J Insect Physiol 29:41–45

Steenstrup JJS (1856) Hectocotyldannelsen hos Octopodslaegterne *Argonauta* og *Tremoctopus*, oplyst ved dagttogelse of lignevde Dannelsen hos Blacksprutterne i Almindelighed. K Danske Vidensk Selsk Skr 5 Raekke 4:187–216. English translation (by WS Dalas): Steenstrup JJ (1857) Hectocotylus-formation in *Argonauta* and *Tremoctopus* explained by observations on a similar formation in the Cephalopoda in general. Ann Mag Nat Hist (2) 20:81–114

Sternlicht M, Griffith DA (1974) The emission and form of spermatophores and the fine structure of adult *Eriophyes sheldoni* Ewing (Acarina, Eriophyoidea). Bull Entomol Res 63:561–565

Stitz H (1901) Der Genitalapparat der Mikrolepidopteren. I. Der männliche Genitalapparat. Zool Jahrb Anat 14:135–176

200

Stockel J (1973) Fonctionnement de l'appareil reproducteur de la femelle de *Sitotroga cerealella* Oliv. (Lep. Gelechiidae). Ann Soc Entomol Fr 9:627–645

Storch V, Ruhberg H (1977) Zur Entstehung der Spermatophore von *Opisthopatus cinctipes* Purcell, 1899 (Onychophora, Peripatopsidae). Zoomorphologie 87:263–276

Straus-Durckheim H (1828) Considérations générales sur l'anatomie comparée des animaux articulés aux quelles on a joint l'anatomie descriptive du Melolonthe vulgaris (hanneton). FG Levrault, Paris Strasbourg Bruxelles, p 434

Strübing H (1955) Spermatophorenbildung bei Fulgoroiden (Hom. Auch.). Naturwiss 42:653

Stuhlmann F (1886) Beiträge zur Anatomie der inneren männlichen Geschlechtsorgane und zur Spermatogenese der Cypriden. Z Wiss Zool 44:536–569

Sturm H (1952) Die Paarung bei *Machilis* (Felsenspringer). Naturwiss 39:308

Sturm H (1955) Beiträge zur Ethologie einiger mitteldeutscher Machiliden. Z Tierpsychol 12:337–363

Sturm H (1956) Die Paarung bei Silberfischen *Lepisma saccharina*. Z Tierphysiol 13:1–12

Suarez SS (1975) The reproductive biology of *Ogilbia cayorum*, a viviparous brotulid fish. Bull Mar Sci 25:143–173

Subramoniam T (1977) Aspects of sexual biology of the anomuran crab *Emerita asiatica*. Mar Biol (Berl) 43:369–377

Subramoniam T (1984) Spermatophore formation in two intertidal crabs *Albunea symnista* and *Emerita asiatica* (Decapoda: Anomura). Biol Bull (Woods Hole) 166:78–95

Sugawara T (1979) Stretch reception in the bursa copulatrix of the butterfly, *Pieris rapae crucivora*, and its role in behavior. J Comp Physiol A Sens Neural Behav Physiol 130:191–200

Suleiman SA, Brown GG (1978) Spermiogenesis in the dog tick *Dermacentor variabilis* (Say). Iowa State J Res 53:93–108

Swailes GE (1971) Reproductive behavior and effects of the male accessory gland substance in the cabbage maggot, *Hylemya brassicae*. Ann Entomol Soc Am 64:176–179

Swammerdam J 1637–1685 *Biblia Naturae (Bybel der Natuure)* was published posthumously, in 1737–1738, Isaak Severinus, Boudowyn Vander and Pieter Vander, Leiden. The German version: *Bibel der Natur*, 1752, Johann Friedrich Gleditschens Buchhandlung, Leipzig. The English translation (by Thomas Flloyd): *The Book of Nature*, 1758, CG Seyfert in Dean Street, Soho, London; the discovery of spermatophores is reported in the Appendix oñ "The anatomy of the Sea Sepia or Cuttle-Fish. Inscribed to the most excellent Francis Redi, physician to the great Duke of Tuscany, a most indefatigable searcher into the miraculous works of Nature". The legend to figure VII, Table LII, and explanations, appear on p LXIII

Swedmark B (1964) The interstitial fauna of marine sand. Biol Rev 39:1–42

Swedmark B (1968) The biology of interstitial Mollusca. Symp Zool Soc London 22:135–149

Szöllösi A (1975) Electron microscope study of spermiogenesis in *Locusta migratoria* (Insects Orthoptera). J Ultrastr Res 50:322–346

Taber S (1977) Semen of *Apis mellifera:* fertility, chemical, and physical characteristics. In: Adiyodi KG, Adiyoidi RG (eds) Advances in invertebrate reproduction, vol 1. Peralam-Kenoth, Karivellur Kerata, India, p 219

Taberly G (1957) Observations sur les spermatophores et leur transfert chez les Oribates. Bull Soc Zool Fr 82:139–145

Takeuchi S, Miyashita K (1975) The process of spermatophore transfer during the mating of *Spodoptera litura* F. Jpn J Appl Entomol Zool 19:41–46

Tan KH (1974) The disruption of neuro-endocrine function and reproduction in the Mediterranean flour moth, *Euphestia kühniella*, by a chemosterilant, hexamethyl-melamine (Hemel). Experientia 30:1403–1404

Tan KH (1977) Chemical induction of mating abnormalities and spermatophore production in the Mediterranean flour moth, *Ephestia kühniella*. Ann Appl Biol 87:1–5

Tatchell RJ (1962) Studies of the male accessory reproductive glands and spermatophore of the tick *Argas persicus* Oken. Parasitology 52:133–142

Tave D, Brown A (1981) A new device to help facilitate manual spermatophore transfer in penaeid shrimps. Aquaculture 25:299–302

Terranova AC, Leopold RA, Degrugillier ME, Johnson JR (1972) Electrophoresis of the male accessory secretion and its fate in the mated female. J Insect Physiol 18:1573–1591

Teuchert G (1968) Zur Fortpflanzung und Entwicklung der Macrodasyoidea (Gastrotricha). Z Morph Tiere 63:343–418

Theis G, Schuster R (1974) Gestielte Tröpfchenspermatophoren bei Calyptostomiden (Acari, Trombidiformes). Mitt Naturwiss Ver Steiermark 104:183–185

Thibout E (1971) Description de l'appareil genital male et formation du spermatophore chez *Acrolepia assectella* (Lepidoptera: Plutellidae). CR Acad Ser D Paris 273:2546–2549

Tinbergen L (1939) Zur Fortpflanzungsethologie von *Sepia officinalis* L. Arch Neerl Zool 3:323–364

Tobe SS, Langley PA (1978) Reproductive physiology of *Glossina*. Ann Rev Entomol 23:283–307

Tobe SS, Loher W (1983) Properties of the prostaglandin synthetase complex in the cricket *Teleogryllus commodus*. Insect Biochem 13:137–141

Traut W (1966) Über die Kopulation bei *Dinophilus gyrociliatus*. Zool Anz 177:402–411

Turner JT, Collard SB, Wright JC, Mitchell DV, Steele P (1979) Summer distribution of pontellid copepods in the neuston of the eastern Gulf of Mexico continental shelf. Bull Mar Sci 29:287–297

Tuzet O, Millot I (1937) Recherches sur la spermiogenèse des ixodes. Bull Biol Fr Belg 71:190–205

Uma K, Subramoniam T (1979) Histochemical characteristics of spermatophore layers of *Scylla serrata* (Forskal) (Decapoda: Portunidae). Int J Invertebr Reprod 1:31–40

Uzzell T (1969) Notes on spermatophore production by salamanders of the *Ambystoma jeffersonianum* complex. Copeia 1969:602–612

Van Goethem JL (1977) Systematic research on the Urocyclinae (Gastropoda, Urocyclidae). Ann Soc R Zool Belg 106:123–132

Vérany JB, Vogt C (1852) Mémoire sur les Hectocotyles et les mâles de quelques céphalopodes. Ann Sci Nat 3 sèr Zool 17:147–191

Verco JC, Cotton BC (1931) The spermatophore of *Sepioteuthis australis* (Quoy & Gaimard). Proc Malacol Soc Lond 19:168–170

Verrell P (1982) The sexual behaviour of the red-spotted newt, *Notophthalmus viridescens* (Amphibia: Urodela: Salamandridae). Anim Behav 30:1224–1236

Vickers RA (1982) Some aspects of reproduction in *Pectinophora scutigera* (Lepidoptera: Gelechiidae). J Aust Entomol Soc 21:63–68

Vistorin HE (1978) Fortpflanzung und Entwicklung der Nicoletiellidae (Labidostomidae); Acari Trombidiformes. Zool Jhrb Syst Oekol Geogr Tiere 105:462–473

Vistorin HE (1981) Anatomische Untersuchungen an Genitalorganen und Drüsen der Nicoletiellidae (Acari, Trombidiformes). Sitzungsberichte der Österr Akad Wiss Mathem-Naturw Kl, Abt I, 190:1–32

Vistorin-Theis G (1975) Entwicklungszyklus der Calyptostomiden (Acari, Trombidiformes). Acarologia 17:683–692

Vistorin-Theis G (1977) Anatomische Untersuchungen an Calyptostomiden (Acari, Trombidiformes). Acarologia 19:242–256

Voss GL (1962) South African cephalopods. Trans R Soc S Afr 36, Part 4:245–272

Wagner R (1834–1835) Lehrbuch der vergleichenden Anatomie. Leopold Voss, Leipzig, S. 311

Wagner R, Leuckart R (1852) Semen. In: Todd RB (ed) Cyclopaedia of anatomy, vol 4. Longman, London, p 472

Waitzbauer J (1983) Licht- und elektronenmikroskopische Untersuchungen der Spermiogenese und Spermatophore von *Hermannia gibba* (Oribatidae, Acari). Acarologia (Paris) 24:95–107

Warren E (1934) On the male genital system and spermatozoa of certain millipedes. Ann Natal Museum Pietermaritzburg 7:351–402

Warren E (1939) The genital system of *Hypoctonus formosus* (Butler) (Thelyphonidae). Ann Natal Mus Pietermaritzburg 9:307–344

Webb GR (1974) The sexual evolution of the polygyrid snails: Pulmonata, Stylommatophora. Gastropodia 1 (9):85–90

Webb GR (1977) On the sexology of *Catinella* (Mediappendix) *avara* (Say) or *C.*(M.) *vermeta* (Say). Gastropodia 1 (10):100–102

Webb M (1963) A reproductive function of the tentacle in the male of *Siboglinum ekmani* Jägersten (Pogonophora). Sarsia 13:45–49

Webber HH (1977) Gastropoda: Prosobranchia. In: Giese AC, Pearse JC (eds) Reproduction of marine invertebrates, vol 4. Academic Press, New York San Francisco London, pp 1–97

Weill R (1927) Recherches sur la structure, la valeur systématique et le fonctionnement du spermatophore de *Sepiola antlantica* d'Orb. Bull Biol Fr Belg 61:59–92

Wells MJ (1964) Hormonal control of sexual maturity in cephalopods. Bull Nat Inst Sci India 27:61–77

Wells MJ (1983) Cephalopods do it differently. New Scientist 100:332–338

Wells MJ, Wells J (1969) Pituitary analogue in the octopus. Nature 222:293–294

Wells MJ, Wells J (1972a) Optic glands and the state of the testis in *Octopus*. Mar Behav Physiol 1:71–83

Wells MJ, Wells J (1972b) Sexual displays and mating of *Octopus vulgaris* Cuvier and *O. cyanea* Gray and attempts to alter performance by manipulating the glandular condition of the animals. Anim Behav 20:293–308

Wells MJ, Wells J (1977) Cephalopoda: Octopoda. In: Giese AC, Pearse JS (eds) Reproduction of marine invertebrates, vol 4. Academic Press, New York San Francisco London, pp 291–336

Wenk P (1965) Über die Biologie blutsaugender Simuliiden (Diptera). II. Schwarmverhalten, Geschlechterfindung und Kopulation. Z Morph Ökol Tiere 55:671–713

Went DF (1982) Egg activation and parthenogenetic reproduction in insects. Biol Rev 57:319–344

Westheide W (1967) Monographie der Gattung *Hesionides* Friedrich und *Microphthalmus* Mecznikow. Ein Beitrag zur Organisation und Biologie psammobionter Polychaeten. Z Morph Oekol Tiere 61:1–159

Westheide W (1969) Spermatodesmen, ein einfacher Spermatophorentyp bei interstitiellen Polychaeten. Naturwiss 12:641–642

Westheide W (1978) Ultrastructure of the genital organs in interstitial polychaetes. I. Structure, development, and function of the copulatory stylets in *Microphthalmus* cf. *listensis*. Zoomorphologie 91:101–108

Westheide W (1979) Unusual granules in the ejaculatory duct of a *Microphthalmus* species (Polychaeta, Annelida). Cell Tissue Res 197:61–68

Westheide W (1980) *Erprobdella octoculata* (Hirudinea) – Spermatophorenübertragung, Kokonablage, Schlüpfen der Jungtiere. Publ Wiss Film, Sekt Biol, Ser 13, Nr 27/E 2562, S. 12

Westheide W (1981) Fortpflanzung bei Egeln (Hirudinea). Publ Wiss Film, Sekt Biol, Ser 14, Nr 6/C 1394, S. 14

Westheide W (1982) Ultrastructure of the genital organs in interstitial polychaetes. III. Penes and ejaculatory ducts in *Hesionides arenaria* (hesionidae). Helg Meeresunters 35:479–488

Westheide W (1984) The concept of reproduction in polychaetes with small body size: Adaptations in interstitial species. Fortschr Zool 29:267–287

Westheide W, Ax P (1965) Bildung und Übertragen von Spermatophoren bei Polychaeten. Untersuchungen an *Hesionides arenarius* Friedrich. Verh Dtsch Zool Ges, Kiel 1964. Zool Anz Suppl 28:196–203

Westheide W, Wawra E (1974) Organisation, Systematik und Biologie von *Microhedyle cryptophthalma* nov. sp. (Gastropoda, Opisthobranchia) aus dem Brandungsstrand des Mittelmeeres. Helgoländer Wiss Meeresunters 26:27–41

Weygoldt P (1966) Mating behaviour and spermatophore morphology in the pseudoscorpion *Dinocheirus tumidus* Banks (Cheliferinea, Chernethidae). Biol Bull (Woods Hole) 130:462–467

Weygoldt P (1969a) Paarungsverhalten und Samenübertragung beim Pseudoscorpion *Withius subruber* Simon (Cheliferidae). Z Tierpsychol 26:230–235

Weygoldt P (1969b) Beobachtungen zur Fortpflanzungsbiologie und zum Verhalten der Geisselspinne *Tarantula marginemaculata* C. L. Koch (Chelicerata, Amblypygi). Z Morph Tiere 64:338–360

Weygoldt P (1972) Spermatophorenbau und Samenübertragung bei Uropygen (*Mastigoproctus brasilianus* C. L. Koch) und Amblypygen (*Charinus brasilianus* Weygoldt und *Admetus pumilio* C. L. Koch) (Chelicerata, Arachnida). Z Morph Tiere 71:23–51

Weygoldt P (1974) Die indirekte Spermatophorenübertragung bei Arachniden. Verh Dtsch Zool Ges 67:308–313

Weygoldt P (1977) Kampf, Paarungsverhalten, Spermatophorenmorphologie und weibliche Genitalien bei neotropischen Geisselspinnen (Amblypygi, Arachnida). Zoomorphologie 86:271–286

Weygoldt P (1978) Paarungsverhalten und Spermatophorenmorphologie bei Geisselskorpionen: *Thelyphonellus amazonicus* Butler und *Typopeltis crucifer* Pocock (Arachnida, Uropygi). Zoomorphologie 89:145–156

Weygoldt P, Weisemann A, Weisemann K (1972) Morphologisch-histologische Untersuchungen an den Geschlechtsorganen der Amblypygi unter besonderer Berücksichtigung von *Tarantula marginemaculata* C. L. Koch (Arachnida). Z Morph Tiere 73:209–247

Whitman CO (1891) Spermatophores as a means of hypodermic impregnation. J Morph 4:361–406

Whitman DW, Loher W (1984) Morphology of male sex organs and insemination in the grasshopper *Taeniopoda eques* (Burmeister). J Morph 179:1–12

Wiebe JP (1968) The reproductive cycle of the viviparous seaperch *Cytomatogaster aggregata* Gibbons. Can J Zool 46:1221–1234

Wigglesworth VB (1972) The principles of insect physiology, 7th edtn. Chapman & Hall, London New York, p 827

Wiles PR (1982) A note on the watermite *Hydrodroma despiciens* feeding on chironomid egg masses. Freshwater Biol 12:83–88

Willey A (1902) Contribution to the natural history of the Pearly *Nautilus:* Zoological results based on material from New Britain, New Guinea, Loyalty Islands and elsewhere collected during the years 1895, 1896, and 1897. Part 6. Cambridge University Press, London New York, pp 691–830

Williams JL (1939) The occurrence of spermatophores and their measurements in some British Lepidoptera. Trans Soc Brit Entomol 6:137–148

Williams JL (1941) The relations of the spermatophore to the female reproductive ducts in Lepidoptera. Entomol News 52:61–65

Williams LW (1910) The anatomy of the common squid, *Loligo pealii* (Lesueur). EJ Brill, Leiden-Holland, p 92

Winkler LR, Ashley LM (1954) The anatomy of the common octopus of northern Washington. Walla Walla Coll Publ Biol Sci 10:1–30

Withycombe CL (1922) Notes on the biology of some British Neuroptera (Planipennia). Trans Entomol Soc Lond 71:501–594

Witte H (1975) Funktionsanatomie der Genitalorgane und Fortpflanzungsverhalten bei den Männchen der Erythraeidae (Acari, Trombidiformes). Z Morph Tiere 80:137–180

Witte H, Storch V (1973) Licht- und elektronmikroskopische Untersuchungen an Hodensekreten und Spermien der trombidiformen Milbe *Abrolophus rubipes* (Trouessart, 1888). Acarologia (Paris) 15:441–450

Wittmann KJ (1982) Untersuchungen zur Sexualbiologie einer mediterranen Mysidacee (Crustacea), *Leptomysis lingvura* G. O. Sars. Zool Anz 209:362–375

Wodinsky J (1973) Ventilation rate and copulation in *Octopus vulgaris*. Mar Biol (Berl) 20:154–164

Wodinsky J (1977) Hormonal inhibition of feeding and death in *Octopus:* Control by optic gland secretion. Science 198:948–951

Wolf E (1905) Die Fortpflanzungsverhältnisse unserer einheimischen Copepoden. Zool Jhrb Syst 22:101–280

Woodring JP (1970) Comparative morphology, homologies, and functions of the male system in oribatid mites (Arachnida: Acari). J Morph 132:425–452

Wülker G (1910) Über japanische Cephalopoden. Beiträge zur Kenntnis der Systematik und Anatomie der Dibranchiaten. Beiträge zur Naturgeschichte Ostasiens (herausgegeben von Dr. F. Doflein). Abh math-phys Klasse der Köngl Bayer Akad Wissensch, Suppl 3, Bd 1. KB Akademie der Wissenschaften, München, S. 71

Yang L-C, Chow Y-S (1978) Spermatophore formation and the morphology of the reproductive system of the diamondback moth, *Plutella xylostella* (L.) (Lepidoptera: Plutellidae). Bull Inst Zool Acad Sin (Taipei) 17:109–116

Young JZ (1967) The visceral nerves of *Octopus*. Phil Trans R Soc [Biol] 253:1–22

Zalisko EJ, Brandon RA, Martan J (1984) Microstructure and histochemistry of salamander spermatophores (Ambystomatidae, Salamandridae, and Plethodontidae). Copeia, in press

Zander CD (1961) Künstliche Befruchtung bei lebendgebärenden Zahnkarpfen. Zool Anz 166:81–87

Zenker W (1854) Monographie der Ostracoden. Arch Naturgesch 20:1–87

Zimmer RL (1967) The morphology and function of accessory reproductive organs in the lophophores of *Phoronis vancouverensis* and *Phoronopsis harmeri*. J Morph 121:159–178

Zissler D (1966) Zur Feinstruktur der Ostracoden-Spermien. Naturwiss 53:561–652

Zissler D (1969 a) Die Spermiohistogenese des Süßwasser-Ostracoden *Notodromas monacha* O.F. Müller. I. Die ovalen und spindelförmigen Spermatiden. Z Zellforsch 96:87–105

Zissler D (1969 b) Die Spermiohistogenese des Süßwasser-Ostracoden *Notodromas monacha* O. F. Müller. II. Die spindelförmigen und schlauchförmigen Spermatiden. Z Zellforsch 96:106–133

Systematic and Species Index

206

210

211

Subject Index

215

Zoophysiology (formerly Zoophysiology and Ecology)

Coordinating Editor: **D.S.Farner**
Editors: **B.Heinrich, K.Johansen, H.Langer, G.Neuweiler, D.J.Randall**

Volume 1: **P.J.Bentley**

Endocrines and Osmoregulation

A Comparative Account of the Regulation of Water and Salt in Vertebrates

1971. 29 figures. XVI, 300 pages. ISBN 3-540-05273-9

"The author... has, with competence and insight, succeeded in the difficult task of covering two fields ... Bentley presents a thoroughly competent synthesis, and the result is a wellintegrated and balanced book. The book follows the zoological point of view, not only in outline, but also in the integration of physiological function with the natural life of the animal. The coherent viewpoint makes the text **readable and interesting,** and a large number of clear tables makes the materials **easily accessible.** The adequate coverage can serve as an introduction to the research literature in both fields treated in this book. If future volumes are of equal quality and value, **the series will be a significant contribution."**
Quarterly Review of Biology

Volume 2: **L.Irving**

Arctic Life of Birds and Mammals

Including Man
1972. 59 figures. XI, 192 pages. ISBN 3-540-05801-X

"The author's intense and unabated interest in arctic biology over the last three decades is reflected in the content and perspective of this volume. His unusually keen insight into the life of arctic birds and mammals (including man), which has led to this competent synthesis, is based on a familiarity conceivable only in a person who has experienced arctic life..."
Quarterly Review of Biology

Volume 3: **A.E.Needham**

The Significance of Zoochromes

1974. 54 figures. XX, 429 pages. ISBN 3-540-06331-5

"Dr. Needham's book is doubly welcome, for it not only considers animal pigments from the points of view of structure and function, but it also presents information and concepts which have never been assembled in one volume before...
The format of the book is very pleasing, with particulary high qualitiy typeface and paper. It was a good idea to preface each chapter with a brief synopsis of its subject matter and to include a conclusion section at the end. There are many tables which collect data not easily found elsewhere: the book is **a valuable and unique contribution** to the literature on pigments." *Nature*

Volume 4/5: **A.C.Neville**

Biology of the Arthopod Cuticle

1975. 233 figures. XVI, 448 pages. ISBN 3-540-07081-8

"...The layout is clear and orderly throughout... As the text has a clear and economical style, as there are numerous figures and electron micrographs and an extensive but selective bibliography, this is an essential work of reference. But the treatment throughout is of a critical review... the book is admirably produced and printed..."
Quarterly J. Exp. Physiology

Volume 6: **K.Schmidt-Koenig**

Migration and Homing in Animals

1975. 64 figures, 2 tables. XII, 99 pages.
ISBN 3-540-07433-3

"The author... has provided a valuable service in collecting together examples of homing and migration in a diversity of animal groups. The plan is an excellent one: each chapter, devoted to a single taxonomic group, is subdivided into examples of field performance in orientation and its experimental analysis..." *The IBIS*

Volume 7: **E.Curio:**

The Etiology of Predation

1976. 70 figures, 16 tables. X, 250 pages.
ISBN 3-540-07720-0

"... It is good because is stimulating, exhaustive and logical. No important aspect of the subject is missed. The author illustrates all his main points with a multiplicity of examples drawn from recent research. The reference list of nearly 700 items is evidence of the thoroughness of the treatment and the marshalling of examples used in explanation. Curio is an enthusiast and conveys the excitement to be found in much of the research on this subject; he also draws pointed attention to the gaps in our knowledge. For all these reasons **the book is a must for ethologists, ecologists, experimental psychologists and university libraries.** As a first treatment of seminal quality the book could well become a reference classic and inspire numerous research projects.
...The illustrations are clear and relevant..."
The Quart. Review Biology

Springer-Verlag Berlin Heidelberg New York Tokyo

Volume 8: **W. Leuthold**

African Ungulates

A Comparative Review of Their Ethology and Behavioral Ecology
1977. 55 figures, 7 tables. XIII, 307 pages.
ISBN 3-540-07951-3

·...Dr. Leuthold displays a masterly command of his subject ... The work is basically a review of published knowledge with an original approach, enlivened by the author's interpretations and based on his intimate first-hand knowledge of the subject. The first chapter, on the application of ethological knowledge to wildlife management, covers an important area... a wealth of references is given so that the chapter provides useful guide to the literature. the illustrations are good and well chosen to demonstrate points made verbally and not first to embellish the text. The book will provide **excellent background reading for undergraduates and research students as well as for anyone seriously interested in African wildlife.** On the whole, **it can be thoroughly recommended.**"
J. Applied Ecology

Volume 9: **E. B. Edney**

Water Balance in Land Arthropods

1977. 109 figures, 36 tables. XII, 282 pages.
ISBN 3-540-08084-8

... Dr. Erdney has provided a wealth of organized information on prior work and ideas for needed research, **all of which make the book a bargain.** The volume should prove useful, not only to those who work in arthopod water relations (it is a must for them), but also those of us interested in invertebrate and general ecology, entomology, comparative physiology, and biophysics." *AWRA Water Res. Bull.*

Volume 10: **H.-U. Thiele**

Carabid Beetles in Their Environments

A Study on Habitat Selection by Adaptations in Physiology and Behaviour

Translated from the German by J. Wieser
1977. 152 figures, 58 tables. XVII, 369 pages.
ISBN 3-540-08306-5

"...Because the book is comparative both in method and interpretation, it is a contribution to systematics as well as to ecology... **a fine synthesis of current knowledge** of homeostatic aspects of ecological relationships of carabids, and it is a fitting tribute to the man to whom it is dedicated: Carl H. Lindroth, who was instrumental in formulating the approaches and techniques that are commonly used in ecological research on these fine beetles. The materials is **well organized** and the text is **easily readable,** **thanks to the clarity of thought and expression** of the author and to the skill of an able translator." *Science*

Volume 11: **M. H. A. Keenleyside**

Diversity and Adaptation in Fish Behaviour

1979. 67 figures, 15 tables. XIII, 208 pages.
ISBN 3-540-09587-X

"... it is important as the first serious attempt by a senior researcher to produce an overview of the discipline. Previous works have all been symposium volumes or collections of papers haphazardly assembled, and Keenleyside has produced a volume that is of substantially greater value than these. In clearly perceiving that the unique and valuable features of fish behavior are its diversity of form and circumstance, he has charted a course that future authors would be wise to follow. The book is well produced, well written, and easy to read. The illustrations are clear and straightforward." *Science*

Volume 12: **E. Skadhauge**

Osmoregulation in Birds

1981. 42 figures. X, 203 pages. ISBN 3-540-10546-8

Contents: Introduction. - Intake of Water and Sodium Chloride. - Uptake Through the Gut. - Evaporation. - Function of the Kidney. - Function of the Cloaca. - Function of the Salt Gland. - Interaction Among the Excretory Organs. - A Brief Survey of Hormones and Osmoregulation. - Problems of Life in the Desert, of Migration, and of Egg-Laying. - References. - Systematic and Species Index. - Subject Index.

Volume 13: **S. Nilsson**

Autonomic Nerve Function in the Vertebrates

1983. 83 figures. XIV, 253 pages. ISBN 3-540-12124-2

Contents: Introduction. - Anatomy of the Vertebrate Autonomic Nervous Systems. - Neurotransmission. - Receptors for Transmitter Substances. - Chemical Tools. - Chromaffin Tissue. - The Circulatory System. - Spleen. — The Alimentary Canal. - Swimbladder and Lung. - Urinary Bladder. - Iris. - Chromatophores. - Concluding Remarks. - References. - Subject Index.

Volume 14: **A. D. Hasler, A. T. Scholz**

Olfactory Imprinting and Homing in Salmon

Investigations into the Mechanism of the Imprinting Process
In collaboration with R. W. Goy
1983. 25 figures. XIX, 134 pages. ISBN 3-540-12519-1

Contents: Olfactory Imprinting and Homing in Salmon: Notes on the Life History of Coho Salmon. Imprinting to Olfactory Cues: The Basis for Home-Stream Selection by Salmon. - Hormonal Regulation of Smolt Transformation and Olfactory Imprinting in Salmon: Factors Influencing Smolt Transformation: Effects of Seasonal Fluctuations in Hormone Levels on Transitions in Morphology, Physiology, and Behavior. Fluctuations in Hormone Levels During the Spawning Migration: Effects on Olfactory Sensitivity to Imprinted Odors. Thyroid Activation of Olfactory Imprinting in Coho Salmon. Endogenous and Environmental Control of Smolt Transformation. - Postscript. - References. - Subject Index.

Springer-Verlag
Berlin
Heidelberg
New York
Tokyo